The IMA Volumes in Mathematics and its Applications

Volume 110

Series Editor
Willard Miller, Jr.

Springer Science+Business Media, LLC

Institute for Mathematics and
its Applications
IMA

The **Institute for Mathematics and its Applications** was established by a grant from the National Science Foundation to the University of Minnesota in 1982. The IMA seeks to encourage the development and study of fresh mathematical concepts and questions of concern to the other sciences by bringing together mathematicians and scientists from diverse fields in an atmosphere that will stimulate discussion and collaboration.

The IMA Volumes are intended to involve the broader scientific community in this process.

Willard Miller, Jr., Professor and Director

* * * * * * * * * *

IMA ANNUAL PROGRAMS

1982–1983	Statistical and Continuum Approaches to Phase Transition
1983–1984	Mathematical Models for the Economics of Decentralized Resource Allocation
1984–1985	Continuum Physics and Partial Differential Equations
1985–1986	Stochastic Differential Equations and Their Applications
1986–1987	Scientific Computation
1987–1988	Applied Combinatorics
1988–1989	Nonlinear Waves
1989–1990	Dynamical Systems and Their Applications
1990–1991	Phase Transitions and Free Boundaries
1991–1992	Applied Linear Algebra
1992–1993	Control Theory and its Applications
1993–1994	Emerging Applications of Probability
1994–1995	Waves and Scattering
1995–1996	Mathematical Methods in Material Science
1996–1997	Mathematics of High Performance Computing
1997–1998	Emerging Applications of Dynamical Systems
1998–1999	Mathematics in Biology
1999–2000	Reactive Flows and Transport Phenomena
2000–2001	Mathematics in Multi-Media

Continued at the back

Christoph Börgers   Frank Natterer
Editors

Computational Radiology and Imaging

Therapy and Diagnostics

With 97 Illustrations

Springer

Christoph Börgers
Department of Mathematics
Tufts University
Medford, MA 02155, USA

Frank Natterer
Institut für Numerische
    und instrumentelle Mathematik
Universität Münster
Einsteinstrasse 62
D-48149 Münster, Germany

Series Editor:
Willard Miller, Jr.
Institute for Mathematics and its
    Applications
University of Minnesota
Minneapolis, MN 55455, USA

Mathematics Subject Classifications (1991): 35R30, 44A12, 65R30, 92C50, 92C55

Library of Congress Cataloging-in-Publication Data
Börgers, Christoph.
    Computational radiology and imaging : therapy and diagnostics /
Christoph Börgers, Frank Natterer.
    p.   cm. — (The IMA volumes in mathematics and its
applications ; v. 110)
    Includes bibliographical references and index.
    ISBN 978-1-4612-7189-5   ISBN 978-1-4612-1550-9 (eBook)
    DOI 10.1007/978-1-4612-1550-9
    1. Diagnostic imaging—Mathematics Congresses.   I. Natterer, F.
(Frank), 1941– .   II. Title.   III. Series.
RC78.7.D53B67   1999
616.07′54′0151—dc21   99-18598

Printed on acid-free paper.

Production managed by A. Orrantia; manufacturing supervised by Jacqui Ashri.
Camera-ready copy prepared by the IMA.

9 8 7 6 5 4 3 2 1

ISBN 978-1-4612-7189-5

FOREWORD

This IMA Volume in Mathematics and its Applications

COMPUTATIONAL RADIOLOGY AND IMAGING: THERAPY AND DIAGNOSTICS

is based on the proceedings of a workshop with the same title. The workshop was an integral part of the 1996–97 IMA program on "MATHEMATICS IN HIGH-PERFORMANCE COMPUTING."

I would like to thank the scientific organizers: Christoph Börgers of Tufts University, Department of Mathematics and Frank Natterer of Universität Münster, Institut für Numerische und instrumentelle Mathematik for their excellent work as organizers of the meeting and for editing the proceedings.

I also take this opportunity to thank the National Science Foundation (NSF) and the Army Research Office (ARO), whose financial support made the workshop possible.

Willard Miller, Jr., Professor and Director

PREFACE

The articles collected in this volume are based on lectures at the IMA Workshop "Computational Radiology and Imaging: Therapy and Diagnostics," March 17–21, 1997. Introductory articles by the editors and organizers of the workshop have been added. The focus is on inverse problems involving electromagnetic radiation and particle beams, with applications to X-ray tomography, nuclear medicine, near-infrared imaging, microwave imaging, electron microscopy, and radiation therapy planning.

For electromagnetic radiation with short wave lengths (for example X-rays, gamma rays, and near-infrared lasers), the particle model is appropriate, and the radiation transport is described by the linear Boltzmann equation. For larger wave lengths (for example microwaves), the wave model is appropriate, and the radiation transport is described by Maxwell's equations. Most of the articles in this volume either make explicit reference to the linear Boltzmann equation, or are implicitly based on limiting forms or approximate solutions of this equation. For example, X-ray tomography is based on transport without scattering, a limit of the linear Boltzmann equation in which the mean free path tends to infinity. Algorithms for near-infrared imaging are frequently based on diffusion, a limit of the linear Boltzmann equation in which the mean free path tends to zero. In radiation therapy planning, the quantity of greatest interest is the dose, the energy deposited per unit mass, a macroscopic quantity that could in principle be derived from the solution of a boundary value problem for the linear Boltzmann equation. In practice, simpler approximate dose calculation methods are often used. One of the articles in this volume, the one by Colton and Monk on microwave imaging, is based on the wave model.

The workshop deliberately did not focus on forward problems, even though their computational solution is often a very challenging task by itself, and the subject of extensive current research.

We thank the IMA for hosting the workshop with hospitality and efficiency, especially Avner Friedman, who was the director of the IMA at the time of the workshop, and Robert Gulliver, who was its associate director. The workshop was part of a year on high performance computing at the IMA, and we also thank the organizers of this year, especially Mitchell Luskin and Ridgway Scott. We thank the extraordinarily capable and helpful technical staff at the IMA, especially Patricia V. Brick. Above all, we thank the participants in the workshop for making it productive and inspiring.

Christoph Börgers
Frank Natterer
May 1998

CONTENTS

Foreword v

Preface vii

INTRODUCTION

The radiation therapy planning problem 1
  Christoph Börgers

Mathematical models for medical imaging 17
  Frank Natterer

ARTICLES BY WORKSHOP SPEAKERS

Tomography through the transport equation 33
  D.S. Anikonov, A.E. Kovtanyuk, and I.V. Prokhorov

A general framework for iterative reconstruction
algorithms in optical tomography, using a finite
element method 45
  Simon R. Arridge and Martin Schweiger

Scattered radiation in nuclear medicine: A case study
on the Boltzmann transport equation 71
  Harrison H. Barrett, Brandon Gallas, Eric Clarkson,
  and Anne Clough

Mathematical aspects of radiation therapy treatment
planning: Continuous inversion versus full discretization
and optimization versus feasibility 101
  Yair Censor

Early results on general vertex sets and truncated
projections in cone-beam tomography 113
  Rolf Clackdoyle, Michel Defrise, and Frédéric Noo

Mathematical problems in microwave medical imaging 137
  David Colton and Peter Monk

Image reconstruction from experimental data in diffusion
tomography.. 157
   Michael V. Klibanov, Thomas R. Lucas, and Robert M. Frank

The Application of the x-ray transform to 3D conformal
radiotherapy... 183
   Robert Y. Levine, Eugene A. Gregerson, and Marcia M. Urie

Optimization methods for radiation therapy plans.................. 229
   Weldon A. Lodwick, Steve McCourt, Francis Newman,
   and Stephen Humphries

Fully three-dimensional reconstruction in electron microscopy........ 251
   Roberto Marabini, Gabor T. Herman, and José M. Carazo

THE RADIATION THERAPY PLANNING PROBLEM

CHRISTOPH BÖRGERS*

Abstract. In this article we describe mathematical aspects of the radiation therapy optimization problem. Various says of formulating the problem are presented and discussed.

1. Introduction. The use of X-rays for cancer therapy began a few days after their discovery. Wilhelm Röntgen announced the discovery of X-rays on December 28, 1895, and Emil Grubbe used them for cancer therapy on January 12, 1896 [40]. X-rays are still the most common form of radiation used for cancer therapy, but beams of electrons, protons, neutrons, and other particles are used as well. The planning of the radiation treatment of a tumor begins with the creation of a three-dimensional image of the tumor and surrounding healthy tissue, using techniques such as computed tomography or MRI. The treatment planning discussed in this article occurs after the imaging is completed. It involves substantial use of computational algorithms.

Radiation therapy planning requires the study of radiation penetrating a background (a portion of a patient's body and the surrounding air, for instance). Both the radiation and the background are, of course, made up of particles. We shall distinguish between the two by referring to *radiation particles* and *background particles*. Background particles can be set in rapid motion as a result of interactions with radiation particles, thereby becoming radiation particles themselves. The transport of the radiation particles through the background is described by a system of coupled Boltzmann transport equations; see for instance Ref. [15], and also Sec. 2 of this article. A solution of this system is a vector of *phase space number densities*, that is, numbers of radiation particles per unit volume in *phase space*, i.e. position-direction-energy space. Different components of this vector correspond to different particle types. Even if the beams aimed at the tumor consist of one particle type only (for instance photons, as in X-rays), interactions between radiation particles and the background will set in motion other types of particles. Careful calculations therefore require consideration of several types of radiation particles in any case.

Interactions of radiation particles with each other are negligible in this context. The relevant transport equations are therefore linear. The speed of the radiation particles is the speed of light (for photons) or a significant fraction of the speed of light. As a result, a steady state is reached in a time that is extremely short in comparison with the times for which the beams are typically turned on during treatment, which are on the order of seconds

---

*Department of Mathematics, Tufts University, Medford, MA 02155, E-mail: borgers@math.tufts.edu

or minutes. The relevant transport equations therefore contain no time derivatives. Information obtained through imaging, such as the locations of soft tissue, bone, or air gaps, yields coefficients in these equations.

Radiation therapy is *fractionated*, that is, delivered in multiple sessions. Furthermore, during a single session, several beam configurations may be used. A radiation therapy plan specifies beam positions, directions, energies, etc., as well as when and how long the specified beams are to be turned on. This can be viewed as specifying a sequence of inflow boundary value problems for a system of steady linear Boltzmann equations.

The full solutions to these boundary value problems are never considered in radiation therapy planning. Of greatest interest is the total *dose*, that is, the amount of energy per unit background mass deposited, during the entire course of the treatment, as a result of excitation and ionization events. In the language of kinetic theory, dose is a *macroscopic* quantity, whereas the solution to a linear Boltzmann equation is a *mesoscopic* quantity. Dose depends on position; to emphasize this dependence, it is often called the *dose distribution*.

Strictly speaking, the dose distribution is not all that matters. Biological effects also depend on the type and energy of radiation used, the fractionation schedule, etc.; see Chapter 17 of Ref. [23] for a discussion of these factors. However, in practice it is usually assumed that for a given type of treatment (for instance, treatment with X-rays of a given energy range, and using a standard fractionation schedule), the effectiveness of a given treatment plan can be predicted from the dose distribution alone.

Computing dose distributions is a matter of computational physics, based on well-understood physical principles. In order to devise a good treatment plan, one must also be able to evaluate the *desirability* of a given dose distribution. This is most typically done by a physician based on experience and intuition, and is not a matter of rigorous science. However, one approach to evaluating the desirability of a given dose distribution is to first estimate, based on clinical data or even radiobiological models, the probabilities $p_1, ..., p_n$ of certain events, such as eradication of the tumor, damage to or destruction of healthy organs, pain relief as a result of tumor size reduction, etc. One can then use a function $\varphi = \varphi(p_1, ..., p_n)$ as the measure of desirability. Obviously φ depends on subjective preferences. Refs. [32] and [38] are basic articles on this sort of approach.

2. Dose calculation. As discussed in the introduction, dose calculation means the computation of macroscopic information related to the solution of an inflow boundary value problem for a system of steady linear Boltzmann equations; see Ref. [28] for a recent survey of this aspect of the problem, and an extensive literature list. To make this more concrete for readers not familiar with the linear Boltzmann equation, we shall outline the derivation of the equation for the special case of a single species of particles moving through a homogeneous, scattering, non-absorbing background; see Sec. 1.4 of Ref. [39].

Consider a particle moving through a background. Assume that the particle experiences collisions with the background at random times, and that the times between collisions are independent of each other, exponentially distributed, with an expected value depending on the pre-collision kinetic energy of the particle. This expected value is called the *mean free time*. Assume further that the collisions cause random direction and energy changes. As a result, the *phase space coordinates* of the particle, i.e. its position $\mathbf{x} \in \mathbb{R}^3$, direction $\omega \in S^2$, and energy $E > 0$, at time t are random. Denote their probability density by $f(\mathbf{x}, \omega, E, t)$. When a particle with pre-collision direction $\omega' \in S^2$ and pre-collision energy $E' > 0$ undergoes a collision, its post-collision direction $\omega \in S^2$ and energy $E > 0$ are random, with probability density

$$(2.1) \qquad \frac{1}{2\pi} p(\omega \cdot \omega', E, E') .$$

This expression depends on the dot product $\omega \cdot \omega'$, but not on ω and ω' individually, reflecting isotropy of scattering. If we define $\mu = \omega \cdot \omega' \in [-1, 1]$ to be the cosine of the angle between the pre- and post-collision directions, the probability density of the pair (μ, E), for a given E', is $p(\mu, E, E')$, without the factor of $1/2\pi$. A particle with kinetic energy $E > 0$ has velocity $v(E)$ and mean free time $\overline{\tau}(E)$. The probability density $p(\mu, E, E')$ is close to zero unless μ is close to one. That is, the deflection experienced by a particle in a single collision is likely to be small; see for instance Ref. [45]. One expresses this by saying that the scattering is *strongly forward-peaked*.

With the notation introduced above,

$$(2.2) \qquad f_t + v(\omega \cdot \nabla)f = Q_0 f ,$$

with

$$(Q_0 f)(\omega, E) = \frac{1}{2\pi} \int_0^\infty \int_{\omega' \in S^2} p(\omega \cdot \omega', E, E') \frac{f(\omega', E')}{\overline{\tau}(E')} d\omega' dE' - \frac{f(\omega, E)}{\overline{\tau}(E)} .$$
$$(2.3)$$

In Eq. (2.3), we have omitted the dependence on $\mathbf{x}$ and t for notational simplicity. The left-hand side in Eq. (2.2) corresponds to the streaming of the particle between collisions. On the right-hand side of Eq. (2.3), the term with the minus sign corresponds to the particle being "lost" by entering a collision, and the integral corresponds to the particle "re-emerging" from a collision with altered direction and energy. Up to now, we have thought of a single particle, and of f as the probability density function of its phase space location. Alternatively, we can think of a very large number of particles, independent of each other, and of f as their phase space number density. This is how we shall think from now on.

It is customary to introduce the independent variable

(2.4)
$$\psi = vf ,$$

called the flux, and the quantity

(2.5)
$$\sigma_s(E) = \frac{1}{v(E)\overline{\tau}(E)} ,$$

called the scattering cross-section. Using this notation, and dropping the time derivative, Eq. (2.2) becomes

(2.6)
$$(\omega \cdot \nabla)\psi = Q\psi ,$$

where

(2.7)
$$Q\psi(\omega, E) = \frac{1}{2\pi} \int_0^\infty \int_{\omega' \in S^2} p(\omega \cdot \omega', E, E')\sigma_s(E')\psi(\omega', E')d\omega'dE' \\ - \sigma_s(E)\psi(\omega, E) .$$

Background inhomogeneity makes σ_s and p functions of $\mathbf{x}$.

Let $\Omega \subseteq \mathbb{R}^3$ be a bounded region with a smooth boundary $\partial\Omega$. Let $\mathbf{n} = \mathbf{n}(\mathbf{x})$, $\mathbf{x} \in \partial\Omega$, denote the exterior unit normal vector field on $\partial\Omega$. A well-posed boundary value problem for $\psi = \psi(\mathbf{x}, \omega, E)$, $(\mathbf{x}, \omega, E) \in \Omega \times S^2 \times \mathbb{R}_+$, is obtained by supplementing Eq. (2.6) with the *inflow boundary condition*

(2.8) $\psi(\mathbf{x}, \omega, E) = g(\mathbf{x}, \omega, E)$ for $\mathbf{x} \in \partial\Omega$, $\omega \in S^2$, $\omega \cdot \mathbf{n}(\mathbf{x}) < 0$, $E > 0$.

For the mathematical theory of inflow boundary value problems for linear Boltzmann equations, see for instance Chapter 21 of Ref. [13]

To illustrate how dose distributions can be obtained from the solutions to boundary value problems for linear Boltzmann equations, let us compute an expression for the time rate at which the particles deposit energy in the background in our simplified setting. The expected amount of energy lost by a particle with pre-collision direction ω' and pre-collision energy E' in a collision is

(2.9) $\overline{\Delta E}(\mathbf{x}, \omega', E') = \dfrac{1}{2\pi} \displaystyle\int_0^\infty \int_{\omega \in S^2} (E' - E)\, p(\mathbf{x}, \omega \cdot \omega', E, E')\, d\omega\, dE .$

The time rate of energy deposition is

(2.10) $d(\mathbf{x}) = \displaystyle\int_0^\infty \int_{\omega' \in S^2} \overline{\Delta E}(\mathbf{x}, \omega', E')\, \sigma_s(\mathbf{x}, E')\, \psi(\mathbf{x}, \omega', E')\, d\omega'dE' ,$

and the energy deposited during a time interval of duration T is

(2.11)
$$D(\mathbf{x}) = T\, d(\mathbf{x}) .$$

As explained earlier, the true equations are a little more complicated, and in particular are coupled systems of linear Boltzmann equations.

In discussions of dose calculation in the Medical Physics literature, the underlying system of linear Boltzmann equations is not usually mentioned. With the codes used in current clinical practice, one typically obtains the dose directly, that is, without first computing the solution of the system of linear Boltzmann equations. There is a wide variety of different algorithms. However, they share the following basic ideas. The incoming radiation is thought of as composed of a finite number of pencil beams, that is, infinitesimally thin, mono-directional, mono-energetic beams. Mathematically, this means approximation of the boundary data by a finite sum of δ-functions. Approximations to the dose distributions due to pencil beams are obtained by laboratory experiments, numerical experiments using Monte Carlo simulation, mathematical analysis, or a combination of these approaches. The overall dose distribution is then obtained by summing such approximations. For discussions of dose calculation methods of this kind, see Refs. [24] and (for electron beams) [22]. There is an extensive literature on the mathematical analysis of pencil beams, starting with work due to Fermi [16]; see Ref. [22] for a survey and references. We studied this subject in Refs. [5]-[7].

In the past, Monte Carlo methods have been too slow for routine clinical use. However, the combination of gains in computer speed and development of faster Monte Carlo methods makes their future widespread clinical use increasingly likely; see for instance Refs. [1], [2], and [30] for Monte Carlo methods for particle transport calculations in general, and [35] for a Monte Carlo method specifically for radiation therapy planning.

Grid-based methods for the linear Boltzmann equation, using finite difference or finite element discretizations of spatial derivatives and, for example, discrete ordinates for the collision operator, are rarely mentioned in the Medical Physics literature. The *phase space evolution methods* (see Refs. [20] and [21]) come close to being such schemes. In general, the use of grid-based deterministic methods requires the development of efficient solvers for linear Boltzmann boundary value problems. This subject has been studied extensively in the Nuclear Engineering literature; see for instance Ref. [27] and references given there. However, most of this work does not apply to the case of strongly forward-peaked scattering. It appears that this is a gap that needs to be filled if grid-based deterministic methods are to become practical for dose calculations; see Refs. [25], [34], and [3] for methods for simplified (one- and two-dimensional) problems.

One might think that deterministic methods, such as finite difference or finite element methods, are not likely to compete well with Monte Carlo methods because of the large number of phase space dimensions (three space and three velocity dimensions). I discussed this argument in detail in Ref. [4], coming to the conclusion that it is not convincing. Therefore the question which of the two families of methods is preferable remains, at least in my view, unsettled.

We conclude this section by mentioning that the unit of dose commonly used in radiation therapy planning is the Gray, abbreviated Gy:

$$(2.12) \qquad 1\text{Gy} = 1\text{J/kg} .$$

For realistic values, see for instance Sec. 6 of Ref. [29]. One of the cases discussed there is a brain tumor, for which a dose of 90Gy was prescribed, with limits on the doses to brainstem and optic nerve of 20Gy and 10Gy, respectively.

3. Realizable dose distributions. We call the mapping from beam intensity distributions to dose distributions the *dose operator*. We call a dose distribution *realizable* if there is a realizable beam intensity distribution generating it. Which beam intensity distributions are realizable depends, of course, on the hardware used to deliver the radiation therapy. The most obvious constraint is that beam intensities must be non-negative. Typically there also is an upper bound on the number of beams that can be used.

Let $\mathcal{R}$ denote the set of realizable dose distributions. The question whether $\mathcal{R}$ is convex will be of interest to us in later sections. If the non-negativity of beam intensities is the only constraint on the treatment plan, then the set of permitted beam intensity distributions is convex, and therefore $\mathcal{R}$, being the image of a convex set under the (linear) dose operator, is convex as well. On the other hand, if there is a bound on the number of beams, but freedom in choosing beam positions and directions, then the set of permitted beam intensity distributions is non-convex, and so is $\mathcal{R}$ in general. We briefly refer to the problem of choosing beam positions and directions in the presence of a bound on the number of beams as the *beam selection problem*. So inclusion of the beam selection problem in the optimization problem makes $\mathcal{R}$ non-convex. The beam selection problem is discussed extensively in Ref. [29].

4. Biological response models. Models attempting to predict the probabilities of certain events, desirable or undesirable, for a given dose distribution, are called *biological response models*. For an introduction to this aspect of the problem, see for instance Sec. 1.1 of Ref. [42] and Refs. [37] and [41]. To illustrate the flavor of these models, we shall consider some simple examples. They are found in the references given above, although our notation is a little non-standard here.

We denote the region occupied by the tumor by Ω_t, the region occupied by healthy tissue by Ω_h, and the region of interest by $\Omega = \Omega_t \cup \Omega_h$. There may be ambiguity about the extent of a tumor; one can model that by not requiring that the intersection of Ω_t and Ω_h be empty.

We first discuss the *tumor control probability* (TCP). Assume that the tumor contains a very large number of small units called clonogens, and that the tumor is eradicated if and only if each clonogen is eradicated.

Denote by ρ the number density of clonogens. Further assume that the deaths of clonogens are independent random events, and that for a given clonogen, the probability of its death only depends on the dose D received by it. Denote this probability by $k(D)$. Suppose now that the tumor region Ω_t is divided into a large number of subregions $\Omega_{t,k}$ of volume V_k, $k = 1, ..., n$. Assume that these subregions are so small that the dose and the clonogen number density in $\Omega_{t,k}$ can be approximated by constants D_k and ρ_k, but so large that $\Omega_{t,k}$ contains many clonogens. Then

$$(4.1) \qquad TCP \approx \prod_{k=1}^{n} k(D_k)^{\rho_k V_k} = \exp \sum_{k=1}^{n} \rho_k V_k \ln k(D_k) \; .$$

A continuous analog of (4.1) is

$$(4.2) \qquad TCP = \exp \int_{\Omega_t} \rho(\mathbf{x}) \ln k(D(\mathbf{x})) \, d\mathbf{x} \; .$$

In the special case of constant D and ρ, this reduces to the obvious formula

$$(4.3) \qquad TCP = k(D)^N \; ,$$

where N denotes the total number of clonogens. So Eq. (4.2) gives the right way of modifying Eq. (4.3) for non-constant D and ρ. To complete the model of the TCP, one has to specify the function $k(D)$. It is always taken to be sigmoidal, as sketched in Fig. 1; compare for instance Fig. 1.18 on p. 37 of Ref. [42].

Lyman [31] proposed a simple formula for *normal tissue complication probabilities* ($NTCP$s). It is not a fundamental model based on radiobiology, but a data fitting scheme. Lyman's model applies to cases when a fraction v, $0 \leq v \leq 1$, of an organ at risk receives a constant dose D, and the rest of the organ receives no dose at all. Several ways of extending this model to the general case of a spatially varying dose have been proposed. The one due to Kutcher and Burman [26] can be shown, after a small amount of algebra, to be equivalent to

$$(4.4) \qquad NTCP = \frac{1}{\sqrt{2\pi}} \int_{-\infty}^{(\langle D \rangle_{L^p} - D_{50})/\sigma} \exp(-t^2/2) \, dt \; ,$$

where $\langle D \rangle_{L^p}$ denotes the L^p-average of the dose over the organ at risk, that is:

$$(4.5) \qquad \langle D \rangle_{L^p} = \frac{\|D\|_{L^p}}{V^{1/p}} \; ,$$

where V is the volume of the organ at risk, and the parameters $p > 0$, $D_{50} > 0$, and $\sigma > 0$ are adjusted to fit experimental data. (The denominator of $V^{1/p}$ in Eq. (4.5) is needed to ensure that $\langle D \rangle_{L^p} = C$ if $D(\mathbf{x}) = C$ for all

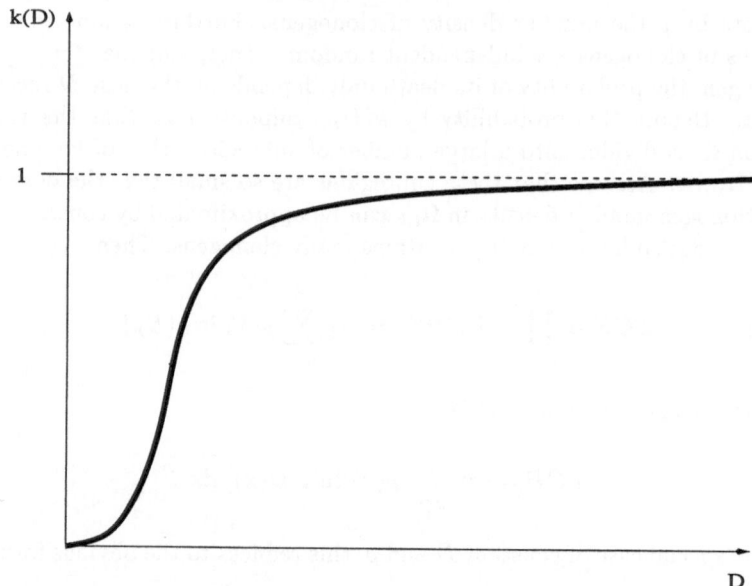

FIG. 1. *Probability of killing a single clonogen with dose D.*

x.) Eq. (4.4) predicts that irradiation at a dose with L^p-average D_{50} leads to a complication with probability 50%; this explains the notation. Table 1 of Ref. [9] suggests values of the parameters p, σ, and D_{50} for various different tissues and organs. The three parameters in Ref. [9] are called n, TD_{50}, and m. These parameters are related to ours as follows: $p = 1/n$, $D_{50} = TD_{50}$, and $\sigma = mTD_{50}$. For example, for the spinal chord, $p = 20$, $D_{50} = 66.5$Gy, and $\sigma = 11.6$Gy, and for the lungs, $p = 1.15$, $D_{50} = 24.5$Gy, and $\sigma = 4.4$Gy. The difference in the values of p reflects that for the spinal chord, even small regions of large dose must be avoided, whereas for the lungs, the average dose is essentially all that matters.

A different approach is described in Ref. [44]. Our presentation of it is close to that of Ref. [41]. Consider an organ at risk occupying a region Ω_o in space (or, more generally, any $\Omega_o \subseteq \Omega_h$). Assume that the organ at risk is made up of a large number of small units called *functional subunits*.

For some organs, such as the spinal chord, it may be appropriate to assume that significant damage to the organ occurs as soon as one of the subunits is destroyed. Organs of this kind are said to have a *serial* structure [42]. The probability of *no* normal tissue complication is then the probability that all subunits survive. This is analogous to the TCP model discussed earlier, where the probability of tumor control is the probability that all clonogens are killed. Following the same arguments that lead to Eq. (4.2), denoting by ρ the number density of subunits, and by $s(D)$ the probability that a single subunit survives irradiation at dose D, we are lead

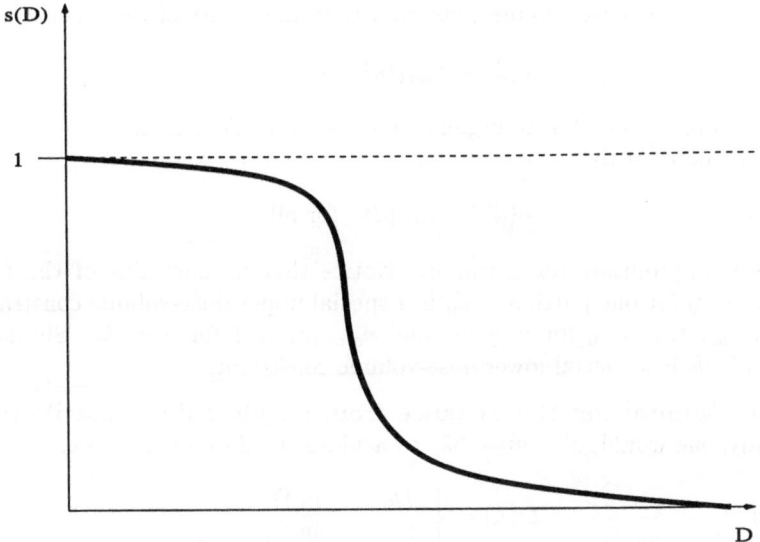

FIG. 2. *Survival probability of a single subunit receiving dose D.*

to the formula

$$(4.6) \qquad NTCP = 1 - \exp \int_{\Omega_o} \rho(\mathbf{x}) \ln s(D(\mathbf{x})) \, d\mathbf{x} \ .$$

It is natural to assume that s is sigmoidal, as shown in Fig. 2.

For other organs, such as the lungs, it may be appropriate to assume that significant damage to the organ occurs only when a certain percentage of the subunits is destroyed. Organs of this kind are said to have a *parallel* structure [42]. The $NTCP$ might, more generally, be a decreasing function of the number of surviving subunits, that is:

$$(4.7) \qquad NTCP = G \left(\int_{\Omega_o} \rho(\mathbf{x}) \, s(D(\mathbf{x})) \, d\mathbf{x} \right) \ ,$$

where G is a decreasing function.

5. Dose-volume constraints. Biological response models offer one way of formulating requirements on the dose distribution. Dose-volume constraints are an alternative approach. For a given dose distribution, and a given region in space, for instance the region occupied by the tumor or by a healthy organ, define $v(d)$ to be the volume fraction that receives a dose $\geq d$. It is clear that v is a decreasing function of d with $v(0) = 1$ and $v(d) = 0$ for sufficiently large d. In the Medical Physics literature, v is called a (differential) *dose-volume histogram*; see for instance Sec. 1.1.9 of

[42]. An *upper dose-volume constraint* is an inequality of the form

(5.1) $$v(d) \leq v_{max}(d) \quad \text{for all } d .$$

This is appropriate for an organ at risk. A *lower dose-volume constraint* is an inequality of the form

(5.2) $$v(d) \geq v_{min}(d) \quad \text{for all } d .$$

This is appropriate for a tumor. Notice that a constraint of the form $v(d_0) \leq v_0$ for one particular d_0 is a special upper dose-volume constraint, with $v_{max}(d) = v_0$ for $d \geq d_0$, and $v_{max}(d) = 1$ for $d < d_0$. Similarly, $v(d_0) \geq v_0$ is a special lower dose-volume constraint.

6. Minimizing the distance from an ideal dose distribution. Ideally, one would, of course, like to achieve the dose distribution

(6.1) $$\hat{D}(\mathbf{x}) = \begin{cases} D_0 & \text{in } \Omega_t , \\ 0 & \text{in } \Omega_h , \end{cases}$$

where D_0 is as large as needed to kill the tumor with certainty, but not very much larger; see for instance Ref. [17]. It is therefore natural to use, as a measure of desirability of a dose distribution D, the quantity

(6.2) $$\varphi(D) = -\|D - \hat{D}\|$$

for some function norm $\| \cdot \|$; see for instance Ref. [19]. The minus sign ensures that larger φ means greater desirability. Of course $\hat{D}$ is not realizable in general, since radiation must pass through healthy tissue to reach a tumor that does not lie at the surface of the patient's body, so in general the maximum of φ is negative.

Assume now that $\| \cdot \|$ is an L^p-norm:

(6.3) $$\|D - \hat{D}\| = \|D - \hat{D}\|_{L^p(\Omega)} = \left[\int_\Omega |D - \hat{D}|^p \, d\mathbf{x} \right]^{1/p}$$

for some $p \geq 1$. Denote the region occupied by the tumor by Ω_t, and the region occupied by the healthy tissue by Ω_h. Then

(6.4) $$\|D - \hat{D}\|_{L^p(\Omega)}^p = \|D - D_0\|_{L^p(\Omega_t)}^p + \|D\|_{L^p(\Omega_h)}^p .$$

Therefore minimizing $\|D - \hat{D}\|_{L^p(\Omega)}$ is equivalent to assuming that $1 - TCP$ is proportional to $\|D - D_0\|_{L^p(\Omega_t)}^p$, $NTCP$ is proportional to $\|D\|_{L^p(\Omega_h)}^p$, and maximizing an expression of the form $\alpha TCP - \beta NTCP$, where α and β are positive weights. For $p > 1$, D approximately constant in Ω_t, and $\|D - D_0\|_{L^p(\Omega_t)}^p$ and small $\|D\|_{L^p(\Omega_h)}^p$, this can, depending on the parameter values, closely resemble the use of the biological response models given by Eqs. (4.2), (4.4), and (4.5); for the TCP, this is illustrated in Fig. 3.

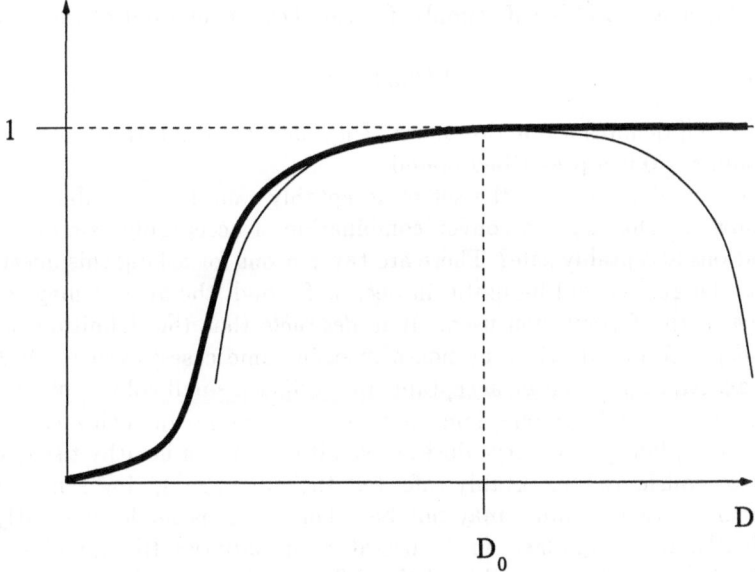

FIG. 3. TCP (bold) and $1 - c|D - D_0|^p$.

It seems to me that the approach of Ref. [11] (see also Ref. [10]) can essentially be viewed as an improvement on minimizing the distance from an ideal dose distribution. To see the similarity, compare for instance Eq. (6.2) with the displayed equation following Eq. (40) in [11]. Notice, however, that exceeding $\hat{D}$ in the tumor is penalized when minimizing the distance from $\hat{D}$, whereas exceeding the lower dose bound in the tumor is not penalized in the approach of [11], unless there is an upper dose bound in the tumor, and that bound is exceeded as well. Similarly, there is no penalty for a positive dose in healthy tissue in the approach of Ref. [11], as long as the upper dose bound in the healthy tissue is not exceeded.

7. Convexity of the set of "acceptably safe" dose distributions. Using biological response models as in Sec. 4, we might call a dose distribution $D \in \mathcal{R}$ *acceptably safe* if it satisfies inequalities of the form

$$(7.1) \qquad NTCP_i \leq \epsilon_i \, ,$$

where the index i labels possible normal tissue complications, and the $\epsilon_i \in (0, 1)$ are prescribed bounds. Alternatively, based on Sec. 5, $D \in \mathcal{R}$ could be called acceptably safe if it satisfies dose-volume constraints

$$(7.2) \qquad v_i(d) \leq v_{max,i}(d) \, ,$$

where the index i labels organs at risk, and the $v_{max,i}$ are prescribed decreasing functions with values between 0 and 1. Based on Sec. 6, we might call $D \in \mathcal{R}$ acceptably safe simply if it satisfies a constraint of the form

$$(7.3) \qquad\qquad \|D|_{\Omega_h}\| \leq \epsilon \,,$$

where $\| \cdot \|$ denotes a function norm, $D|_{\Omega_h}$ denotes the restriction of D to Ω_h, and $\epsilon > 0$ is a prescribed bound.

Let us denote by $\mathcal{A}$ the set of acceptably safe dose distributions. Is $\mathcal{A}$ convex? That is, is a convex combination of acceptably safe dose distributions acceptably safe? There are two reasons for asking this question. First, the answer will be useful in Sec. 8. Second, the answer may reveal a flaw in the formulation used. It is *desirable* that the definition of *acceptably safe* permit $\mathcal{A}$ to be non-convex in some cases, even when $\mathcal{R}$ is convex. Namely, it can be acceptable to sacrifice a small volume of healthy tissue, but not a large one; lung tissue is an example. In such a case, two treatment plans giving large doses to small volumes of healthy tissue may each be considered acceptably safe, but their average, giving a moderate dose to a larger volume, may not be. This point is made, explicitly or implicitly, in several places in the literature on radiation therapy planning; see for instance the discussion of the difference between the prostate and lung cases in Ref. [33], or p. 1296 of Ref. [41].

For the remainder of the section, we assume that $\mathcal{R}$ is convex, in particular that the beam selection problem is not included in the optimization problem. If the definition of $\mathcal{A}$ is then based on $NTCP$ models as defined by Eqs. (4.4) and (4.5), $\mathcal{A}$ is assured to be convex, regardless of the parameter choices. The same holds if its definition is based on (7.3), regardless of the choice of norm, or on an improved inequality along the lines of Ref. [11] (compare the discussion at the end of Sec. 6). On the other hand, if the definition is based on (7.2), then $\mathcal{A}$ is assured *not* to be convex, except in the trivial case when all $v_{max,i}$ are constant functions. If an $NTCP$ model of the form (4.6) underlies the definition of $\mathcal{A}$, convexity of $\mathcal{A}$ is assured provided that $\ln s(D)$ is a concave (that is, concave-down) function of D. Whether or not this is the case cannot be deduced from the general qualitative shape in Fig. 2. Finally, if an $NTCP$ model of the form (4.7) underlies the definition of $\mathcal{A}$, then convexity of $\mathcal{A}$ is certainly not assured, since the function s is not everywhere concave.

8. Uniqueness of solutions to the optimization problem. We conclude with a discussion of conditions that imply uniqueness of the solution to the radiation therapy optimization problem. From a practical point of view, this is important because the choice of optimization algorithm depends on it. For instance, algorithms such as simulated annealing have been proposed for radiation therapy optimization because of the possibility of multiple local optima [43]. As suggested by Niemierko in Ref. [36] and shown by Deasy in Ref. [14], insight into the issue of uniqueness of solutions can be gained from elementary convexity considerations. We shall present a variation on Deasy's argument, and point out an issue arising in this context that seems important and not yet well-understood [8].

The issue of uniqueness of the optimal treatment plan can be divided into two parts as follows. The first question is when, and in which sense, the optimal realizable dose distribution is unique. The second question is to which extent, for a given optimal realizable dose distribution, the treatment plan generating it is unique. This amounts to studying the nullspace of the dose operator.

To discuss the first question, let us assume that our goal is to determine a dose distribution $D \in \mathcal{A}$ (see Sec. 7) with maximal TCP. This corresponds to the choice $\varphi = TCP$ if $D \in \mathcal{A}$, and $\varphi = 0$ otherwise. Other choices of φ could be discussed similarly. Of course, maximizing TCP means the same as maximizing $q(TCP)$ if q is a strictly increasing function. If $\mathcal{A}$ is convex, then uniqueness of the optimal dose distribution in $\mathcal{A}$ depends on concavity properties of $q(TCP)$ as a functional of the dose distribution D. Strict concavity rules out multiple local maxima. Non-strict concavity rules out multiple local maxima with different TCP values.

Two sources of non-convexity of $\mathcal{A}$ have already been discussed. First, $\mathcal{R}$ is non-convex if beam selection is included in the optimization problem, as discussed in Sec. 3. But even if $\mathcal{R}$ is convex, $\mathcal{A}$ can be non-convex, and ought to be non-convex at least in some cases because of dose-volume constraints; see Sec. 7.

Let us assume now that $\mathcal{A}$ is convex for our problem. Using Eq. (4.2),

$$(8.1) \qquad \ln TCP = \int_{\Omega_t} \rho(\mathbf{x}) \ln k(D(\mathbf{x})) \, d\mathbf{x} \ .$$

If k were a concave function, then (8.1) would be a concave functional of D. Although k is not concave everywhere (compare Fig. 1), it is concave where k is close to 1. This implies uniqueness of the locally optimal TCP value at least among dose distributions for which the minimum tumor dose is not too low.[1] Things are simpler if we maximize the minimum tumor dose instead of the TCP over $\mathcal{A}$:

$$(8.2) \qquad \min_{\mathbf{x} \in \Omega_t} D(\mathbf{x})$$

is a concave functional of D.

---

[1] Deasy [14] proposed using the notion of *quasi-concavity* instead of concavity. A function $g = g(\mathbf{z})$, $\mathbf{z} \in \mathbf{R}^l$, is called quasi-concave if $g(\theta \mathbf{z}_1 + (1-\theta)\mathbf{z}_2) \geq \min(g(\mathbf{z}_1), g(\mathbf{z}_2))$ whenever $\mathbf{z}_1 \neq \mathbf{z}_2$ and $0 < \theta < 1$, and strictly quasi-concave if strict inequality holds. Strict quasi-concavity rules out multiple local maxima. Deasy argued that the TCP, as a function of beam weights, should be strictly quasi-concave. His reasoning, however, seems based on the incorrect assumption that strict monotonicity in each coordinate direction implies strict quasi-concavity ([14], p. 1159). A counterexample is $g(u, w) = (1 + u^2)(1 + w^2)$, $u \geq 0$ and $w \geq 0$. This function is strictly increasing in u and w, but not quasi-concave since $g(1/2, 1/2) = 25/16 < \min(g(1, 0), g(0, 1)) = \min(2, 2) = 2$.

We are currently studying the second question [8]. Making greatly simplifying assumptions, including the absence of scattering, the dose operator can be modeled as the dual exponential X-ray transform; see Refs. [12] and [29]. The question raised then reduces to studying the nullspace of this transform. However, the question should also be posed with discrete sets of permitted beam positions and directions, and with more realistic dose operators. Even when the nullspace of the dose operator is trivial, one may ask whether the dose operator has singular values that are *nearly* zero. If the answer is yes, then there may be beam intensity distributions that are significantly different from the optimal one(s), but generate nearly optimal dose distributions. Among these beam intensity distributions, one could then try to find a particularly simple one.

9. Acknowledgments. While writing this article, I had helpful discussions with Martin Altschuler, Yair Censor, Todd Quinto, and Jim Satterthwaite. My work is supported in part by NSF grant DMS-9626696.

REFERENCES

[1] A. F. BIELAJEW, *Photon Monte Carlo Simulation*, Lecture Notes, National Research Council of Canada, Report PIRS-0393, available on the internet at http://ehssun.lbl.gov/egs/epub/course.html, 1993.

[2] A. F. BIELAJEW AND D. W. O. ROGERS, *Electron Monte Carlo Simulation*, Lecture Notes, National Research Council of Canada, Report PIRS-0394, available on the internet at http://ehssun.lbl.gov/egs/epub/course.html, 1993.

[3] C. BÖRGERS, *A fast iterative method for computing particle beams penetrating matter*, J. Comp. Phys., **133**, 323–339, 1997.

[4] C. BÖRGERS, *Complexity of Monte Carlo and deterministic dose-calculation methods*, Phys. Med. Biol., **43**, 517-528, 1998.

[5] C. BÖRGERS AND E. W. LARSEN, *The transversely integrated scalar flux of a narrowly focused particle beam*, SIAM J. Appl. Math, **50**, No. 1, 1–22, 1995.

[6] C. BÖRGERS AND E. W. LARSEN, *Asymptotic derivation of the Fermi pencil beam approximation*, Nucl. Sci. Eng., **123**, No. 3, 343–357, 1996.

[7] C. BÖRGERS AND E. W. LARSEN, *On the accuracy of the Fokker-Planck and Fermi pencil beam equations for charged particle transport*, Med. Phys., **23**, 1749–1759, 1996.

[8] C. BÖRGERS AND E. T. QUINTO, *Nullspace and conditioning of the mapping from beam weights to dose distributions in radiation therapy planning*, in preparation.

[9] C. BURMAN, G. J. KUTCHER, B. EMAMI, AND M. GOITEIN, *Fitting of normal tissue tolerance data to analytic functions*, Int. J. Rad. Onc. Biol. Phys., **21**, 123–135, 1991.

[10] Y. CENSOR, *Mathematical aspects of radiation therapy treatment planning: Continuous inversion versus full discretization and optimization versus feasibility*, in this volume.

[11] Y. CENSOR, M. D. ALTSCHULER, AND W. D. POWLIS, *On the use of Cimmino's simultaneous projection method for computing a solution of the inverse problem in radiation therapy treatment planning*, Inverse Problems, **4**, 607–623, 1988.

[12] A. M. CORMACK AND E. T. QUINTO, *The mathematics and physics of radiation dose planning using X-rays*, Contemporary Mathematics, **113**, 41–55, 1990.

[13] R. DAUTRAY AND J.-L. LIONS, *Mathematical Analysis and Numerical Methods for Science and Technology*, vol. 6, Springer-Verlag, 1993.

[14] J. O. DEASY, *Multiple local minima in radiotherapy optimization problems with dose-volume constraints*, Med. Phys., **24**, No. 7, 1157–1161, 1997.

[15] J. J. DUDERSTADT AND W. R. MARTIN, *Transport Theory*, John Wiley & Sons, 1979.

[16] E. FERMI, result reported in B. Rossi and K. Greisen, *Cosmic ray theory*, Rev. Mod. Phys., **13**, 240, 1941.

[17] M. GOITEIN, *Causes and consequences of inhomogeneous dose distributions in radiation therapy*, Int. J. Rad. Onc. Biol. Phys., **12**, 701–704, 1986.

[18] M. GOITEIN AND A. NIEMIERKO, *Biologically based models for scoring treatment plans*, presentation to the Joint U.S./Scandinavian Symposium on Future Directions of Computer-Aided Radiotherapy, San Antonio, 1988.

[19] T. HOLMES AND T. R. MACKIE, *A comparison of three inverse treatment planning algorithms*, Phys. Med. Biol., **39**, 91–106, 1994.

[20] J. J. JANSSEN, D. E. J. RIEDEMAN, M. MORAWSKA-KACZYŃSKA, P. R. M. STORCHI, AND H. HUIZENGA, *Numerical calculation of energy deposition by high-energy electron beams: III. Three-dimensional heterogeneous media*, Phys. Med. Biol., **39**, 1351–1366, 1994.

[21] J. J. JANSSEN, E. W. KOREVAAR, R. M. STORCHI, AND H. HUIZENGA, *Numerical calculation of energy deposition by high-energy electron beams: III-B. Improvements to the 6D phase space evolution model*, Phys. Med. Biol., **42**, 1441–1449, 1997.

[22] D. JETTE, *Electron beam dose calculations*, in Radiation Therapy Physics, A. R. Smith (ed.), 95–121, Springer-Verlag, Berlin, 1995.

[23] H. E. JOHNS AND J. R. CUNNINGHAM, *The Physics of Radiology*, fourth edition, Charles C. Thomas, 1983.

[24] F. M. KHAN, *The Physics of Radiation Therapy*, second edition, Williams & Wilkins, 1994.

[25] K. M. KHATTAB AND E. W. LARSEN, *Synthetic acceleration methods for linear transport problems with highly anisotropic scattering*, Nucl. Sci. Eng., **107**, 217–227, 1991.

[26] G. J. KUTCHER AND C. BURMAN, *Calculation of complication probability factors for non-uniform normal tissue irradiation: the effective volume method*, Int. J. Rad. Onc. Biol. Phys., **16**, 1623–1630, 1989.

[27] E. W. LARSEN, *Diffusion-synthetic acceleration methods for discrete-ordinate problems*, Transport Theory Stat. Phys., **13**, 107–126, 1984.

[28] E. W. LARSEN, *Tutorial: The nature of transport calculations used in radiation oncology*, Transport Theory Stat. Phys., **26**, No. 7, 739, 1997.

[29] R. Y. LEVINE, E. A. GREGERSON, AND M. M. URIE, *The application of the X-ray transform to 3D conformal radiotherapy*, in this volume.

[30] I. LUX AND L. KOBLINGER, *Monte Carlo Particle Transport Methods: Neutron and Photon Calculations*, CRC Press, 1991.

[31] J. T. LYMAN, *Complication probability as assessed from dose-volume histograms*, Rad. Res., **104**, S-13–S-19, 1985.

[32] R. MOHAN, G. S. MAGERAS, B. BALDWIN, L. J. BREWSTER, G. J. KUTCHER, S. LEIBEL, C. M. BURMAN, C. C. LING, AND Z. FUKS, *Clinically relevant optimization of 3-D conformal treatments*, Med. Phys., **19**, 933–944, 1992.

[33] R. MOHAN, X. WANG, A. JACKSON, T. BORTFELD, A. L. BOYER, G. J. KUTCHER, S. A. LEIBEL, Z. FUKS, AND C. C. LING, *The potential and limitations of the inverse radiotherapy technique*, Radiother. Oncol., **32**, 232–248, 1994.

[34] J. E. MOREL AND T. A. MANTEUFFEL, *An angular multigrid acceleration technique for S_n equations with highly forward-peaked scattering*, Nucl. Sci. Eng., **107**, 330–342, 1991.

[35] H. NEUENSCHWANDER, T. R. MACKIE, AND P. J. RECKWERDT, *MMC – A high-performance Monte Carlo code for electron beam treatment planning*, Phys.

Med. Biol., **40**, 543, 1995.

[36] A. NIEMIERKO, *Optimization of intensity modulated beams: Local or global optimum?*, Med. Phys., **23**, 1072, 1996.

[37] A. NIEMIERKO AND M. GOITEIN, *Calculation of normal tissue complication probability and dose-volume histogram reduction schemes for tissues with a critical element structure*, Radiother. Oncol., **20**, 166–176, 1991.

[38] A. NIEMIERKO, M. URIE, AND M. GOITEIN, *Optimization of 3D radiation therapy with both physical and biological end points and constraints*, Int. J. Rad. Onc. Biol. Phys., **23**, 99–108, 1992.

[39] G. C. POMRANING, *Linear Kinetic Theory and Particle Transport in Stochastic Mixtures*, Series on Advances in Mathematics for Applied Sciences Vol. 7, World Scientific, Singapore, 1991.

[40] *Radiology Centennial, Inc.*, A Century of Radiology, http://www.xray.hmc.psu.edu/rci/centennial.html, 1993.

[41] C. RAPHAEL, *Mathematical modelling of objectives in radiation therapy treatment planning*, Phys. Med. Biol., **37**, No. 6, 1293–1311, 1992.

[42] S. WEBB, *The Physics of Three-Dimensional Radiation Therapy*, IOP Publishing, Bristol and Philadelphia, 1993.

[43] S. WEBB, *Optimization by simulated annealing of three-dimensional conformal treatment planning for radiation fields defined by multi-leaf collimators: II. Inclusion of two-dimensional modulation of X-ray intensities*, Phys. Med. Biol., **37**, 1689–1704, 1992.

[44] H. R. WITHERS, J. M. G. TAYLOR, AND B. MACIEJEWSKI, *Treatment volume and tissue tolerance*, Int. J. Rad. Onc. Biol. Phys., **14**, 751–759, 1988.

[45] C. D. ZERBY AND F. L. KELLER, *Electron transport theory, calculations, and experiments*, Nucl. Sci. Eng., **27**, 190–218, 1967.

MATHEMATICAL MODELS FOR MEDICAL IMAGING

FRANK NATTERER*

Abstract. In this article we describe the mathematical models used in medical imaging, their limitations, and the pertinent mathematical methods and problems. In many cases these problems have not yet been solved satisfactorily. In the description of the mathematical models we restrict ourselves to those features which are important for the mathematical scientist. More about the physical details, the assumptions made, the limitations can be found in [26].

Key words. Tomography, Inverse Problems

AMS(MOS) subject classifications. 92C55, 86A22, 44A12

1. Transmission CT. This is the original and simplest case of CT. In transmission tomography one probes an object with non diffracting radiation, e.g. x-rays for the human body. If I_0 is the intensity of the source, $a(x)$ the linear attenuation coefficient of the object at point x, L the ray along which the radiation propagates, and I the intensity past the object, then

$$(1.1) \qquad I = I_0 \, e^{-\int_L a(x)dx}.$$

In the simplest case the ray L may be thought of as a straight line. Modeling L as strip or cone, possibility with a weight factor to account for detector inhomogeneites may be more appropriate. In (1.1) we neglegt the dependence of a from the energy (beam hardening effect) and other non linear phenomena (e.g. partial volume effect), see [26], [28].

The mathematical problem in transmission tomography is to determine a from measurements of I for a large set of rays L. If L is simply the straight line connecting the source x_0 with the detector x_1, (1.1) gives rise to the integrals

$$(1.2) \qquad \ln \frac{I}{I_0} = -\int_{x_0}^{x_1} a(x)dx$$

where dx is the restriction to L of the Lebesgue measure in $\mathbb{R}^n$. We have to compute a in a domain $\Omega \subseteq \mathbb{R}^n$ from the values of (1.2) where x_0, x_1 run through certain subsets of $\partial\Omega$.

*Institut für Numerische, und instrumentelle Mathematik, Universität Münster, Einsteinstrasse 62, D-48149 Münster, Germany.
E-mail: nattere@math.uni-muenster.de

For $n = 2$, (1.2) is simply a reparametrization of the Radon transform

$$(1.3) \qquad (Ra)(\theta, s) = \int_{x \cdot \theta = s} a(x) dx$$

where $\theta \in S^{n-1}$. Thus our problem is in principle solved by Radon's inversion formula

$$(1.4) \qquad a = R^* K g, \qquad g = Ra$$

where

$$(1.5) \qquad (R^* g)(x) = \int_{S^{n-1}} g(\theta, x \cdot \theta) d\theta$$

is the so-called backprojection and

$$(1.6) \qquad K = \frac{1}{2}(2\pi)^{1-n} \begin{cases} (-1)^{(n-2)/2} \, H \frac{\partial^{n-1}}{\partial s^{n-1}} & , \quad n \text{ even} \\ (-1)^{(n-1)/2} \, \frac{\partial^{n-1}}{\partial s^{n-1}} & , \quad n \text{ odd} \end{cases}$$

with H the Hilbert transform. In fact the numerical implementation of (1.4) leads to the filtered backprojection algorithm which is the standard algorithm in commercial CT scanners [26], [32], [40], [20], [29].

For $n = 3$, the relevant integral transform is the x-ray transform

$$(Pa)(\theta, x) = \int_{\mathbb{R}^1} a(x + s\theta) ds$$

where $\theta \in S^{n-1}$ and $x \in \theta^\perp$. P admits a similar inversion formula as R, to wit

$$(1.7) \qquad a = P^* K g \qquad g = Pf$$

with K very similar to (1.6) and

$$(P^* g)(x) = \int_{S^{n-1}} g(\theta, E_\theta x) d\theta$$

where E_θ is the orthogonal projection onto $\theta^\perp$. Unfortunately, (1.7) is not as useful as (1.4). The reason is that (1.7) requires g for all θ and $y \in \theta^\perp$, i.e. (1.2) has to be available for all $x_0, x_1 \in \partial\Omega$. This is not practical. Also, it is not necessary for unique reconstruction of a. In fact it can be shown that a can be recovered uniquely from (1.2) with sources x_0 on a circle surrounding supp(a) and $x_1 \in S^{n-1}$. Unfortunately the determination of a in such an arrangement is, though uniquely possible, highly unstable. The condition of stability is the following: Each plane meeting supp(a) must contain at least one source. This condition is obviously violated for sources on a circle. Cases in which the condition is satisfied include the helix and a pair of orthogonal circles. A variety of inversion formulae has been derived for 3D [46], [14].

Methods completely independent of inversion formulas have become known as ART (algebraic reconstruction technique). In ART one discretisizes the problem and solves the resulting linear system iteratively [10], [19], [20], [32].

If scatter is to be included, a transport model is more appropriate. Let $u(x, \theta)$ be the density of the particles at x travelling (with speed 1) in direction θ. Then,

$$\theta \cdot \nabla u(x, \theta) + a(x) u(x, \theta) = \int_{S^{n-1}} \eta(x, \theta, \theta') u(x, \theta') d\theta'$$

(1.8a)
$$+ \delta(x - x_0) \, .$$

Here, $\eta(x, \theta, \theta')$ is the probability that a particle at x travelling in direction θ is scattered in direction θ'. Again we neglect dependence on energy. δ is the Dirac δ-function modeling a source of unit strength. (2.1a) holds in a domain Ω of $\mathbb{R}^n$ ($n = 2$ or 3), and $x_0 \in \partial\Omega$. Since no radiation comes in from outside we have

(1.8b) $$u(x, \theta) = 0 \, , \quad x \in \partial\Omega \, , \quad \nu_x \cdot \theta \le 0$$

where ν_x is the exterior normal on $\partial\Omega$ at $x \in \partial\Omega$. (1.1) is now replaced by

(1.8c) $$I(x_1, x_0, \theta) = I_0 u(x_1, \theta) \, , \quad x_1 \in \partial\Omega \, , \quad \nu_{x_1} \cdot \theta \ge 0 \, .$$

The problem of recovering a from (1.8) in much harder. An explicit formula for a such as (1.4) has not become known and is unlikely to exist. Nevertheless numerical methods have been developed for special choices of η [2], [6]. The situation gets even more difficult if one takes into account that, strictly speaking, η is object dependent and hence not known in advance. (1.8) is a typical example of an inverse problem for a partial differential equation. In an inverse problem one has to determine the differential equation - in our case a, η - from information about the solution - in our case (1.8c). More on inverse problems can be found in section 3.

2. Emission CT. In emission tomography one determines the distribution f of radiating sources in the interior of an object by measuring the radiation outside the object in a tomographic fashion [12]. Let again $u(x, \theta)$ be the density of particles at x travelling in direction θ with speed 1, and let a be the attenuation distribution of the object. (This is the quantity which is sought for in transmission CT). Then,

(2.1a) $$\theta \cdot \nabla u(x, \theta) + a(x) u(x, \theta) = f(x) \, .$$

This equation holds in the object region $\Omega \subseteq \mathbb{R}^n$ for each $\theta \in S^{n-1}$. Again there exists no incoming radiation, i.e.

(2.1b) $\qquad\qquad u(x, \theta) = 0 \quad , \quad x \in \partial\Omega , \quad \nu_x \cdot \theta \le 0$

while the outgoing radiation

(2.1c) $\qquad\qquad u(x, \theta) = g(x, \theta) , \quad x \in \partial\Omega , \quad \nu_x \cdot \theta \ge 0$

is measured and hence known. (2.1) again constitutes an inverse problem for a transport equation. For a known, (2.1a-b) is readily solved to yield

$$u(x, \theta) = \int\limits_{x-\infty\cdot\theta}^{x} e^{-\int\limits_{y}^{x} a ds} f(y) dy .$$

Thus (2.1c) leads to the integral equation

(2.2) $\qquad\qquad g(x, \theta) = \int\limits_{x-\infty\cdot\theta}^{x} e^{-\int_{y}^{x} a ds} f(y) dy$

for f. Apart from the exponential factor, (2.2) is identical - up to notation - to the integral equation in transmission CT. Except for very special cases - e.g. a constant in a known domain [45], [35] no explicit inversion formulas are available. Numerical techniques have been developed but are considered to be slow. Again the situation becomes worse if scatter is taken into account. This can be done by simply adding the scattering integral in (1.8a) to the right had side of (2.1a).

What we have described to far is called SPECT (= single particle emission CT). In PET (= positron emission tomography) the sources eject the particles pairwise in opposite directions. They are detected in coincidence mode, i.e. only events with two particles arriving at opposite detectors at the same time are counted. (2.1b) has to be replaced by

$$g(x, \theta) = \int\limits_{x-\infty\cdot\theta}^{x} e^{-\int\limits_{y}^{x} a ds - \int\limits_{x-\infty\cdot\theta}^{y} a ds} f(y) dy$$

(2.3) $\qquad\qquad = e^{-\int\limits_{x-\infty\cdot\theta}^{x} a ds} \int\limits_{x-\infty\cdot\theta}^{x} f(y) dy .$

Thus PET is even closer to the case of transmission CT. If a is known, we simply have to invert the x-ray transform. Inversion formulas which can make use of the data collected in PET are available. Their numerical implementation is presently under consideration.

The main problems in emission CT are

1. Unknown attenuation
2. Noise
3. Scatter

For the attenuation problem, the ideal mathematical solution would be a method for determining f and a in (9) simultaneously. Under strong assumption on a (e.g. a constant in a known region [21], a affine distortion of a prototype [33], a close to a known distribution [8]) encouraging results have been obtained. Theoretical results based on the transport formulation have been obtained, even for models including scatter [36]. But a clinically useful way of determining a from the emission data has not yet been found.

Noise and scatter are stochastic phenomena. Thus, besides models using integral equations, stochastic models have been set up for emission tomography [42]. These models are completely discrete. We subdivide the reconstruction region into m pixels or voxels. The number of events in pixel/voxel j is a Poisson random variable φ_j whose mathematical expection $f_j = E\varphi_j$ is a measure for the activity in pixel/voxel j. The vector $f = (f_1, \ldots, f_n)$ is the sought - for quantity. The vector $g = (g_1, \ldots, g_n)$ of measurements is considered as realization of the random variable $\gamma = (\gamma_1, \ldots, \gamma_n)$ where γ_i is the number of events detected in detector i. The model is determined by the (n, m)-matrix $A = (a_{ij})$ whose elements are

$a_{ij} = P$ (event in pixel/voxel j detected in detector i)

where P denotes probability. We have $E(\gamma) = Af$. f is determined from g by the maximum liklihood method. A numerical method for doing this is the EM (= expection maximation) algorithm. In its basic form it reads

$$f^{k+1} = f^k A^* \frac{g}{Af^k}, \qquad k = 0, 1, \ldots$$

where division and multiplication are to be understood component wise. The problem with the EM-algorithm is that it is only semi-convergent (compare 4), i.e. noise is amplified at high iteration numbers. This is known as the checkerboard effect. Various suggestions have been made to get rid of this effect. The most exciting and interesting ones use "prior information" and attempt to maximize "posterior likelihood". Thus f is assumed to have a prior probability distribution, called a Gibbs-Markov random field $\pi(f)$, which gives preference to certain functions f. Most prior π simply add a penalty term to the likelihood function to account for correction between neighboring pixels and do not use biological information. However if π is carefully chosen so that piecewise constant functions f with smooth boundaries forming the region of constancy are preferred then the noise amplification at high iteration numbers can be avoided. The question remains as to whether this conclusion will remain valid for function f which are assigned low probability by π - or, more to the point - whether "real" emission densities f will be well-resolved by this Bayesian method. An ROC study (i.e. double-blind trials where radiologists are to find lesions from images produced by two different algorithms) concluded that maximum likelihood methods were superior to the filtered backprojection algorithm in certain clinical applications [13], [16], [38]. The same type of study is needed to determine whether or not Gibbs priors will improve the maximum likelihoof reconstruction (stopped short of convergence to avoid noise amplification) on real data.

3. Ultrasound CT. X-rays travel along straight lines. For other sources of radiation, such as ultrasound and microwaves, this is no longer the case. The paths are no longer straight, and their exact shape depends on the internal structure of the object. We can no longer think in terms of simple projections and linear integral equations. More sophisticated non linear models have to be used.

In the following we consider an object $\Omega \subseteq \mathbb{R}^n$ with refractive index n. We assume $n = 1$ outside the object. The object is probed by a plane wave

$$e^{-ikt}u_\theta(x) , \quad u_\theta(x) = e^{ikx\cdot\theta}$$

with wave number $k = \frac{2\pi}{\lambda}$, λ the wave length, travelling in the direction θ. The resulting wave $e^{-ikt}u(x)$ satisfies the reduced wave equation

$$(3.1a) \qquad \Delta u + k^2(1 + f)u = 0 , \quad f = n^2 - 1 ,$$

plus suitable boundary conditions at infinity. The inverse problem to be solved is now the following. Assume that

$$(3.1b) \qquad g(x,\theta) = u(x) , \quad \theta \in S^{n-1}$$

is known outside Ω. Determine f inside Ω!

Uniqueness and stability of the inverse problem (3.1) has recently been settled. However, stability is only logarithmic, i.e. a data error of size δ results in a reconstruction error $1/\log(1/\delta)$, see section 4.

Numerical methods for (3.1) are mostly based on linearizations, such as the Born and Rytov approximation. In order to derive the Born approximation, one rewrites (3.1a) as

$$(3.2) \qquad u(x) = u_\theta(x) - k^2 \int_\Omega G(x - y)f(y)u(y)dy$$

where G is an appropriate Green's function. For $n = 3$, we have

$$(3.3) \qquad G(x) = \frac{e^{ik|x|}}{4\pi|x|} .$$

The Born approximation is now obtained by assuming $u \sim u_\theta$ in the integral in (3.2). With this approximation, (3.1b) reads

$$(3.4) \qquad g(x,\theta) = u_\theta(x) - k^2 \int_\Omega G(x - y)u_\theta(y)f(y)dy , \quad x \notin \Omega .$$

This is a linear integral equation for f, valid for all x outside the object and for all measured directions θ.

Numerical methods based on (3.3) - and a similar equation for the Rytov approximation - have become known as diffraction tomography. Unfortunately, the assumptions underlying the Born- and Rytov approximations are not satisfied in medical imaging. Thus, the reconstructions of f obtained from (3.3) are very poor. However, we may use (3.3) to get some encouraging information about stability. For $|x|$ large, (3.3) assumes the form

$$(3.5) \qquad g(x,\theta) = u_\theta(x) - \frac{k^2}{4\pi|x|} e^{ik|x|} \int e^{ik(\theta - \frac{x}{|x|})\cdot y} f(y) dy$$

$$= u_\theta(x) - \frac{k^2}{4\pi|x|} e^{ik|x|} (2\pi)^{3/2} \hat{f}(k(\frac{x}{|x|} - \theta))$$

with $\hat{f}$ the Fourier transform of f. (3.5) determines $\hat{f}$ within a ball of radius $\sqrt{2}k$ from the data (3.1b) in a completely stable way. We conclude that the stability of the inverse problem (3.1) is much better than logarithmic. If the resolution is restricted to spatial frequencies below $\sqrt{2}k$ - which is perfectly reasonable from a physical point of view - then we can expect (3.1) to be perfectly stable.

So far we considered plane wave irradiation at fixed frequency, and we worked in the frequency domain. Time domain methods are conceivable as well. We start out from the wave equation

$$(3.6a) \qquad \frac{\partial^2 u}{\partial t^2} = c^2 \Delta u$$

with the propagation speed c assumed to be 1 outside the object. With x_0 a source outside the object we consider the initial conditions

$$(3.6b) \qquad u(x,0) = 0 , \quad \frac{\partial u}{\partial t}(x,0) = \delta(x - x_0) .$$

We want to determine c inside the object from knowing

$$(3.6c) \qquad g(x_0, x_1, t) = u(x_1, t) , \quad t > 0$$

for many sources x_0 and receivers x_1 outside the object. In the one dimensional case, the inverse problem (3.6) can be solved by the famous Gelfand-Levitan method [9] in a stable way. It is not clear how Gelfand-Levitan can be extended to dimensions two and three. The standard methods use sources and receivers on all of the boundary of the object. This is not practical in medical imaging. However, for reduced data sets, comparable to those in 3D X-ray tomography (compare (1.1)), we do not know how to use Gelfand-Levitan, nor do we know anything about stability.

Of course one can always solve the nonlinear problem (3.6) by a Newton type method. Such methods have been developed [17], [27], [7]. They suffer from excessive computing time and from their apparent inability to handle large wave numbers k. More efficient methods based on initial value techniques have been developed in [34]. A Riccati approach can be found in [11].

4. Optical tomography. Here one uses NIR (= near infra-red) lasers for the illumination of the body. The process is now described by the transport equation

$$\frac{\partial u}{\partial t}(x,\theta,t) + \theta \cdot \nabla u(x,\theta,t) + a(x)u(x,\theta,t)$$

(4.1a) $$= b(x) \int_{S^{n-1}} \eta(\theta \cdot \theta')u(x,\theta',t)d\theta' + f(x,\theta,t)$$

for the density $u(x,\theta,t)$ of the particles at $x \in \Omega$ flying in direction $\theta \in S^{n-1}$ at time t. a and b are the sought - for tissue parameters. The scattering kernel η is assumed to be known. The source term f is under the control of the experimenter. Together with the initial and boundary conditions

$$u(x,\theta,0) = 0 \quad \text{in } \Omega \times S^{n-1} ,$$

(4.1b)

$$u(x,\theta,t) = 0 \quad \text{on } \partial\Omega \times S^{n-1} \times \mathbb{R}^1 , \quad v_x \cdot \theta \leq 0$$

(4.1a) has a unique solution under natural conditions on a, b, η and f. As in (1.1) we pose the inverse problem. Assume that we know the outward radiation

(4.1c) $\quad g(x,\theta,t) = u(x,\theta,t) \quad \text{on } \partial\Omega \times S^{n-1} \times \mathbb{R}^1 , \quad v_x \cdot \theta \geq 0 ,$

can we determine one or both the quantities a, b?

There are essentially three methods for illuminating the object, i.e. for choosing the source term f in (4.1a). In the stationary case one puts $f = \delta(x - x_0)$ where $x_0 \in \partial\Omega$ is a source point. u is considered stationary, too. A second possibility is the light flash $f = \delta(x - x_0)\delta(t)$. Finally one can also use time harmonic illumination, in which case $f = \delta(x - x_0)\,e^{i\omega t}$. This case reduces to the stationary case with a replaced by $a + i\omega$. In all three cases, the data function g of (4.1c) is measured at $x \in \partial\Omega$, possibly averaged over one or both of the variables θ, t.

Light tomography is essentially a scattering phenomenon. This means that the scattering integral in (4.1a) is essential. It can no longer be treated merely as a perturbation as in x-ray CT. Thus the mathematical analysis and the numerical methods are expected to be quite different from what we have seen in other types of tomography.

The mathematical theory of the inverse problem (4.1) is in a deplorable state. There exist some Russian papers on uniqueness [3]. General methods have been developed, too, but apparently they have been applied to 1D problems only [37]. Nothing seems to be known about stability. The numerical methods which have become known are of the Newton type, either applied directly to the transport equation or to the so-called diffusion approximation [4]. The diffusion approximation is an approximation to the transport equation by a parabolic differential equation. Since inverse problems for parabolic equations are severely ill-posed, this approach is questionable. Higher order approximations [15] are hyperbolic, making the inverse problem much more stable.

As an alternative to the transport equation one can also model light tomography by a discrete stochastical model [41]. In the 2D case, break up the object into a rectangular arrangement of pixels labelled by indices i, j with $a \leq i \leq b$ and $c \leq j \leq d$. Attach to each pixel the quantities f_{ij}, b_{ij}, r_{ij}, ℓ_{ij} meant to denote the probability of a forward, backward, rightward or leftward transition out of the pixel i, j with respect to the direction used to get into this pixel. For each pair of boundary pixels i, j and i', j' let $P_{ij,i'}$ be the probability that a particle that enters the object at pixel i, j will eventually leave the object at pixel i', j'. The problem is to determine the quantities f_{ij}, b_{ij}, r_{ij}, ℓ_{ij} from the values of $P_{ij,i'j'}$ for all boundary pixels. Preliminary numerical tests show that this is possible, at least in principle. However, the computations are very time consuming. More seriously, they reveal a very high degree of instability.

5. Electrical impedance tomography. Here, the sought-for quantity is the electrical impedance σ of an object Ω. Voltages are applied via electrodes on $\partial\Omega$, and the resulting currents at these electrodes are measured [5]. With u the potential in Ω, we have

$$\mathrm{div}(\sigma \cdot \nabla u) = 0 \quad \text{in} \quad \Omega$$

(5.1)

$$u = g \quad , \quad \sigma\frac{\partial u}{\partial \nu} = f \quad \text{on} \quad \partial\Omega .$$

Knowing many voltage - current pairs g, f on $\partial\Omega$, we have to determine σ from (5.1).

Uniqueness for the inverse problem (5.1) has recently been settled [23]. Unfortunately, the stability properties are very bad. Numerical methods based on Newton's method, linearization, simple backprojection, layer stripping have been tried [43]. All these methods suffer from the severe ill-posedness of the problem. There seems to be no way to improve stability by purely mathematical means.

6. Magnetic resonance imaging (MRI). The physical phenomena exploited here is the precession of the spin of a proton in a magnetic field of strength H about the direction of that field. The frequency of this precession is the Larmor frequency γH where γ is the gyromagnetic ratio. By making the magnetic field H space dependent in a controled way the local magnetization $M_0(x)$ (together with the relaxation times $T_1(x)$, $T_2(x)$) can be imaged. In the following we derive the imaging equations [22].

The magnetization $M(x, t)$ caused by a magnetic field $H(x, t)$ satisfies the Bloch equation

$$(6.1) \qquad \frac{\partial M}{\partial t} = \gamma M \times H - \frac{1}{T_2}(M_1 e_1 + M_2 e_2) - \frac{1}{T_1}(M_3 - M_0)e_3 \ .$$

Here, M_i is the i-th component of M, and e_i is the i-th unit vector $i = 1, 2, 3$. The significance of T_1, T_2, M_0 become apparent if we solve (6.1) with the static field $H = H_0 e_3$ and with initial values $M(x, 0) = M^0(x)$. We obtain with $\omega_0 = \gamma H_0$

$$\begin{aligned}
M_1(x, t) &= e^{-t/T_2}(M_1^0 \cos(\omega_0 t) + M_2^0 \sin(\omega_0 t)) \\
(6.2) \qquad M_2(x, t) &= e^{-t/T_2}(-M_1^0 \sin(\omega_0 t) + M_2^0 \cos(\omega_0 t)) \\
M_3(x, t) &= e^{-t/T_1} M_3^0 + (1 - e^{-t/T_1})M_0
\end{aligned}$$

Thus the magnetization rotates in the $x_1 - x_2$-plane with Larmor frequency ω_0 and returns to the equilibrium condition $(0, 0, M_0)$ with speed controlled by T_2 in the $x_1 - x_2$-plane and by T_1 in the x_3-direction.

In an MRI scanner one generates a field

$$H(x, t) = (H_0 + G(t) \cdot x)e_3 + H_1(t)(\cos(\omega_0 t)e_1 + \sin(\omega_0 t)e_2)$$

where G and H_1 are under control. In the jargon of MRI, $H_0 e_3$ is the static field, G the gradient, and H_1 the radio frequency (RF) field. The input G, H_1 produces in the detecting system of the scanner the output signal

$$(6.3) \qquad S(t) = -\frac{d}{dt} \int_{\mathbb{R}^3} M(x, t)B(x)dx$$

where B characterizes the detecting system. Depending on the choice of H_1 various approximation to S can be derived.

(i) H_1 is constant in the small interval $[0, \tau]$ and $\gamma \int_0^\tau H_1 dt = \frac{\pi}{2}$ (Short $\frac{\pi}{2}$

pulse). In that case,

$$S(t) = \int_{\mathbf{R}^3} M_0(x) e^{-i\gamma \int_0^t G(t')dt' \cdot x - t/T_2(x)} \, dx \; .$$

Choosing G constant for $\tau \leq t \leq \tau + T$ and zero otherwise we get for $T \ll T_2$

$$S(t) = \int_{\mathbf{R}^3} M_0(x) e^{-i\gamma(t-\tau)G \cdot x} dx$$

(6.4)

$$= (2\pi)^{3/2} \hat{M}_0(\gamma(t-\tau)G)$$

where $\hat{M}_0$ is the 3D Fourier transform of M_0. From here we can proceed in several ways. We can either use (6.4) to determine the 3D Fourier transform $\hat{M}_0$ of M_0 and to compute M_0 by an inverse 3D Fourier transform. This requires $\hat{M}_0$ to be known on a Cartesian grid, what can be achieved by a proper choice of the gradients or by interpolation. We can also evoke the central slice theorem to obtain the 3D Radon transform RM_0 of M_0 by a series of 1D Fourier transforms. M_0 is recovered in turn by inverting the 3D Radon transform.

(ii) H_1 is the shaped pulse

$$H_1(t) = \phi(t\gamma G) e^{i\gamma G x_3 t}$$

where ϕ is a smooth positive function supported in $[0, \tau]$. Then, with x', G' the first to components of x, G, respectively, we have

(6.5) $$S(t) = \int_{\mathbf{R}^2} M_0'(x', x_3) e^{-i\gamma \int_0^t G'(t')dt' \cdot x' - t/T_2(x', x_3)} dx'$$

where

$$M_0'(x', x_3) = \int M_0(x', y_3) Q(x_3' - y_3) dy_3$$

with a function Q essentially supported in a small neighborhood of 0. (6.5) is the 2D analogue of (6.4). So we have again the choice between Fourier imaging (i.e. computing the 2D Fourier transform from (6.5) and doing an inverse 2D Fourier transform) and projection imaging (i.e. doing a series of 1D Fourier transforms in (6.5) and inverting the 2D Radon transform).

Some of the mathematical problems of MRI, e.g. interpolation in Fourier space, are common to other techniques in medical imaging. An interesting mathematical problem occurs if the magnets to not produce suffiently homogeneous fields. It calls for the reconstruction of a function from integrals over slightly curved manifolds. Even though this is a problem of classical integral geometry it has not yet found a satisfactory solution. Another problem in MRI is incomplete sampling in Fourier space [31].

7. Vector tomography. If the domain under consideration contains a moving fluid, then the Doppler shift can be used to measure the velocity $u(x)$ of motion. Assume the time harmonic signal $e^{i\omega_0 t}$ is transmitted along the oriented line L. This signal is reflected by particles travelling with speed ν in the direction of L as $e^{i(\omega_0 - k\nu)t}$, where $k = 2\omega_0/c$, c the speed of the probing signal, i.e. $k\nu$ is the Doppler shift. Let $S(L, \nu)$ be the Lebesgue measure of those particles on L which move with speed $< \nu$, i.e. $u(x) \cdot e_L < \nu$, e_L the tangent vector on L. Then the total response is

$$g(L, t) = \int\limits_{-\infty}^{+\infty} e^{i(\omega_0 - k\nu))t} S(L, \nu) d\nu .$$

Thus S can be recovered from g by a Fourier transform. The problem is to recover u from S.

Not much is known about uniqueness. However, the first moment of S,

$$(7.1) \qquad \int\limits_{-\infty}^{+\infty} \nu S(L, \nu) d\nu = \int\limits_{L} u(x) \cdot e_L dx = (Ru)(L)$$

is similar to the Radon transform. Inverting R is a problem of vector tomography. Mathematically this belongs to the recently developed field of integral geometry of tensor fields [39]. One can show that curl u can be computed from Ru, and an inversion formula similar to the Radon inversion formula exists. Numerical simulations are given in [25], [44].

8. Tensor tomography. As an immediate extension of transmission CT to non-isotropic media we consider a matrix valued attenuation $a(x) = (a_{ij}(x))$, $i, j = 1, \ldots, n$. We solve the vector differential equation

$$(8.1) \quad \frac{du(t)}{dt} = -a(x(t))u(t) , \quad x(t) = (1 - t)x_0 + tx_1 , \quad 0 \le t \le 1$$

for the vector valued function $u(t) = (u_i(t))_{i=1,\ldots,n}$. Let a be defined in a convex domain Ω, and let $x_0, x_1 \in \partial\Omega$. Then, $u(x_1) = U(x_0, x_1)u(x_0)$ with a nonlinear map $U(x_0, x_1)$ depending on a. The problem is to recover a in Ω from the knowledge of $U(x_0, x_1)$ for $x_0, x_1 \in \partial\Omega$. For $n = 1$ we regain (1.1). Applications of (7.1) for $n > 1$ have become known in photoelasticity [1], but applications to medicine are not totally out of question.

In a further extension we let a depend on the direction $\xi = (x_1 - x_0)/|x_1 - x_0|$. Such problems occur in the polarization of harmonic electromagnetic and elastic waves in anisotropic media. In linearized form these problems give rise to the transverse x-ray transform

$$(8.2) \qquad (Ja)(x,\theta,\omega) = \int \omega^T a(x + t\theta)\omega dt , \quad \omega \perp \theta$$

and to the longitudenal x-ray transform

$$(8.3) \qquad (Ja)(x,\theta) = \int \theta^T a(x + t\theta)\theta dt .$$

(8.2) can easily be reduced to the $(n - 1)$-dimensional x-ray transform in the plane $H_{\omega,s} = \{y : y \cdot \omega = s\}$. We only have to introduce the function $a_\omega(y) = \omega^T a(y)\omega$. Then, (8.2) provides for $x \in H_{\omega,s}$ all the line integrals of a_ω on $H_{\omega,s}$. For (8.3), the situation is not so easy. We decompose a in its solenoidal and potential part, i.e.

$$a = a_1 + \nabla a_2 \quad \text{div } a_1 = 0 \quad a_2 = 0 \quad \text{on } \partial\Omega .$$

It can be shown that a_1 can be recovered uniquely from (8.3), but a_2 is completely undetermined. This is reminiscent of vector tomography in 1.7.

9. Magnetic source imaging. Here one wants to find the electric currents inside the body from measurements of the induced magnetic field outside the patient. These fields may be caused by epileptic fits or by cardiac infarction. The appropriate mathematical model relating the electric field J inside the body Ω to the magnetic field B is the Biot-Savart law

$$(9.1) \qquad B(x) = \frac{\mu_0}{4\pi} \int_\Omega J(y) \times \frac{x - y}{|x - y|^3} dy .$$

Here, $J = J^i + \sigma E$ with J^i the impressed source current and σE the Ohmic current. Since σE depends on J^i, (9.1) is a nonlinear equation for J^i.

In general the relation abetween J^i and σE is quite complicated and requires extensive computations [18]. If the body is modeled as a spherically symmetric conductor, this relation can be made explicit. In that case, (9.1) reduces to the linear integral equation

$$(9.2) \quad B(x) = \frac{\mu_0}{4\pi} \int_\Omega \left\{ \frac{y \times Q(y)}{F(x,y)} - \frac{(y \times Q(y)) \cdot x}{F^2(x,y)} \nabla_x F(x,y) \right\} dy ,$$

$$F(x,y) = |x - y|(|x - y||x| + (x - y) \cdot x)$$

for the dipol moment Q. Obviously, (9.2) is not uniquely solvable for Q. However it makes sense to compute the generalized solution (4.2), which is called lead field solution in the biomagnetism literature.

Besides non-uniqueness, the main problem in magnetic source imaging is again instability. One tries to overcome this difficulties either by assuming only few (1-3) magnetic dipoles or by strong regularization in a continuous dipole distribution.

10. Electric source imaging. Here one wants to find the electric potential u generated by electric sources inside the body (typically in the heart) on a surface Γ_1 ("epicardial") close to the sources from the potential on the surface Γ_0 of the body Ω [24]. Thus we have

$$\Delta u = 0 \quad \text{between} \quad \Gamma_1 \text{ and } \Gamma_0$$

(10.1)
$$\frac{\partial u}{\partial \nu} = 0 \quad \text{on } \Gamma_0$$

The problem is to recover u on Γ_1 from the values of u on a part of Γ_0. The degree of ill-posedness of this inverse problem depends on how close Γ_1 is to the sources. In an idealized setting this can be analysed quantitatively [30]. Let Γ_1, Γ_0 be concentric circles with radius $r_1 < r_0$, respectively, and let all the sources lie inside a third concentric circle with radius $r_2 < r_1$. Then, the reconstruction error $\varepsilon(\delta)$ on Γ_1 for a data error δ on Γ_0 is

(10.2)
$$\varepsilon(\delta) = M_2^{1-c}\delta^c \,,$$

$$c = \frac{\log(r_1/r_2)}{\log(r_0/r_2)} \quad , \quad M_2 = \underset{|x| = r_2}{\text{Max}} \ |u(x)| \,.$$

In the terminology of ill-posed problems the problem is only modestly ill-posed if c is not too small, i.e. if the sources are not too close to Γ_1.

Unfortunately, in practice Γ_1 is always very close to the sources, i.e. c is small. Thus the inverse problem (10.1) behaves like a severely ill-posed problem, and regularization is mandatory. Let $A : L_2(\Gamma_1) \to L_2(\Gamma_0)$ be the forward operator, which associates with each potential f on Γ_1 the function $g = u$ on Γ_0 where u is the solution of (10.1) with $u = f$ on Γ_1. Then, the Tikhonov-Phillips regularization is obtained by minimizing

$$\|Af - g\|_{L_2(\Gamma_0)}^2 + \alpha^2\|f - f_0\|_{L_2(\Gamma_1)}^2 \,.$$

A possible choice for f_0 is $f_0 = 0$. However, since measurements are done for many time instants, one can also use an approximation to f computed from earlier timeframes.

REFERENCES

[1] Aben, H.K.: Integrated Photoelasticity, *Valgus*, Talin 1975 (Russian).

[2] Anikonov, D.S. - Prokhorov, I.V. - Kovtanyuk, E.E.: Investigation of Scattering and Absorbing Media by the Methods of X-ray Tomography, *J. Inv. Ill-Posed Problems*, **1**, 259-281, (1993).

[3] Anikonov, D.S.: Uniqueness of Simultaneous Determination of two Coefficients of the Transport Equation, *Soviet Math. Dokl.* **30**, 149-151, (1984).

[4] Arridge, S.R. - Van der Zee, P. - Cope, M. - Delpy, D.T.: Reconstruction Methods for Infrared Absorption Imaging, *Poc. SPIE* **1431**, 204-215, (1991).

[5] Barber, D.C.: Electrical Impedance Tomography, in: *The Biomedical Engineering Handbook, Bronzino, J. D. (ed.)*, CRC Press, Boca Raton, Flo., 1995.

[6] Bondarenko, A.- Antyufeev, V.: X-Ray Tomography in Scattering Media, *Institute of Mathematics*, Novosibirsk, Russia (1990).

[7] Borup, D.T. - Johnson, S.A. - Kim, W.W. - Berggren, M.J.: *Nonperturbative diffraction tomography via Gauss-Newton iteration applied to the scattering integral equation*, Ultrasonic Imaging **14**, 69-85 (1992).

[8] Bronnikov, A.V.: Degradation Transform in Tomography, *Pattern Recognition Letters* **15**, 527-592, (1994).

[9] Burridge, R.: The Gelfand-Levitan, the Marchenko, and the Gopinath-Sondhi Integral Equations of Inverse Scattering Theory, Regarded in the Context of Inverse Impulse-Response Problems, *Wave Motion*, **2**, 305-323, (1980).

[10] Censor, Y.: Finite Series-Expansion Reconstruction Methods, *Proc. IEEE* **71**, 409-419, (1983).

[11] Chen, Y.: *Inverse scattering via Heisenberg's uncertainty principle*, Inverse Problems **13**, 253-282 (1997).

[12] Ell, P.J. - Holman, B.L. (eds.): Computed Emission Tomography. Oxford University Press (1982).

[13] Geman, S. - McClure, D.: Statistical Methods for Tomographic Image Reconstruction, ISI Tokio session, *Bull. Int. Statist. Inst.*, **LII(4)**, 5-21, 1987.

[14] Grangeat, P.: Mathematical Framework of Cone Beam 3D Reconstruction via the First Derivative of the Radon Transform, in: Herman, G.T. et al. (eds.): Mathematical Methods in Tomography, Springer (1991).

[15] Gratton, E. et al.: A novel approach to laser tomography, *Bioimaging*, **1**, 40-46 (1993).

[16] Green, P.J.: Bayesian Reconstructions from Emission Tomography Data Using a Modified EM Algorithm, *IEEE Transactions on Medical Imaging*, **9**(1), 84-93, März 1990.

[17] Gutman, S. - Klibanov, M.V.: Regularized Quasi-Newton Method for Inverse Scattering Problems, *Mathl. Comput. Modelling* **18**, No. 1, 5-31, Pergamon Press Ltd. (1993).

[18] Hämäläinen, M.S. - Sarvas, J.: Realistic Conductivity Model of the Human Head for Interpretation of Neuromagnetic Data, *IEEE Trans. Biomed. Eng.* **36**, 165-171, (1989).

[19] Herman, G.T. - Lent, A.: Iterative Reconstruction Algorithms, *Comput. Biol. Med.* **6**, 273-294, (1976).

[20] Herman, G.: Image Reconstruction From Projections. The Fundamentals of Computerized Tomography. Academic Press 1980.

[21] Hertle, A.: The Identification Problem for the Constantly Attenuation Radon Transform, *Math. Z.* **197**, 13-9, (1988). Problems **9**, 579-617 (1993).

[22] Hinshaw, W.S. - Lent, A.H.: *An Introduction to NMR Imaging: From the Bloch Equation to the Imaging Equation*, Proc. IEEE **71**, 338-350 (1983).

[23] Isakov, V.: *Uniqueness and Stability in Multi-Dimensional Inverse Problems*, Inverse Problems **9**, 579-617 (1993).

[24] Johnson, C.R. - MacLeod, R.S.: Inverse Solutions for Electric and Potential Field Imaging, in: Barbour, R.L. - Carolin (eds.): *Physiological Imaging, Spec-*

troscopy, and *Early Detection Diagnostic Methods 1987*, 130–139 (1993).

[25] Juhlin, P.: Doppler Tomography, *Proc. IEEE Eng. Med. Biol.*, San Diego, CA, October 28–81, (1993).

[26] Kak, A.C. - Slaney, M.: Principle of Computerized Tomography Imaging. IEEE Press 1987.

[27] Kleinman, R.E. - van den Berg, P.M.: A Modified Gradient Method for Two-Dimensional Problems in Tomography, *J. Comp. Appl. Math.*, **42**, 17–35, (1992).

[28] Krestel, E. (ed.): Imaging Systems for Medical Diagnostics. Siemens Aktiengesellschaft (1990).

[29] Louis, A.K.: *Medical Imaging: State of the Art and Future Development*, Inverse Problem **8**, 709–738 (1992).

[30] Lavrentiev, M.M. - Romanov, V.G. - Shishatskij, S.P.: Ill-Posed Problems of Mathematical Physics and Analysis, *Translation of Mathematical Monographs* **64**, AMS 1986.

[31] Liang, Z.-P. - Boada, F.E. - Constable, R.T. - Haacke, E.M. - Lauterbur, P.C. - Smith, M.R.: Constrained Reconstruction Methods in MR Imaging, Rev Magn. Reson. Med. **4**, 67–185, (1992).

[32] Natterer, F.: The Mathematics of Computerized Tomography. Wiley-Teubner 1986.

[33] Natterer, F.: Determination of Tissue Attenuation in Emission Tomography of Optically Dense Media, *Inverse Problems* **9**, 731–736 (1993).

[34] Natterer, F. - Wübbeling, F.: *A propagation - backpropagation algorithm for ultrasound tomography*, Inverse Problems **11**, 1225–1232 (1995).

[35] Palamodov, V.: An Inversion Method for Attenuated X-Ray Transform in Space, submitted to *SIAM J. Appl. Math.*.

[36] Romanov, V.G.: Conditional Stability Estimates for the Problem of Recovering of Absorption Coefficients and Right Hand Side of Transport Equations (Russian), to appear in *Siberia Math. J.*

[37] Sanchez, R. - McCormick, N.J.: General Solutions to Inverse Transport Problems, *J. Math. Phys.* **22**, 847–855, (1981).

[38] Setzepfandt, B.: ESNM: Ein rauschunterdrückendes EM - Verfahren für die Emissionstomographie. Thesis, Fachbereich Mathematik der Universität Münster, Germany 1992.

[39] Sharafutdinov, V. A.: Integral Geometry of Tensor Fields. Nauka, Novosibirsk (1993).

[40] Smith, K.T. - Solmon, D.C. - Wagner, S.L.: *Practical and Mathematical Aspects of the Problem of Reconstructing Objects From Radiographs*, Bull AMS **83**, 1227–1270 (1977).

[41] Singer, J.R. - Grünbaum, F.A. - Kohn, P. - Zubelli, J.R.: *Image Reconstruction of the Interior of Bodies that Diffuse Radiation*, Science **248**, 990-993 (1990).

[42] Shepp, L.A. - Vardi, Y.: *Maximum Likelihood Reconstruction for Emission Tomography*, IEEE Trans. Med. Im. **1**, 115-122 (1982).

[43] Somersalo, E. - Cheney, M. - Isaacson, D. - Isaacson, E.: Layer Stripping: A Direct Numerical Method for Impedance Imaging, *Inverse Problems* **7**, 899-926 (1991).

[44] Sparr, G. - Stråklén, K. - Lindström, K. - Persson, W.: *Doppler tomography for vector fields*, Inverse Problems **11**, 1051-1061 (1995).

[45] Tretiak, O.J. - Metz, C.: The Exponential Radon Transform, *SIAM J. Appl. Math.*, **39**, 341-354, (1980).

[46] Tuy, H.K.: An Inversion Formula for Cone-Beam Reconstruction. *SIAM J. Appl. Math.* **43**, 546-552 (1983).

TOMOGRAPHY THROUGH THE TRANSPORT EQUATION

D.S. ANIKONOV*, A.E. KOVTANYUK*†, AND I.V. PROKHOROV*‡

Introduction. This paper is a written version of Anikonov's report at Workshop 6, "Computational Radiology and Imaging: Therapy and Diagnostics", March 17-21, 1997, IMA, Minneapolis. The contents of the report is the following.

1. Mathematical models of passing radiation through a medium. Setting of tomography problems.

2. Methods for suppression of influence of particles scattering in a medium.

3. Construction of the heterogeneity indicator for radiography.

4. Comparison of two mathematical models in tomography.

5. Testing of the algorithms.

1. Mathematical models of passing radiation through medium. The transport equation is the mathematical model created specially for process of passing radiation through a medium. In the steady-state case it has a form:

$$\omega \cdot \nabla_r f(r, \omega, E) + \mu(r, E) f(r, \omega, E) =$$

$$(1.1) \qquad \frac{1}{4\pi} \int_{E_1}^{E_2} \int_{\Omega} k(r, \omega, \omega', E, E') f(r, \omega', E') d\omega' dE' + J(r, \omega, E)$$

The variable $r \in G$, G is a bounded convex domain in 3-dimensional space R^3, vectors $\omega \in \Omega$, (Ω is an unit sphere in R^3), numeral variable $E \in [E_1, E_2]$. The function $f(r, \omega, E)$ is interpreted as the density of the particles flux at the point r in direction ω with energy E. The function $\mu(r, E)$ is the coefficient of full attenuation of the radiation and $k(r, \omega, \omega', E, E')$ is the indicatrix of scattering. Also $\mu(r, E)$ is in proportion to the density of the probability of a particle interaction within medium G at point $r \in G$ for a particle with energy E. Often $\mu(r, E)$ is called the

---

*Institute of Applied Mathematics, FEB Russian Academy of Science, Radio 7, Vladivostok, 690041, Russia; E-mail: anik@ipm.marine.su

† E-mail: ankov@ipm.marine.su

‡ E-mail: prh@ipm.marine.su

macro cross-section of all interactions. The value $1/\mu(r, E)$ is equal to one mean free path. The function $k(r, \omega, \omega', E, E')$ is the macro cross-section of scattering at point $r \in G$ and is in proportion to the probability density of the transformation of a particle from state (ω', E') to (ω, E). Finally $J(r, \omega, E)$ is the density of radiation due to internal sources.

The equation (1.1) describes α, β, γ rays, neutron motion and photon migration. The concrete case of radiation is provided by a choice of coefficient classes.

The important partial case of (1.1) is its monoenergetic variant:

$$(1.2) \quad \omega \cdot \nabla_r f(r, \omega) + \mu(r) f(r, \omega) = \frac{1}{4\pi} \int_\Omega k(r, \omega, \omega') f(r, \omega') d\omega' + J(r, \omega)$$

Equation (1.1) transforms into (1.2) if the indicatrix depends on an energy by the factor $\delta(E - E')$ where δ is Dirac's function. This indicatrix corresponds to Rayleigh scattering. And studying of equation (1.2) is a preliminary stage of research for other kinds of scattering, for instance, for Compton's scattering and appearance of pairs: electron, positron.

Besides (1.1) and (1.2) there exist more simple cases: sphere symmetrical, cylinder symmetrical and plane parallel cases in which the function f and the coefficients depend on less number of variables. For example in sphere symmetrical case f depends only on $|r|$ and one angle between vectors r and ω.

In addition we note the transport equation in diffuse approximation. It corresponds to the isotropic case: $k = k(r)$, $J = J(r)$. If the coefficients $\mu(r)$ and $k(r)$ are almost equal and $\mu(r) > k(r)$ the transport equation has the following form in the diffuse version:

$$(1.3) \qquad div\left(\frac{1}{3\mu(r)}\nabla\varphi(r)\right) - (\mu(r) - k(r))\,\varphi(r) + J(r) = 0$$

where

$$\varphi(r) = \frac{1}{4\pi}\int_\Omega f(r, \omega)d\omega$$

The equation (1.3) may be deduced from (1.2) by using P_1-approximation which is the roughest of all P_N-approximations for a solution of equation (1.2).

We research equation (1.2) mainly. Let us denote $d(r, \omega)$ $(r, \omega) \in \overline{G} \times \Omega$ the length of the intersection of a ray $L_{r,\omega} = \{r + t\omega, t \geq 0\}$ and set $\overline{G}$. It is evident that for any $(r, \omega) \in \overline{G} \times \Omega$ the points $\xi = r - d(r, -\omega)\omega$ and $\eta = r + d(r, \omega)\omega$ belong to $\partial G = \overline{G} - G$.

The set of pairs (ξ, ω) is denoted as Γ^- and the set of pairs (η, ω) is designated as Γ^+. It is clear that $\Gamma^\pm \subset \partial G \times \Omega$.

We impose the boundary condition for equation (1.2):

$$(1.4) \qquad\qquad f(\xi, \omega) = h(\xi, \omega), \quad (\xi, \omega) \in \Gamma^-,$$

where the function $h(\xi, \omega)$ is the density of the input flux at the boundary of G. Besides that, we consider function $H(\eta, \omega)$, $H(\eta, \omega) = f(\eta, \omega)$, $(\eta, \omega) \in \Gamma^+$ which is the density of the output flux on ∂G.

The following partition

$$\overline{G} = \bigcup_{i=1}^{p} \overline{G}, \quad G_i \bigcap G_j = \emptyset, \quad i \neq j, \quad i,j = 1, ..., p, \quad G_o = \bigcup_{i=1}^{p} G_i$$

is used where the boundaries ∂G_i of domains G_i are assumed to be piecewise smooth. The nonnegative coefficients μ, k, J are smooth within each G_i and may be discontinuous on the surface $\partial G_o = \bigcup_{i=1}^{p} \partial G_i$. Also we call the domains G_i as zones, inclusions or heterogeneities.

It is well known that the problem of determination of function $f(r, \omega)$ from equation (1.2) and boundary condition (1.4) with known μ, k, J, h has the unique solution, if

$$\frac{1}{4\pi} \int_{\Omega} k(r, \omega, \omega') d\omega' \leq \mu(r), \quad (r, \omega) \in G \times \Omega.$$

This problem is usually called the forward problem. Its solution is a sum of Heumann series in which n-th term corresponds to the density of particles after n scattering acts. The quantity

$$S(r) = (\mu(r))^{-1} \left(\frac{1}{4\pi} \int_{\Omega} k(r, \omega, \omega') d\omega' \right)$$

means the ratio of the number of scattering particles at the point r to the number of all interacting particles at the same point r. For example if $S(r) = 0.9$ then scattering constitutes 90% of all interactions at the point r.

Problem 1. *It is required to determine the coefficients μ, k, J from equation (1.2) and boundary condition (1.4) if only the functions $h(\xi, \omega)$ and $H(\eta, \omega)$ are known.*

From physical point of view this problem is directed to determination of the internal structure of unknown medium G when the intensity on the boundary of G is known. It gives us rights to call the problem 1 a X-tomography problem.

Problem 2. *It is required to determine the surface ∂G_o of discontinuity of coefficients μ, k, J from equation (1.2) and boundary condition (1.4) if the only function $H(\eta, \omega)$, $(\eta, \omega) \in \Gamma^+$ is known.*

This problem differs from the problem 1 because it requires less known data and less information is sought for. The problem 2 like the problem 1 may be considered as a X-tomography problem, because the surface ∂G_o is interpreted as the boundaries between various materials in medium G.

2. Methods for suppression of influence of particle scattering in a medium. More than ten years ago the author of this report found that using the external sources with singularities might suppress the influence of scattering and internal sources on output intensity $H(\eta, \omega)$ [3]. Now we have justified three versions of research on this way.

a). Let vector ω be characterized by spherical angles: the latitude θ, $0 \le \theta \le \pi$ and the longitude γ, $0 \le \gamma < 2\pi$, $\omega = \omega(\theta, \gamma)$, $\omega(\theta, \gamma) = (\sin\theta\cos\gamma, \sin\theta\sin\gamma, \cos\theta)$. So the angle $\theta = \pi/2$ corresponds to the horizontal plane. Hereafter we designate $\omega_o = \omega(\pi/2, \gamma)$.

It was proved: if

$$(2.1) \qquad \lim_{\theta \to \frac{\pi}{2}} \left| \frac{\partial}{\partial \theta} h(\xi, \omega(\theta, \gamma)) \right| = \infty, \ (\xi, \omega) \in \Gamma^-$$

then

$$\int_0^{d(\eta, -\omega_o)} \mu(\eta - \tau\omega_o) d\tau = \ln \lim_{\theta \to \frac{\pi}{2}} \frac{\frac{\partial}{\partial \theta} h(\eta - d(\eta, -\omega(\theta, \gamma))\omega(\theta, \gamma), \omega(\theta, \gamma))}{\frac{\partial}{\partial \theta} H(\eta, \omega(\theta, \gamma))},$$

$$(2.2) \qquad\qquad\qquad (\eta, \omega) \in \Gamma^+$$

As the integrals in the left side of (2.2) are known then it is possible to determine function $\mu(r)$ using the inverse Radon transformation.

After determination $\mu(r)$ other coefficients may be determined also, but it requires additional assumptions. For instance let $k = k(r)$, $J = J(r)$ and surfaces ∂G_i, $i = 1, ..., p$ be strictly convex. In this case $k(r)$ and $J(r)$ for $r \in \partial G$ can be determined. Also we may determine $k(r)$ and $J(r)$ for all $r \in G$ by using two irradiations of medium G with the essentially different output densities $h_1(\xi, \omega)$ and $h_2(\xi, \omega)$ [1,2].

b). Let $h(\xi, \omega)$ be discontinuous at $\omega_o = \omega(\pi/2, \gamma)$ [4,10]. The value of the jump of $h(\xi, \omega(\theta, \gamma))$ is designated as $[h(\xi, \omega_o)]$. More exactly

$$[h(\xi, \omega_o)] = \lim_{\epsilon \to 0} (h(\xi, \omega(\pi/2 + \epsilon, \gamma)) - h(\xi, \omega(\pi/2 - \epsilon, \gamma)))$$

If

$$(2.3) \qquad\qquad [h(\xi, \omega_o)] \ne 0, \ (\xi, \omega_o) \in \Gamma^-$$

then

$$\int_0^{d(\eta, -\omega_o)} \mu(\eta - \tau\omega_o) d\tau = \ln \frac{[h(\eta - d(\eta, -\omega_o)\omega_o, \omega_o)]}{[H(\eta, \omega_o)]},$$

$$(2.4) \qquad\qquad\qquad (\eta, \omega_o) \in \Gamma^+$$

c). We assume the function h besides ξ, ω also depends on a parameter $\lambda \in [0, \lambda_o]$ which signifies multiple irradiation of medium G by external sources with density $h(\xi, \omega, \lambda)$ [4,11].

If

(2.5)
$$\lim_{\lambda \to 0} \left| \frac{\partial}{\partial \lambda} h(\xi, \omega_o, \lambda) \right| = \infty, \quad (\xi, \omega_o) \in \Gamma^-$$

then

$$\int\limits_0^{d(\eta, -\omega_o)} \mu(\eta - \tau \omega_o) d\tau = \ln \lim_{\lambda \to 0} \frac{\frac{\partial}{\partial \lambda} h(\eta - d(\eta, -\omega_o)\omega_o, \omega_o, \lambda)}{\frac{\partial}{\partial \lambda} H(\eta, \omega_o, \lambda)},$$

(2.6)
$$(\eta, \omega_o) \in \Gamma^+$$

The further research of items b) and c) is similar to one in item a).

The external sources satisfying to conditions (2.1), (2.3) and (2.5) have been founded. To do that it is enough to put obstacles of the specific properties on the way of an arbitrarily radiation flux [5].

3. Construction of the heterogeneity indicator under medium radiography. For simplicity we consider the cases in which $f(r, \omega)$ is Holder continuous for each $(r, \omega) \in G \times \Omega$. We intend to define the integro-differential operator on the output desity [6-8]

(3.1)
$$V(r) = \left| \nabla_r \int\limits_\Omega f(r + d(r, \omega)\omega, \omega) d\omega \right|, \quad r \in G_o$$

Function $V(r)$ has been defined at all points $r \in G_o$. It is continuous at each $r \in G_o$. Formerly analogous integro-differential operators but only applied to Radon transformations were considered at [14-16].

Let $r_o \in \partial G_o$ and r_o be a common point of ∂G_i and ∂G_j, $i < j$. The surfaces ∂G_i and ∂G_j are supposed to be smooth at r_o. We designate the jump of an arbitraraly function $\chi(r, \omega)$, $(r, \omega) \in \overline{G} \times \Omega$ by the following formula:

$$[\chi(r_o, \omega)] = \lim_{r' \to r_o} \chi(r', \omega) - \lim_{r'' \to r_o} \chi(r'', \omega), \quad r' \in G_i, r'' \in G_j$$

Let us introduce

$$m(r_o, \omega) = (-[\mu(r_o)]f(r_o, \omega) + [N(r_o, \omega)] + [J(r_o, \omega)])$$

(3.2)
$$\exp(-\int\limits_0^{d(r_o, \omega)} \mu(r_o + \tau\omega) d\tau)$$

where $N(r,\omega)$ is the integral of collision

$$N(r,\omega) = \frac{1}{4\pi} \int\limits_{\Omega} k(r,\omega,\omega')f(r,\omega')d\omega'$$

Due to equation (1.2) $m(r_o,\omega)$ can be presented in the form

(3.3) $$m(r_o,\omega) = [\omega \cdot \nabla_r f(r_o,\omega)] \exp(- \int\limits_{0}^{d(r_o,\omega)} \mu(r_o + \tau\omega)d\tau).$$

We denote P_{r_o} the tangent plane to surfaces ∂G_i and ∂G_j at point $r_o \in \partial G_i \cap \partial G_j$; $\Omega_{r_o} = \{\omega :| \omega - r_o |= 1, \omega \in P_{r_o}\}$ and

(3.4) $$M(r_o) = \int\limits_{\Omega_{r_o}} m(r_o,\omega)d\omega$$

The next formula has been proved [8]

(3.5) $$V(r) = 2|M(r_o)|| \ln |r - r_o|| + O(1)$$

where $r \in G_i$ or $r \in G_j$, r_o is the point from $\partial G_i \cap \partial G_j$ nearest to r and $O(1)$ means a bounded function. The next statement follows from equality (3.5).

Theorem 3.1 *Function $V(r) \to \infty$ when $r \to r_o \in \partial G_i \cap \partial G_j$ if $M(r_o) \neq 0$.*

This fact allows us to name function $V(r)$ as the indicator of heterogeneities, because $V(r) = \infty$ only on ∂G_o. The uniqueness theorem for problem 2 easly follows from theorem 3.1 and a constructive method is given for determination of surface ∂G_o.

We remark the condition $M(r_o) \neq 0$ is satisfied if $m(r_o,\omega) \neq 0$, $r_o \in \partial G_o$, $\omega \in \Omega$. For instance it is valid when either $[\mu(r_o)] > 0$ and $[k(r_o,\omega, \omega')] < 0$, $[J(r_o,\omega)] \leq 0$ or $[\mu(r_o)] < 0$ and $[k(r_o,\omega,\omega')] > 0$, $[J(r_o,\omega)] \geq 0$. The next statement has been proved as an invertion of the previous theorem.

Theorem 3.2 *Let coefficients μ, k and J be piecewise constant and surface ∂G_o is smooth. If $m(r,\omega) = 0$ for each $(r,\omega) \in \overline{G} \times \Omega$ then problem 2 has infinite set of solutions.*

Remark 1. *Condition $m(r,\omega) = 0$ is an essential requirement only for $r = r_o$, $r_o \in \partial G_o$ because for all the rest $r(r \in G_o)$ equality $m(r,\omega) = 0$ always takes place.*

Remark 2. *The problem of determination of ∂G_o is not to have the unique solution in the case $m(r,\omega) = 0$ even if function $h(\xi,\omega)$ is known in addition to $H(\eta,\omega)$.*

Remark 3. *It is not difficult to give examples where condition $m(r, \omega) = 0$, $(r,\omega) \in \overline{G} \times \Omega$ is satisfied [8].*

The proved theorems give us some justification to name $|m(r_o, \omega)|$ as the measure of visibility of a heterogeneity at point r_o in direction ω. Analogously we name inequality $|m(r_o, \omega)| > 0$, $r_o \in \partial G_o$, $\omega \in \Omega$ as the condition of distinguishability of ∂G_o at point r_o and direction ω. We have the grounds to consider that the media in wich $m(r, \omega) = 0$ for all $(r, \omega) \in G \times \Omega$ are invisible for tomography [13].

4. Comparison of two mathematical models in tomography. Here we compare the problem (1.2), (1.4) and a boundary problem for the equation corresponding to diffuse approximation

$$(4.1) \qquad div\left(\frac{1}{3\mu(r)}\nabla\varphi(r)\right) - (\mu(r) - k(r))\varphi(r) + J(r) = 0$$

where $\varphi(r)$ is interpreted as the density of radiation in point $r \in G$.

One of common assumptions which is being added to (4.1) is the requirement of a sewing together of normal derivatives of $\varphi(r)$ on surface ∂G_o inside G. To satisfy that it is required $\varphi(r) \in C^1(\overline{G}_i)$, $i = 1, ..., p$ i.e. $\varphi(r)$ has continuous derivatives bounded up to ∂G_i. Probably this assumption appears from needs of mathematical apparatus traditionally used in research of similar problems. At least using of the same requirement is essential for the proofs.

The density of radiation at point $r \in G$ corresponding to the transport equation is given by the equality

$$(4.2) \qquad \varphi(r) = \frac{1}{4\pi}\int_\Omega f(r, \omega)d\omega,$$

where $f(r, \omega)$ is the solution of problem (1.2), (1.4).

The next formula has been proved for the function $\varphi(r)$, defined in (4.2) [9]

$$(4.3) \qquad |\nabla\varphi(r)| = |\ln|r - r_o|||\widetilde{M}(r_o)| + O(1),$$

$$\widetilde{M}(r_o) = \int_{\Omega_{r_o}}[\omega \cdot \nabla_r f(r_o, \omega)]d\omega.$$

Here $r_o \in \partial G_o$, $r \in G_o$, r_o is the nearest point from ∂G_o to point r.

Since $|\ln|r - r_o|| \to \infty$ when $r \to r_o$, function $\varphi(r)$ has bounded derivatives in G_i if and only if $\widetilde{M}(r_o) = 0$, $r_o \in \partial G_o$. The equality $\widetilde{M}(r_o) = 0$ has to be interpreted as an additional special condition connecting together functions μ, k, J, h. The character of this connection is not quite clear still, but there are invisible media under radiography among the cases satisfying to equality $\widetilde{M}(r_o) = 0$ (see theorem 3.2).

Thus a contradiction appears between two mathematical models of the same natural process. In the case when this contradiction is absent we have a chance to consider the medium invisible for radiography from the transport equation description. As far as the transport equation is more exact model than the diffuse approximation, the last would be improved. For the time being we can suggest only some preliminary variants of the compromise [9].

5. Testing of the algorithms. For preparating tests we calculate the forwards problem (1.2),(1.4) taking any known functions μ, k, J and h. Our purpose is to receive the values of the output density $H(\eta, \omega)$, $(\eta, \omega) \in \Gamma^+$. Then using our algorithms we are to reconstruct function μ or surface ∂G_o.

We use Monte-Carlo method for calculation $H(\eta, \omega)$ which is presented as a sum of Neumann series. The main difficulty of the tests is calculation of $H(\eta, \omega)$. It usually takes us 20 hours or more with using personal computer "Pentium-100". Nevertheless the exactness of calculation $H(\eta, \omega)$ has not been quite enough. Of course time of calculation is essentially less and exactness is much higher in the spherical symmetry case.

In all tests the domain G is the unit ball. Hearafter we use the following designation: q - amount of trajectories used in Monte-Carlo method for calculation of one value $H(\eta, \omega)$, n - amount of the terms of Neumann series taken for calculation, $S(r)$ - the ratio (in per cents) of amount of scattering particles to amount of all particles interacting in the same point, A - the maximum number of the next property: the density of the flux passed throgh the medium decreases in A times.

Test 1. (Figure 1.) Reconstruction of the function $\mu(r)$ by using formula (2.4).

Let inclusions G_i be the following

$$G_1 = \{r = (r_1, r_2, r_3) : c_1(r) = \left(\frac{r_1}{0,2}\right)^2 + \left(\frac{r_2 + 0,2}{0,2}\right)^2 + \left(\frac{r_3}{0,2}\right)^2 < 1\}$$

$$G_2 = \{r = (r_1, r_2, r_3) : c_2(r) = \left(\frac{r_1}{0,8}\right)^2 + \left(\frac{r_2}{0,4}\right)^2 + \left(\frac{r_3}{0,4}\right)^2 < 1, c_1(r) > 1\}$$

$$G_3 = \{r = (r_1, r_2, r_3) : c_3(r) = r_1^2 + r_2^2 + r_3^2 < 1, c_2(r) > 1\}$$

$$\mu(r) = k(r) = \begin{cases} 2 \cdot c_1(r), & r \in G_1 \\ 2 \cdot c_2(r), & r \in G_2 \\ 2\left(1 - 0,75\frac{r_1^2}{c_3(r)}\right), & r \in G_3 \end{cases}$$

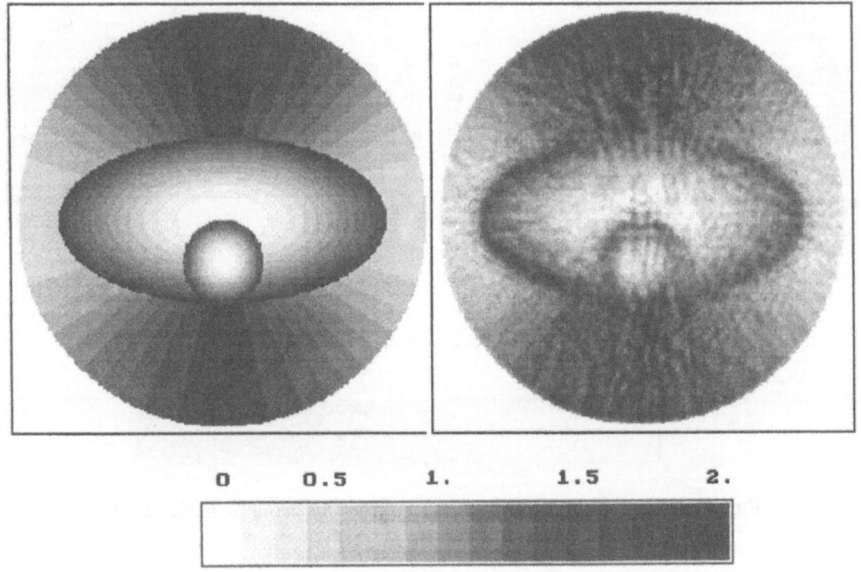

$$0 \qquad 0.5 \qquad 1. \qquad 1.5 \qquad 2.$$

FIG. 1. *Reconstruction of the function $\mu(r)$ by using formula (2.4).*

$$q = 2000, \ n = 10, \ S = 100\%, \ A \simeq 50$$

$$h(\eta, \omega, r) = \begin{cases} 1, & \theta \leq \frac{\pi}{2} \\ 0,5, & \theta > \frac{\pi}{2} \end{cases}$$

Test 2. (Figure 2.) Reconstruction of the function $\mu(r)$ by using formula (2.6).

$$G_1 = \{r = (r_1, r_2, r_3) : c_1(r) = \left(\frac{r_1}{0,1}\right)^2 + \left(\frac{r_2}{0,1}\right)^2 + \left(\frac{r_3}{0,1}\right)^2 < 1\}$$

$$G_2 = \{r = (r_1, r_2, r_3) : c_2(r) = \left(\frac{r_1 - 0.5}{0,2}\right)^2 + \left(\frac{r_2}{0,2}\right)^2 + \left(\frac{r_3}{0,2}\right)^2 < 1\}$$

$$G_3 = \{r = (r_1, r_2, r_3) : c_3(r) = \left(\frac{r_1}{0,9}\right)^2 + \left(\frac{r_2}{0,9}\right)^2 + \left(\frac{r_3}{0,9}\right)^2 < 1, |r| < 1\}$$

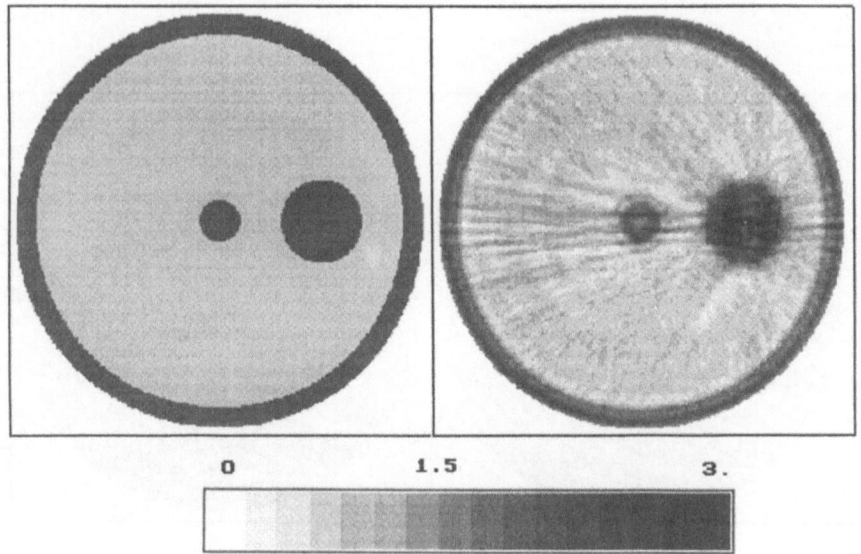

FIG. 2. *Reconstruction of the function $\mu(r)$ by using formula (2.6).*

$$G_4\{r = (r_1, r_2, r_3) : |r| < 1, r \notin \overline{G}_1 \cup \overline{G}_2 \cup \overline{G}_3\}$$

$$\mu(r) = \begin{cases} 3, & r \in G_1 \cup G_2 \cup G_3 \\ 1, & r \in G_4 \end{cases}$$

$$k(r) = \begin{cases} 2.7, & r \in G_1 \cup G_2 \cup G_3 \\ 1, & r \in G_4 \end{cases}$$

$$h(\eta, \omega, \lambda) = \begin{cases} 0, & \lambda + |\omega_3| \le 0 \\ (\lambda + |\omega_3|)^{0,01}, & \lambda + |\omega_3| > 0 \end{cases}$$

$$q = 2000, \, n = 10, \, S = 90\%, A \simeq 50.$$

Test 3. (Figure 3.) Reconstruction of the indicator $V(r)$ by using formula (3.1).

$$G_1 = \{r = (r_1, r_2, r_3) : r_1^2 + r_2^2 + r_3^3 < (0,2)^2\}$$

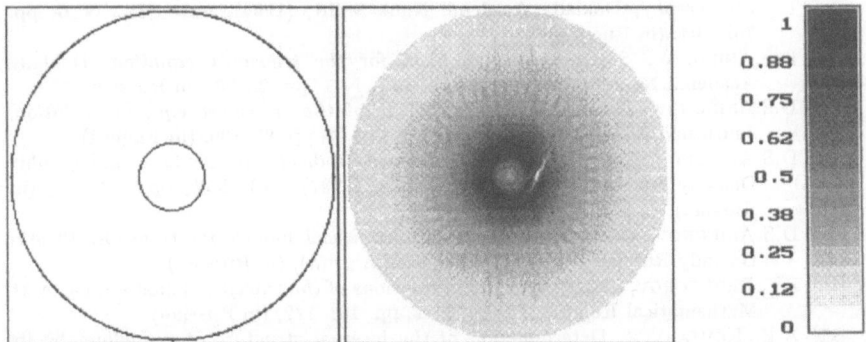

FIG. 3. *Reconstruction of the indicator V(r) by using formula (3.1).*

$$G_2 = G - \overline{G}_1$$

$$\mu(r) = \begin{cases} 10, & r \in G_1 \\ 11, & r \in G_2 \end{cases}$$

$$k(r) = \begin{cases} 10, & r \in G_1 \\ 9, & r \in G_2 \end{cases}$$

$$h(\eta, \omega) = 1, \; q = 20000, \; n = 80, \; S \sim 90\% \; A \simeq 500000000$$

The results of the numerical experiments are shown on the figures. As a rule more dark calours correspond to more large values of reconstructed quantities.

REFERENCES

[1] D.S.ANIKONOV, *Many-dimensional inverse problems for the transport theory*, Differential equations, (1984), **vol. 20, N 5**, pp. 817–824, (in Russian).

[2] D.S.ANIKONOV, *Uniqueness of the joint determination of two coefficients of the transport equation*, Doklady Academii Nauk SSSR, (1984), **vol. 277, N 4**, pp. 777–780, (in Russian).

[3] D.S.ANIKONOV, *Uniqueness of the determination of a coefficient of the transport equation for special type of sources*, Sov. Math. Dokl., (1985), **vol. 32**, pp. 511–515, (in English).

[4] D.S.ANIKONOV, I.V.PROKHOROV, AND A.E.KOVTANYUK, *Investigation of scattering and absorbing media by the methods of X-ray tomography*, Journ. Inv. Ill-Posed Problems, (1993), **vol. 1**, pp. 259–281, (in English).

[5] D.S.ANIKONOV, *Using singularities of the transport equation solution in X-tomography*, Doklady Academii Nauk SSSR, (1994), **vol. 335, N 6**, pp. 702–704, (in Russian).

[6] D.S.ANIKONOV, *Stephan type problem for the transport equation*, Doklady Academii Nauk SSSR (1994), **vol. 338, N 1**, pp. 25–28, (in Russian).

[7] D.S.ANIKONOV, *Formula for the gradient of the transport equation solution*, Journ.Inv.Ill-Posed Problems, (1996), **vol. 4**, pp. 85–100, (in English).

[8] D.S.ANIKONOV, *Construction of heterogeneity indicator for medium radiography*, Doklady Russian Academy of Science, (1997), **vol. 357**, pp. 324–327, (in Russian).

[9] D.S.ANIKONOV, *Comparison of two mathematical models for transport theory*, Doklady Russian Academy of Science,(in print), (in Russian).

[10] I.V. PROKHOROV, *Some properties of solutions of the transport equation*, Far East Mathematical Reports, (1996), **N 2**, pp. 161–172, (in Russian).

[11] A.E. KOVTANYUK, *Determination of the internal structure of a medium by its multiple irradiation*, Far East Mathematical Reports, (1995), **N 1**, pp. 101–118, (in Russian).

[12] D.S.ANIKONOV, *Integro-differential operator in tomography problem*, Journ. Inv. Ill-Posed Problems, (in print), (in English).

[13] D.S.ANIKONOV, V.G.NAZAROV AND I.V.PROKHOROV, *Visible and invisible media in tomography*, Doklady Russian Academy of Science, (1997), **vol. 1, N 5**, pp. 599–603, (in Russian).

[14] E.I.VAYNBERG, I.A.KAZAK AND M.L.FANGOYZ, *X-ray computer tomography through method of inverse projection with a double differentiation filtering*, Defectoscopy, (1985), **N 2**, pp. 31–39, (in Russian).

[15] F.NATTERER, *The mathematics of computerized tomography*, (1986), (in English).

[16] A.I.KATSEVICH AND A.G.RAMM, *A method for finding discontinuities of functions from the tomographic data*, Lectures in Applied Mathematics, (1993), **vol. 30**, pp. 115–123.

A GENERAL FRAMEWORK FOR ITERATIVE RECONSTRUCTION ALGORITHMS IN OPTICAL TOMOGRAPHY, USING A FINITE ELEMENT METHOD

SIMON R. ARRIDGE* AND MARTIN SCHWEIGER†

Abstract. In this paper we present several schemes for solving the inverse problem in Optical Tomography. We first set the context of Optical Tomography and discuss alternative photon transport models and measurement schemes. We develop the inverse problem as the optimisation of an objective functions and develop three classes of algorithms fors its solution : Newton methods, linearised methods, and gradient methods. We concentrate on the use numerical methods based on Finite Elements, and discuss how efficient methods may be developed using adjoint solutions. A taxonomy of algorithms is given, with an analysis of their spatial and temporal complexity.

Key words. Optical Tomography, Diffusion, Inverse Problems, Finite Elements.

1. Introduction. By Optical Tomography we mean the methodology of using light in a narrow wavelength band in the near-infrared ($\sim$700nm-1000nm), to *transilluminate* tissue, and to use the resulting measurements of intensity on the tissue boundary to reconstruct a map of the optical properties within the tissue. This quite complex field is relatively new, yet has attracted considerable interest from theoreticians, experimental scientists, and clinicians. Recent developments can be found in several review articles in the recent Royal Society meeting [29], and other journals [14, 3].

In this paper we present several schemes for solving the inverse problem in Optical Tomography. We make the assumption that the forward problem is governed by a diffusion equation, and that a Finite Element Method (FEM) is used for its solution. Then classical optimisation algorithms may be developed to determine the mesh parameters that have maximum posterior probability given the data. Our aim herein is to analyse the efficiency and complexity of different methods.

The paper is organized as follows. In §2 we give a short summary of the physical and clinical context for this subject. In §3 we define the inverse problem as an optimisation problem. In §4 we define the governing differential equation of the forward model, and in §5 summarize the FEM approach for its solution. In §6 we develop three classes of algorithm for the inverse problem : Newton methods, linearised methods, and gradient methods; these are all developed in the frequency-domain version of the problem. One of the most intriguing aspects of Optical Tomography is the rich variety of measurement types that can be employed, including steady-state, intensity modulated, and time-resolved instruments, as well as other more exotic measurements that are generated in post-processing of

---

*Department of Computer Science, University College London, WC1E 6BT, UK.

†Department of Medical Physics, University College London, WC1E 6JA, UK. The work of the second author was supported by Action Research Grant A/P/0503.

the acquired data. Therefore in §7 we develop the treatment of §6 for other measurement domains that are of current interest. In §8 we summarize our treatment and in §9 present conclusions.

2. Background. Optical Tomography is potentially a powerful tool to noninvasively obtain spatially resolved data of the optical parameters of tissue, from which physiologically relevant information such as local oxygenation can be calculated. Primary applications of this new imaging modality are the monitoring of cerebral blood and tissue oxygenation of newborn and preterm infants [13, 38] to prevent death or permanent brain damage caused by asphyxiation during birth, functional mapping of brain activation during physical or mental exercise [36], and imaging of the breast to detect tumours [15].

The principal problem that arises in Optical Tomography is the dominance of scattering, which causes light to propagate diffusely in tissue and thus prohibits the application of direct reconstruction methods using the Radon transform. Various reconstruction methods, including *ad hoc* backprojection [11, 37] and semi-analytic [26] schemes have been proposed, but increasingly attention is turning to iterative, model-based reconstruction methods [6, 18, 30].

Data acquisition for Optical Tomography is not limited to steady-state attenuation measurements. Time-of-flight systems using an ultra-short pulsed laser as a light source and time-resolved detectors allow the boundary measurement of the temporal intensity response function with a resolution of a few picoseconds [12]. Frequency domain systems use a radio frequency modulated light source and measure the phase shift and modulation amplitude of the transmitted light [10]. While the data acquisition equipment for the steady-state, temporal and frequency domain are quite different, the methods are related from the theoretical viewpoint : the time-of-flight method includes the steady-state case via the integral over time of the detected light, and is related to the frequency domain case via its Fourier transform. In this paper we formally develop the problem in the Fourier domain, and then consider the temporal domain.

3. Variational approach to the inverse problem. Optical Tomography is an example of an optimisation problem over two independent spaces, $(X^{(\mu)}, X^{(\kappa)})$, illustrated in Figure 1, where the absorption parameter μ and diffusion parameter κ arise from the underlying forward model, discussed below, in §4.

Let Ω be the domain under consideration, with surface $\partial\Omega$. We consider S source positions $\vec{\zeta}_j \in \partial\Omega$ $(j = 1...S)$ and M_j measurement positions $\vec{\xi}_{j,i} \in \partial\Omega$ for the jth source $(i = 1...M_j)$, resulting in a total number of measurements $M_{TOT} = \sum_{j=1}^{S} M_j$, with $M_{UNIQ} \leq M_{TOT}$ distinct measurement positions. The forward problem is non-linear and is represented by :

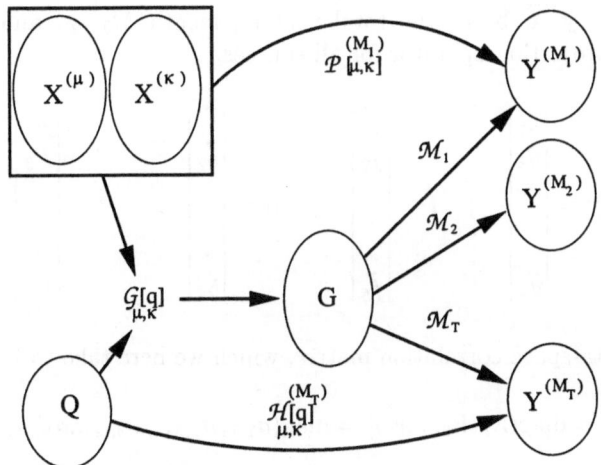

FIG. 1. *Spaces and operators used in Optical Tomography.* $(X^{(\mu)}, X^{(\kappa)})$ *are the solution spaces,* Q *is the space of sources,* G *is the space of solutions to the governing equation,* $Y^{\mathcal{M}_d}$ *are the data spaces.* $\mathcal{G}$ *is the Green's function operating on a source, and* $\mathcal{M}_d$ *are the measurement operators, operating on the solutions to the governing equation to give the data.* $\mathcal{P}$ *is the forward operator that maps the solution directly to the data. For completeness, the operator* $\mathcal{H}^{\mathcal{M}_d}$ *is defined as the Dirichlet-to-Neumann map for data type* $\mathcal{M}_d$.

$$(3.1) \qquad \vec{y} = \vec{\mathcal{P}}[\mu, \kappa]$$

Under the usual assumptions of a system corrupted by multivariate Gaussian noise, and a maximum-likelihood approach to the solution, we may define the inverse problem as the optimisation of an objective function

$$(3.2) \qquad \Psi = \frac{1}{2} \sum_{j=1}^{S} \sum_{i=1}^{M_j} \left(\frac{y_{j,i} - \mathcal{P}_{j,i}[\mu, \kappa]}{\sigma_{j,i}} \right)^2 = \sum_{j=1}^{S} \psi_j$$

which we write in vector form as

$$(3.3) \qquad \Psi = \frac{1}{2} \left(\vec{y} - \vec{f} \right)^{\mathrm{T}} R^{-2} \left(\vec{y} - \vec{f} \right) = \frac{1}{2} \vec{b}^{\mathrm{T}} \vec{b}$$

where $y_{j,i}$ is the data for the ith measurement from source j with standard deviation $\sigma_{j,i}$, $f_{j,i}$ is the modelled data for this source-detector pair, $\mathcal{P}_{j,i}[\mu, \kappa]$ is the scalar operator mapping $(X^{(\mu)}, X^{(\kappa)})$ to $f_{j,i}$ at position $\vec{\xi}_{j,i}$, and $b_{j,i} = \sigma_{j,i}^{-1}(y_{j,i} - f_{j,i})$ is the *residual data* for this measurement. We use subscripted vectors $\vec{y}_j, \vec{f}_j, \vec{\sigma}_j$ to represent all the relevant quantities for a single source j. Thus $\mathcal{P}_j[\mu, \kappa]$ is the *projection operator* $\mathcal{P}_j : (X^{(\mu)}, X^{(\kappa)}) \to$ Y for source j, $\vec{f}_j$ is the *projection data* obtained by sampling $\mathcal{P}_j$ at the measurement positions $\left\{ \vec{\xi}_{j,1}, \vec{\xi}_{j,2} \ldots \vec{\xi}_{j,M_j} \right\}$, and $\vec{b}_j = R_j^{-1} \left(\vec{y}_j - \vec{f}_j \right)$.

We use $\vec{\mathbf{y}}$, $\vec{\mathbf{f}}$, $\vec{\mathbf{b}}$, as the total vectors, length M_{TOT}, and $\vec{\mathcal{P}}$ as the combined projection operator for all sources.

$$(3.4) \qquad \vec{\mathbf{y}} = \begin{vmatrix} \vec{y}_1 \\ \vec{y}_2 \\ \cdot \\ \cdot \\ \cdot \\ \vec{y}_S \end{vmatrix} \;\; ; \;\; \vec{\mathbf{f}} = \begin{vmatrix} \vec{f}_1 \\ \vec{f}_2 \\ \cdot \\ \cdot \\ \cdot \\ \vec{f}_S \end{vmatrix} \;\; ; \;\; \vec{\mathbf{b}} = \begin{vmatrix} \vec{b}_1 \\ \vec{b}_2 \\ \cdot \\ \cdot \\ \cdot \\ \vec{b}_S \end{vmatrix} \;\; ; \;\; \vec{\mathcal{P}} = \begin{vmatrix} \mathcal{P}_1 \\ \mathcal{P}_2 \\ \cdot \\ \cdot \\ \cdot \\ \mathcal{P}_S \end{vmatrix}$$

R is the data-space correlation matrix, which we here take to be

$$(3.5) \qquad \mathrm{R} = \mathrm{diag}(\mathrm{R}_1, \mathrm{R}_2, ... \mathrm{R}_S) = \mathrm{diag}(\sigma_{1,1}, \sigma_{1,2}, ...\sigma_{j,i}, ...\sigma_{S,M_s})$$

We will assume that $(X^{(\mu)}, X^{(\kappa)}) = (L^{\infty}(\Omega), L^{\infty}(\Omega))$, which restricts μ and κ to be within strictly positive lower and upper bounds. In Transport theory G is taken to be $L^1(\Omega)$, but in the theory of elliptic PDEs it is usually considered to be a Sobolev space $H^1(\Omega)$ with $Q = H^{\frac{1}{2}}(\Omega)$ and $Y = H^{-\frac{1}{2}}(\Omega)$. Note that we use the term "Dirichlet-to-Neumann" operator for the mapping from Q to Y because of the large number of results available in the literature on PDEs [17], although this is certainly a misnomer since it more typical to use interior sources and Robin boundary conditions for this problem. The nature of the spaces resulting from the different measurement operators used in Optical Tomography has to the best of our knowledge not been established and is an important topic for further research.

We assume that $\mu(\vec{r})$ and $\kappa(\vec{r})$ are expressed in a basis :

$$(3.6) \qquad \kappa(\vec{r}) = \sum_{k=1}^{N_\kappa} \kappa_k v_k^{(\kappa)}(\vec{r})$$

$$(3.7) \qquad \mu(\vec{r}) = \sum_{k=1}^{N_\mu} \mu_k v_k^{(\mu)}(\vec{r})$$

REMARK 2.1. In principle different bases $\{v_k^{(\kappa)}(\vec{r}), v_k^{(\mu)}(\vec{r})\}$ can be used for κ and μ, or different limits to the order of summation (N_κ, N_μ). This is potentially useful since it will appear that the obtainable resolution for κ is much lower than for μ.

We further assume that the Fréchet derivatives of $\mathcal{P}$ exist, $\mathcal{P}'_j$: $(X^{(\mu)}, X^{(\kappa)}) \to Y$, and $\mathcal{P}''_j$: $(X^{(\mu)}, X^{(\kappa)}) \times (X^{(\mu)}, X^{(\kappa)}) \to Y$, and that these are representable in the basis as the Jacobian and Hessian matrices respectively.

To solve the optimisation problem, we consider the *gradient* of the objective function which in component form is $\Psi'_{x,k} = \frac{\partial \Psi}{\partial x_k}$ where x is either μ or κ. We can now write :

$$(3.8) \qquad \frac{\partial \Psi}{\partial x_k} = \sum_{j=1}^{S} \sum_{i=1}^{M_j} \left(\frac{y_{j,i} - \mathcal{P}_{j,i}[\mu, \kappa]}{\sigma_{j,i}^2} \right) \left(-\frac{\partial \mathcal{P}_{j,i}[\mu, \kappa]}{\partial x_k} \right)$$

$$(3.9) \qquad = -\vec{\mathcal{P}'}^{\mathrm{T}} \mathsf{R}^{-2} (\vec{y} - \vec{f})$$

$$(3.10) \qquad = -\vec{\mathcal{P}'}^{\mathrm{T}} \mathsf{R}^{-1} \vec{b} = \vec{z}$$

where $\vec{z}$ is a $(N_\mu + N_\kappa)$ vector representing the gradient of the objective function in the chosen basis.

Two classes of optimisation scheme will be discussed

1. Newton methods. These methods seek a zero of Ψ' by an iterative method, using a Taylor expansion around the current estimate $\vec{x}^{(n)}$

$$(3.11) \qquad \Psi'(\vec{x}^{(n+1)}) = \Psi'(\vec{x}^{(n)}) + \Psi''(\vec{x}^{(n)}) \Delta \vec{x}^{(n)} = 0$$

$$(3.12) \qquad \Rightarrow \Delta \vec{x}^{(n)} = -\left(\Psi''(\vec{x}^{(n)}) \right)^{-1} \Psi'(\vec{x}^{(n)})$$

These methods imply an explicit linearisation of the inverse problem and become computationally intractable when the number of dimensions of the inverse search space is large.

2. Gradient methods. These methods require only knowledge of the gradient $\vec{z}$, and are generally preferable for an optimisation problem over a large numbers of dimensions

4. Forward model. Light transport in biological tissue in the near-infrared (NIR) wavelength range is dominated by scattering, and the un-scattered component at a penetration depth of several centimetres is negligible.

We assume here that the propagation model is the diffusion approximation to the radiative transfer equation which in the frequency domain is :

$$(4.1) \qquad -\nabla \cdot \kappa(\vec{r}) \nabla \hat{\Phi}(\vec{r}, \omega) + \mu(\vec{r}) \hat{\Phi}(\vec{r}, \omega) + \frac{\imath \omega}{c} \hat{\Phi}(\vec{r}, \omega) = \hat{q}_0(\vec{r}, t),$$

and in the time domain :

$$(4.2) \qquad -\nabla \cdot \kappa(\vec{r}) \nabla \Phi(\vec{r}, t) + \mu(\vec{r}) \Phi(\vec{r}, t) + \frac{1}{c} \frac{\partial \Phi(\vec{r}, t)}{\partial t} = q_0(\vec{r}, t),$$

where Φ is the isotropic photon density, q_0 is an isotropic source distribution and c is the speed of light in the medium. For a detailed discussion of the physical and mathematical models employed in this field, refer to the recent review articles by Hebden and Arridge [14, 3]. The model is characterised by the two spatially varying functions $\mu(\vec{r})$ (absorption)[1] and $\kappa(\vec{r})$

---

[1]In the optics literature the absorption parameter is usually denoted μ_a, with the scattering parameter denoted μ_s', and the diffusion parameter given by $\frac{1}{3(\mu_a + \mu_s')}$. We will use μ to denote μ_a throughout, to avoid a proliferation of subscripts.

(diffusion), which gives rise to the dual search-space nature of the optimization problem defined in 3.2.

For both 4.1 and 4.2 the boundary measurement $\Gamma(\xi)$ at $\xi \in \partial\Omega$ is related to $\Phi(\vec{r})$ by

$$(4.3) \qquad \Gamma(\xi) = -c\kappa(\xi)\vec{\hat{n}} \cdot \nabla\Phi(\xi),$$

where $\vec{\hat{n}}$ is the outer normal of $\partial\Omega$ at ξ. We use the Robin-type boundary condition

$$(4.4) \qquad \Phi(\xi) + \frac{\kappa}{\alpha}\vec{\hat{n}} \cdot \nabla\Phi(\xi) = 0,$$

where α is a term to incorporate boundary reflections as a result of a refractive index mismatch at $\partial\Omega$ [22, 35]. A collimated source incident at $\zeta \in \partial\Omega$ is commonly represented by a diffuse point source $q_0(\vec{r}) = \delta(\vec{r} - \vec{r}_s)$ where $\vec{r}_s$ is located at a depth of one scattering length below the surface.

5. Finite element approach. Although the above approach is general, and can be applied to analytical forward models, the most successful approaches use a numerical approach to solve the Forward problem. Here we assume it is an FEM approach.

The domain Ω is divided into P elements, joined at D vertex nodes. The solution Φ is approximated by the piecewise linear function $\Phi^h(\vec{r}) = \sum_i^D \Phi_i u_i(\vec{r}) \in \mho^h$, where $\mho^h$ is a finite dimensional subspace spanned by basis functions $u_i(\vec{r})$, $i = 1 \ldots D$ chosen to have limited support. The problem of solving for Φ^h becomes one of sparse matrix inversion for which standard methods such as Cholesky decomposition or conjugate gradient solvers are readily available. The advantage of the FEM approach is its versatility which makes it applicable to complex geometries and highly inhomogeneous parameter distributions.

As developed in [7, 35], the diffusion equation in the FEM framework is expressed, in the frequency domain as :

$$(5.1) \qquad (K(\kappa) + C(\mu) + \alpha A + \imath\omega B)\,\vec{\Phi}(\omega) = \vec{Q}(\omega)$$

and in the time-domain as :

$$(5.2) \qquad (K(\kappa) + C(\mu) + \alpha A)\,\vec{\Phi}(t) + B\frac{\partial\vec{\Phi}(t)}{\partial t} = \vec{Q}(t)$$

where the *system matrices* K, C, A and B have entries given by :

$$(5.3) \qquad K_{ij} = \int_\Omega \kappa(\vec{r})\nabla u_i(\vec{r}) \cdot \nabla u_j(\vec{r})d^n\vec{r}$$

$$(5.4) \qquad C_{ij} = \int_\Omega \mu(\vec{r})u_i(\vec{r})u_j(\vec{r})d^n\vec{r}$$

$$(5.5) \qquad B_{ij} = \frac{1}{c}\int_\Omega u_i(\vec{r})u_j(\vec{r})d^n\vec{r}$$

$$(5.6) \qquad A_{ij} = \int_{\partial\Omega} u_i(\vec{r})u_j(\vec{r})d(\partial\Omega)$$

Using the basis expansion of $(\kappa(\vec{r}), \mu(\vec{r}))$ given in 3.6 we can further express K, C as :

$$(5.7) \qquad \mathsf{K}(\kappa) = \sum_{k}^{N_\kappa} \kappa_k \mathsf{V}_k^{(\kappa)}$$

$$(5.8) \qquad \mathsf{C}(\mu) = \sum_{k}^{N_\mu} \mu_k \mathsf{V}_k^{(\mu)}$$

Where $\mathsf{V}_k^{(\kappa)}, \mathsf{V}_k^{(\mu)}$ represent *basis system matrices* whose entries are given by :

$$(5.9) \qquad \mathsf{V}_{k,ij}^{(\kappa)} = \int_\Omega v_k^{(\kappa)}(\vec{r}) \nabla u_i(\vec{r}) \cdot \nabla u_j(\vec{r}) d^n \vec{r}$$

$$(5.10) \qquad \mathsf{V}_{k,ij}^{(\mu)} = \int_\Omega v_k^{(\mu)}(\vec{r}) u_i(\vec{r}) u_j(\vec{r}) d^n \vec{r}$$

To relate the FEM approach to the forward model, we define :

$$(5.11) \qquad \mathcal{P}_j = \mathcal{M}[\Phi_j]$$
$$(5.12) \qquad \mathcal{P}'_j = \mathcal{M}'[\Phi_j]$$

where $\vec{\Phi}_j$ is the solution to 5.1 for the jth source, and $\mathcal{M} : \mathrm{G} \to \mathrm{Y}$ is a *measurement operator*. In discrete mode, we define also $\vec{\mathcal{M}}_j(\vec{\xi}_j)$: $\mathbb{R}^D \to \mathbb{R}^{M_j}$ where $\left\{\vec{\xi}_j\right\}, j = 1 \ldots M_j$ is the set of measurement positions for source j. The equivalent definitions become :

$$(5.13) \qquad \vec{f}_j = \vec{\mathcal{M}}_j(\vec{\xi}_j)[\vec{\Phi}_j]$$
$$(5.14) \qquad \vec{f}_j' = \vec{\mathcal{M}}_j(\vec{\xi}_j)'[\vec{\Phi}_j]$$

Henceforward we will use $\mathcal{M}$ for both the operator and its discrete representation, with the distinction being clear by the parameter type.

REMARK 5.1. The simplest measurement operator to consider is the *boundary operator* $\mathcal{B}$ which implements 4.3. In the discrete case this is a $M_j \times D$ matrix. In §7 we consider other types of measurement giving rise to a multiplicity of data types.

REMARK 5.2. If $\mathcal{M}$ is linear then $\frac{\partial}{\partial x_k}$ and $\mathcal{M}$ commute leading to $\mathcal{M}'[\vec{\Phi}] = \mathcal{M}\left[\frac{\partial \vec{\Phi}}{\partial x_k}\right]$. If the measurement is non-linear, a more complex relationship develops. We return to this in §7.3.

6. Frequency-domain case. We will first consider the frequency domain, i.e. 5.1, then extend our results to the time-domain, and to simplifications of the time-domain problem through the use of integral transforms. We will introduce the problem Jacobian as a convenient tool with which to compare algorithms.

6.1. Derivation of the problem Jacobian. We define the Jacobian $J = R^{-1}\vec{f}\,'$ of the forward problem as a $M_{TOT} \times N_{TOT}$ matrix where $M_{TOT} = \sum M_j$ and $N_{TOT} = N_\mu + N_\kappa$ with the following structure

(6.1)
$$
\begin{vmatrix} \vec{b}_1 \\ \vec{b}_2 \\ \cdot \\ \cdot \\ \cdot \\ \vec{b}_S \end{vmatrix} = \begin{vmatrix} J_{1,(\mu)} & J_{1,(\kappa)} \\ J_{2,(\mu)} & J_{2,(\kappa)} \\ \cdot & \cdot \\ \cdot & \cdot \\ \cdot & \cdot \\ J_{S,(\mu)} & J_{S,(\kappa)} \end{vmatrix} \begin{vmatrix} \Delta\mu \\ \Delta\kappa \end{vmatrix}
$$

where we define J_j as the $M_j \times N_{TOT}$ matrix that is the Jacobian for the sub-objective function ψ_j and $J_{j,(\mu)}, J_{j,(\kappa)}$ as the sub-matrices corresponding to the basis coefficients with respect to the two variables μ, κ. We further establish the notation that the vector corresponding to the ith row of J_j is the *Photon Measurement Density Function* $\vec{\rho}_{(j,i)}^{\mathrm{T}}$, as defined in [2, 5] which has the structure :

(6.2)
$$
\vec{\rho}_{(j,i)}^{\mathrm{T}} = \left[\vec{\rho}_{(j,i),(\mu)}^{\mathrm{T}} \; \vec{\rho}_{(j,i),(\kappa)}^{\mathrm{T}} \right]
$$

we will also sometimes use a single index $m = \sum_{j'}^{j-1} M_{j'} + i$ for a row of J whereby $\vec{\rho}_m^{\mathrm{T}} = \vec{\rho}_{(j,i)}^{\mathrm{T}}$ and the matrix components are related by

(6.3)
$$
J_{(x)m,k} = \rho_{m,(x)_k} = J_{j,(x)i,k} = \rho_{(j,i),(x)_k}
$$

To construct J, consider the expansion of 5.7 and differentiate 5.1 with respect to each basis coefficient :

(6.4)
$$
(K(\kappa) + C(\mu) + \alpha A + \imath\omega B)\frac{\partial\vec{\Phi}(\omega)}{\partial x_k} = \frac{\partial\vec{Q}(\omega)}{\partial x_k} - V_k^{(x)}\vec{\Phi}(\omega)
$$

In the following, we will make the assumption that the source term is constant, and does not need to be modified during the optimisation process, so that the term $\frac{\partial\vec{Q}(\omega)}{\partial x_k} = 0$. Solving 6.4 and applying the discrete measurement operator $\vec{\mathcal{M}}_j(\vec{\xi}_j)$ yields two vectors $\vec{d}_{j,k}, \vec{e}_{j,k}$ of length M_j, given by :

(6.5)
$$
\vec{d}_{j,k} = R_j^{-1}\vec{\mathcal{M}}_j(\vec{\xi}_j)\left[\frac{\partial\vec{\Phi}_j(\omega)}{\partial\mu_k}\right]
$$

(6.6)
$$
\vec{e}_{j,k} = R_j^{-1}\vec{\mathcal{M}}_j(\vec{\xi}_j)\left[\frac{\partial\vec{\Phi}_j(\omega)}{\partial\kappa_k}\right]
$$

which are located into the Jacobian as in Figure 2.

We can thus state the direct method, Algorithm 1, to construct the Jacobian. The cost is $S\cdot(1+N_{TOT})$ FEM solves and $S\cdot N_{TOT}$ sparse matrix

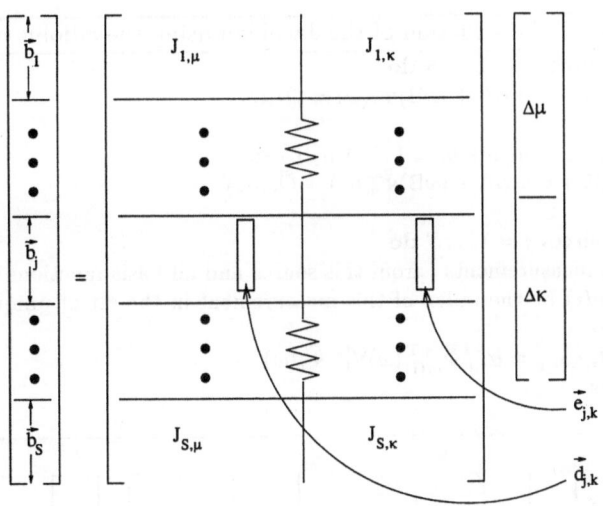

FIG. 2. *Terms in the Jacobian calculated by the Basis System Matrices*

Algorithm 1 Construction of the Jacobian using the direct method

---

for all sources $j = 1 \ldots S$ **do**
  Solve $(\mathsf{K} + \mathsf{C} + \alpha\mathsf{A} + \imath\omega\mathsf{B})\vec{\Phi}_j(\omega) = \vec{Q}_j(\omega)$
  for all basis functions $v_k^{(x)}$ **do**
    Solve $(\mathsf{K} + \mathsf{C} + \alpha\mathsf{A} + \imath\omega\mathsf{B})\frac{\partial\vec{\Phi}_j(\omega)}{\partial x_k} = -\mathsf{V}_k^{(x)}\vec{\Phi}_j(\omega)$
    apply measurement $\mathsf{R}_j^{-1}\vec{\mathcal{M}}_j(\vec{\xi}_j)\left[\frac{\partial\vec{\Phi}_j(\omega)}{\partial x_k}\right]$
  end for
end for

---

vector multiplications. In the DC case ($\omega = 0$), the system matrix $(\mathsf{K} + \mathsf{C})$ is solved using Cholesky decomposition and forward and back substitution, and in the complex case by using the Biconjugate Gradients method [33].

An alternative scheme utilizes the solution to the *Adjoint problem*

$$(6.7) \qquad (\mathsf{K} + \mathsf{C} + \alpha\mathsf{A} - \imath\omega\mathsf{B})\vec{\Phi}_m^+(\omega) = \vec{Q}_m^+(\omega)$$

where $\vec{Q}_m^+(\omega)$ is the *adjoint source* [5]. Using this solution, we have from [5] that

$$(6.8) \qquad \rho_{(j,i),(x)_k} = \sigma_{j,i}^{-1}\vec{\Phi}_{m(i)}^{+\mathrm{T}}(\omega)\mathsf{V}_k^{(x)}\vec{\Phi}_j(\omega)$$

which leads to Algorithm 2, as illustrated in Figure 3. It has the cost of $(S + M_{UNIQ})$ solutions of the forward problem and $M_{TOT} \cdot N_{TOT}$ sparse matrix quadratic forms.

6.2. Algorithms involving the Jacobian explicitly. The prototypical algorithm for Newton-methods is the Levenberg-Marquardt algo-

---

Algorithm 2 Construction of the Jacobian using the adjoint method

---

for all sources $j = 1 \ldots S$ **do**
    Solve $(\mathsf{K} + \mathsf{C} + \alpha\mathsf{A} + \imath\omega\mathsf{B})\vec{\Phi}_j(\omega) = \vec{Q}_j(\omega)$
end for
for all measurements $m = 1 \ldots M_{UNIQ}$ **do**
    Solve $(\mathsf{K} + \mathsf{C} + \alpha\mathsf{A} - \imath\omega\mathsf{B})\vec{\Phi}_m^+(\omega) = \vec{Q}_m^+(\omega)$
end for
for all sources $j = 1 \ldots S$ **do**
    for all measurements i from this source and all basis functions $v_k^{(x)}$ **do**
        let $m(i)$ be the index of this measurement in the set of unique measurements
        set $J_{j,(x)_{i,k}} = \sigma_{j,i}^{-1}\vec{\Phi}_{m(i)}^{+T}(\omega)\mathsf{V}_k^{(x)}\vec{\Phi}_j(\omega)$
    end for
end for

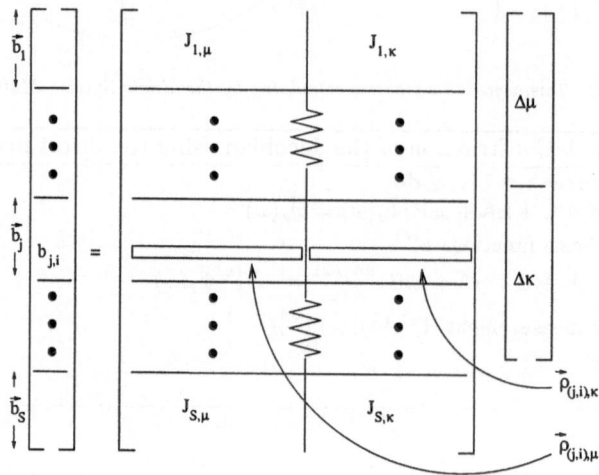

FIG. 3. *Terms in the Jacobian calculated by the Photon Measurement Density Functions*

rithm. For our purposes we state it in the form of Algorithm 3. In Optical tomography this was discussed in some detail in [34, 27].

From 3.11 we require to solve a linear step :

$$(6.9) \qquad \Delta\vec{x} = -\left(\Psi''(\vec{x})\right)^{-1}\Psi'(\vec{x})$$

Using 3.9 we have

$$(6.10) \qquad \Psi'(\vec{x}) = -\vec{\mathcal{P}'}^{T}\mathsf{R}^{-1}\vec{b} = -\mathsf{J}^{T}\vec{b}$$

$$(6.11) \qquad \Psi''(\vec{x}) = \vec{\mathcal{P}'}^{T}\mathsf{R}^{-2}\vec{\mathcal{P}'} - \mathsf{R}^{-1}\vec{\mathcal{P}''}\vec{b}$$

$$(6.12) \qquad = \mathsf{J}^{T}\mathsf{J} - \mathsf{H}\vec{b}$$

Therefore we can represent 6.9 as

$$(6.13) \qquad \Delta \vec{x} = \left(\mathsf{J}^T \mathsf{J} - \mathsf{H} \vec{b} \right)^{-1} \mathsf{J}^T \vec{b}$$

In the Levenberg-Marquardt method the Hessian H is ignored and a control parameter is introduced. In the under-determined case, the number of rows of J is less than the number of columns, i.e $M_{TOT} < N_{TOT}$ so that the $N_{TOT} \times N_{TOT}$ matrix $\mathsf{J}^T \mathsf{J}$ is certainly rank-deficient, whereas in the over-determined case $M_{TOT} > N_{TOT}$ so that $\mathsf{J}^T \mathsf{J}$ may be invertible. Nevertheless, in practical inverse problems even the over-determined case is usually illconditioned so that both cases require stabilization via the Levenberg-Marquardt method :

$$(6.14) \qquad \Delta \vec{x} = \left(\mathsf{J}^T \mathsf{J} + \lambda \mathsf{I} \right)^{-1} \mathsf{J}^T \vec{b}$$

REMARK 6.1. In Marquardt's original paper [20] it is noted to rescale J by it's column sums $J_{i,j} \to J_{i,j} / \sum_k^{M_{TOT}} J_{k,j}$ so that the scaled matrix has its greatest eigenvalue equal to 1. This gives a consistent way to derive the control parameter λ. In general we may incorporate other pre-conditioning matrices on J.

REMARK 6.2. λ exerts a tradeoff between a value of $\lambda = 0$ which is the Newton-Kantarovitch scheme, to a steepest-descent scheme as λ becomes very large.

Algorithm 3 The Levenberg-Marquardt algorithm

1: set $n = 0$, set $\lambda = \lambda_0$, set $\vec{x}^{(0)}$ to an initial state, evaluate $\Psi^{(0)}$
2: **while** end condition invalid **do**
3:      evaluate J, $\vec{b}$.
4:      solve 6.14 for $\Delta \vec{x}^{(n)}$
5:      evaluate Ψ
6:      **if** $\Psi < \Psi^{(n)}$ **then**
7:          set $n = n + 1$, set $\vec{x}^{(n+1)} = \vec{x}^{(n)} + \Delta \vec{x}^{(n)}$
8:          go to 2
9:      **else**
10:          set $\lambda = c \cdot \lambda$ where c is a constant
11:          go to 4
12:      **end if**
13: **end while**

6.3. Algorithms involving the Jacobian implicitly. The principal difficulty with the Levenberg-Marquardt algorithm is the explicit formation and inversion of the matrix J. Suppose that we rearrange 3.11 :

$$(6.15) \qquad \Psi'(\vec{x}) \quad = \quad -\Psi''(\vec{x}) \Delta \vec{x}$$

$$(6.16) \qquad \Rightarrow \quad \mathsf{J}^T \vec{b} \quad = \quad \mathsf{J}^T \mathsf{J} \Delta \vec{x}$$

$$(6.17) \qquad \Rightarrow \quad \vec{b} \quad = \quad \mathsf{J} \Delta \vec{x}$$

We obtain a linear matrix problem (which may be ill-posed). Among the methods for solving such problems, the Kacmarcz method, commonly known as ART (Algebraic Reconstruction Technique) is a popular method in Medical Imaging [24]. Using the single-indexing of the Jacobian rows as introduced in Eq. 6.3, and allowing for the well-known improvement by non-sequential ordering of rows (see for example [23]), we may state ART as Algorithm 4.

Algorithm 4 ART with row permutations

$\Delta \vec{x} = 0$
for L sweeps **do**
  for $m = 1 \ldots M_{\text{TOT}}$ **do**
    $m' =$ next index from permutation set $\pi(1, 2, \ldots, M_{\text{TOT}})$
    $\Delta \vec{x} \mathrel{+}= \lambda \vec{\rho}_{m'} \left(\frac{b_{m'} - \vec{\rho}_{m'}^{\text{T}} \Delta \vec{x}}{\|\vec{\rho}_{m'}\|^2} \right)$
  end for
end for
$\vec{x} \mathrel{+}= \Delta \vec{x}$

ART is also formulated in terms of subsets. To fix our ideas, a natural choice for the size of each subset is M_j which corresponds to considering the projection $\mathcal{P}_j$ from each source in turn. We are still free to choose the source order in an non-sequential manner, leading to the form given in Algorithm 5, where W is a $M_j \times M_j$ filter.

Algorithm 5 Block-ART

$\Delta \vec{x} = 0$
for L sweeps **do**
  for sources $j = 1 \ldots S$ **do**
    $j' =$ next index from permutation set $\pi(1, 2, \ldots, S)$
    $\Delta \vec{x} \mathrel{+}= \mathsf{J}_{j'}^{\text{T}} \mathsf{W}_{j'} \left(\vec{b}_{j'} - \mathsf{J}_{j'} \Delta \vec{x} \right)$
  end for
end for
$\vec{x} \mathrel{+}= \Delta \vec{x}$

Variant 1. In the standard Block-ART algorithm, $\mathsf{W}_{j'}$ is a diagonal matrix whose entries are the norms of the rows of $\mathsf{J}_{j'}$:

$$(6.18) \qquad W_{j',ii} = \left\| \rho_{j_i'} \right\|^{-2}$$

This is equivalent to the ART algorithm, except that $\Delta \vec{x}$ is updated only after each row of sub-block j' has been back-projected from the same value of $\vec{b}_{j'} - \mathsf{J}_{j'} \Delta \vec{x}$

Variant 2. Another possibility is to regard each sub-block $\mathsf{J}_{j'}$ as an underdetermined linear problem :

$$(6.19) \qquad \vec{b}_{j'} - \mathsf{J}_{j'} \Delta \vec{x} = \mathsf{J}_{j'} \vec{x}$$

which is solved by the under-determined version of the Moore-Penrose generalised inverse :

(6.20)
$$\vec{x} = J_{j'}^T \left(J_{j'} J_{j'}^T\right)^{-1} \left(\vec{b}_{j'} - J_{j'} \Delta\vec{x}\right)$$

which implies that

(6.21)
$$W_{j'} = \left(J_{j'} J_{j'}^T\right)^{-1}$$

For practical problems 6.21 may still need to be regularized :

(6.22)
$$W_{j'} = \left(J_{j'} J_{j'}^T + \lambda I\right)^{-1}$$

Variant 3. A suggestion by Natterer and Wübbeling for ultrasound tomography [25], is to use $W = I$, the identity matrix. This algorithm becomes a steepest descent algorithm in the *sub*-objective functions ψ_j defined in 3.2 which we will refer to as "Block Steepest-Descent". The disadvantage of the slow convergence of a steepest descent algorithm in the full objective function Ψ is offset by decomposition into subproblems, without imposing the computational burden of matrix construction or inversion stages. We will show below that this variant can be implemented very efficiently.

The advantage of ART over other algorithms for the linearised problem, such as truncated SVD [8], arises from the use of the adjoint method for the Jacobian, described in Algorithm 2. By precomputing the forward vectors $\{\vec{\Phi}_j(\omega)\}, j = 1 \ldots S$, and the adjoint vectors $\{\vec{\Phi}_m^+(\omega)\}, j = 1 \ldots M_{UNIQ}$ the vectors $\vec{\rho}_{m'}^T$ (in ART), or matrices $J_{j'}$ (in Block-ART) can be generated efficiently on the fly, thus justifying our statement that the Jacobian is only represented *implicitly*.

6.4. Algorithms involving the gradient. Whenever the number of parameters of the optimisation problem is large, explicit calculation and inversion of the linearised step is regarded as intractable. Instead, gradient methods require only the computation of the gradient $\vec{z}$ defined in 3.10, the proto-typical example being Conjugate gradients, which is stated in Algorithm 6.

The gradient $\vec{z}$ can be calculated using either of Algorithms 1 or 2 for generating the Jacobian. However an alternative is to generate it directly using the adjoint scheme.

Consider that

(6.23)
$$\vec{z} = -J^T \vec{b}$$

(6.24)
$$= -\sum_{j=1}^{S} J_j^T \vec{b}_j$$

(6.25)
$$= -\sum_{j=1}^{S} \sum_{i=1}^{M_j} \vec{\rho}_{(j,i)}^T b_{j,i}$$

But from Eqs. 6.2 and 6.8 we have that $\vec{\rho}_{(j,i)}^{\mathrm{T}}$ is a vector in two parts, whose k^{th} components are given by $\sigma_{j,i}^{-1}\vec{\Phi}_{m(i)}^{+\mathrm{T}}(\omega)V_k^{(\mu)}\vec{\Phi}_j(\omega)$ and $\sigma_{j,i}^{-1}\vec{\Phi}_{m(i)}^{+\mathrm{T}}(\omega)V_k^{(\kappa)}\vec{\Phi}_j(\omega)$ respectively. Therefore we may write

$$(6.26) \qquad z_{(x)_k} = -\sum_{j=1}^{S}\sum_{i=1}^{M_j}\sigma_{j,i}^{-1}\vec{\Phi}_{m(i)}^{+\mathrm{T}}(\omega)b_{j,i}V_k^{(x)}\vec{\Phi}_j(\omega)$$

But $\vec{\Phi}_m^+(\omega)$ is the solution to :

$$(6.27) \qquad (K + C + \alpha A - \imath\omega B)\vec{\Phi}_m^+(\omega) = \vec{Q}_m^+(\omega)$$

so we form a vector

$$(6.28) \qquad \vec{v}_j^+(\omega) = \sum_{i=1}^{M_j}\frac{b_{j,i}}{\sigma_{j,i}}\vec{Q}_{m(i)}^+(\omega)$$

and solve

$$(6.29) \qquad (K + C + \alpha A - \imath\omega B)\vec{\eta}_j^+(\omega) = \vec{v}_j^+(\omega)$$

leading to :

$$(6.30) \qquad z_{(x)_k} = -\sum_{j=1}^{S}\vec{\eta}_j^{+\mathrm{T}}(\omega)V_k^{(x)}\vec{\Phi}_j(\omega)$$

The cost is $2 \cdot S$ FEM solves, and $S \cdot N_{TOT}$ sparse matrix vector multiplies.

REMARK 6.3. This method is as example of the method of *adjoint differentiation*.

REMARK 6.4. The above scheme is particularly efficient if there are more detectors than sources, or if each detector is only used for one source (in which case $M_{UNIQ} = M_{TOT}$).

In nonlinear conjugate gradient methods a set of conjugate search directions is generated to find the minimum of the objective function. At each iteration step a one-dimensional line minimisation along the current search direction is performed. Conjugate gradient methods are well-established in nonlinear optimisation. See for example [9]. A brief outline is given in Algorithm 6.

Different choices for the weighting term β exist. Here we used the Polak-Ribière method. Note that this requires restarting of the conjugate gradient method whenever $\beta < 0$ to guarantee convergence. A variety of line search methods can be used, either utilizing gradient information, such as the secant method, or using only function evaluations such as the quadratic fit method. Often an exact line search is too computationally

---

Algorithm 6 Nonlinear conjugate-gradient method

Set search direction $\vec{d}^{\,(0)} = -\vec{z}(\vec{x}^{\,(0)})$
Set residual $\vec{p}^{\,(0)} = \vec{d}^{\,(0)}$
Define termination criterion ε
Set iteration counter $n = 0$
repeat
  Find $\gamma^{(n)}$ that minimises $\Psi(\vec{x}^{\,(n)} + \gamma^{(n)}\vec{d}^{\,(n)})$
  $\vec{x}^{\,(n+1)} = \vec{x}^{\,(n)} + \gamma^{(n)}\vec{d}^{\,(n)}$
  $\vec{p}^{\,(n+1)} = -\vec{z}(\vec{x}^{\,(n+1)})$
  $\beta^{(n+1)} = \max\left(\dfrac{\vec{p}^{\,(n+1)^{\mathrm{T}}}(\vec{p}^{\,(n+1)} - \vec{p}^{\,(n)})}{\vec{p}^{\,(n)^{\mathrm{T}}}\vec{p}^{\,(n)}}, 0\right)$
  $\vec{d}^{\,(n+1)} = \vec{p}^{\,(n+1)} + \beta^{(n+1)}\vec{d}^{\,(n)}$
  $n = n + 1$
until $\|\vec{z}(\vec{x}^{\,(n)})\| < \varepsilon$

---

expensive due to the large number of function or derivative computations. Inexact line search methods such as Armijo's Rule define the bounds for acceptable step lengths which guarantee convergence. There the line search is terminated when the step length is within the valid range.

As an alternative to Conjugate-Gradients, the Block Steepest-Descent referred to above is also very efficient. Here the gradient of each sub-objective function ψ_j is used to take a steepest descent step. This method is given in Algorithm 7.

Algorithm 7 Block Steepest-Descent Algorithm

Set $\Delta\vec{x} = 0$
for L sweeps **do**
  for $j = 1 \ldots S$ **do**
    $j' = $ next index from permutation set $\pi(1, 2, \ldots, S)$
    $\vec{b}_{j'} = \mathsf{R}_{j'}^{-1}\left(\vec{y}_{j'} - \vec{f}_{j'}\right)$
    $\vec{v}_{j'}^{+}(\omega) = \sum_{i=1}^{M_{j'}} \sigma_{j',i}^{-1} b_{j',i} \vec{Q}_{m(i)}^{+}(\omega)$
    Solve $(\mathsf{K} + \mathsf{C} + \alpha\mathsf{A} - \imath\omega\mathsf{B})\vec{\eta}_{j'}^{+}(\omega) = \vec{v}_{j'}^{+}(\omega)$
    for $k = $ basis functions **do**
      $\Delta\vec{x}\; += \; \lambda\vec{\eta}_{j}^{+\mathrm{T}}(\omega)\mathsf{V}_{k}^{(x)}\vec{\Phi}_{j}(\omega)$
    end for
  end for
end for

---

7. Time-dependent problem.

7.1. Fully time-dependent problem. A time resolved data acquisition system provides the temporally resolved flux intensity for each source-detector pair (j, i). In principle the entire data set could be used to form the objective functions and its derivatives. Some work along these lines has been done [21, 31].

In this approach, the objective function 3.2 becomes :

$$(7.1) \qquad \Psi^{temporal} = \frac{1}{2} \sum_{j=1}^{S} \sum_{i=1}^{M_j} \sum_{t=1}^{T_{j,i}} \left(\frac{y_{j,i}(t) - \mathcal{P}_{j,i}[\mu, \kappa](t)}{\sigma_{j,i}(t)} \right)^2$$

and the Jacobian $J(t)$ becomes a $M_{TOT}^T \times N_{TOT}$ matrix where

$$M_{TOT}^T = \sum_{j=1}^{S} \sum_{i=1}^{M_j} T_{j,i} \ .$$

Due to the high degree of noise in the temporal measurements, it is usually thought that many time samples $T_{j,i}$ will be required for each source-detector pair, leading to a highly over-determined problem.

The forward problem may be solved using a finite-differencing scheme in time [7] which effectively increases the time for the forward problem solution by $T_{MAX} = \frac{1}{\Delta t}(T_{final} - T_{initial})$. Here $T_{MAX} > T_{j,i}$ is dictated by stability and accuracy considerations, so that even if only a small number of time-samples are used in the problem specification 7.1, the computation time of the problem is still increased by an order of magnitude[2]. Typical values of T_{MAX} are $100 \to 1000$.

Proceeding in the same manner as 6.4, we can write :

$$(7.2) \qquad \left(K(\kappa) + C(\mu) + \alpha A + B \frac{\partial}{\partial t} \right) \frac{\partial \vec{\Phi}(t)}{\partial x_k} = \frac{\partial \vec{Q}(t)}{\partial x_k} - V_k^{(x)} \vec{\Phi}$$

giving rise to an obvious extension to Algorithm 1, with cost $T_{MAX} \cdot S \cdot (1 + N_{TOT})$ FEM solves and $T_{MAX} \cdot S \cdot N_{TOT}$ sparse matrix vector multiplications.

The adjoint method may also be used, but the product terms $\vec{\Phi}_{m(i)}^{+T}(\omega) V_k^{(x)} \vec{\Phi}_j(\omega)$ become convolutions

$$(7.3) \qquad \int_{T_{initial}}^{T_{final}} \vec{\Phi}_{m(i)}^{+T}(t - t') V_k^{(x)} \vec{\Phi}_j(t') dt'$$

where $\vec{\Phi}_m^{+}(t)$ is the solution to the adjoint problem :

$$(7.4) \qquad \left(K(\kappa) + C(\mu) + \alpha A - B \frac{\partial}{\partial t} \right) \vec{\Phi}_m^{+}(t) = \vec{Q}_m^{+}(t)$$

which prorogates backward in time from the initial condition $\vec{\Phi}_m^{+}(t) = 0$ for $t > T_{final}$. Finally the gradient can be determined in a more efficient

---

[2]It is worth noting that in Finite Differencing models where operator-splitting methods can be employed, Δt can be taken arbitrarily large without compromising stability, so that the computation time becomes reasonable.

scheme using the full mechanism of adjoint differentiation :

(7.5)
$$z_{(x)_k} = -\sum_{j=1}^{S} \int_{T_{initial}}^{T_{final}} \vec{\eta}_j^{+^{\mathrm{T}}}(t) \mathsf{V}_k^{(x)} \vec{\Phi}_j(t) dt$$

where $\vec{\eta}_j^{+}(t)$ is the solution to the adjoint problem :

(7.6)
$$\left(\mathsf{K}(\kappa) + \mathsf{C}(\mu) + \alpha\mathsf{A} - \mathsf{B}\frac{\partial}{\partial t} \right) \vec{\eta}_j^{+}(t) = \vec{\nu}_j^{+}(t)$$

where $\vec{\nu}_j^{+}(t)$ is formed from the data using

(7.7)
$$\vec{\nu}_j^{+}(t) = \sum_{i=1}^{M_j} \frac{b_{j,i}(t)}{\sigma_{j,i}(t)} \vec{Q}_{m(i)}^{+}(t)$$

Whichever solution approach is used, the problem is, at best, increased T_{MAX} in computational cost.

7.2. Problem reduction using temporal transforms. The time-domain problem may be reduced in size by applying a transformation to the data space to reduce the size of the vector $\vec{b}(t)$, and the Jacobian $J(t)$. In general any spatio-temporal operator might be used, such as the singular vectors of $J(t)$ or the singular vectors of the Dirichlet-to-Neumann operator $\mathcal{H}$, as suggested for an equivalent problem in Electrical Impedance Tomography [16]. However we will restrict ourselves to operators that can be computed in an efficient manner. Let us consider a *temporal operator* $\mathcal{T}$ with general form

(7.8)
$$\mathcal{T}[h(t)] = \int_{-\infty}^{\infty} f(t)h(t)dt$$

from which we can derive further measurement operators

(7.9)
$$\mathcal{M}_{\mathcal{T}}[\cdot] = \mathcal{M}\mathcal{T}[\cdot]$$

To simplify notation, we make the following definition

DEFINITION 7.1. *we will use the notation* $\mathcal{T}[H]$ *to mean* $\mathcal{T}[h(t)]$ *with* $H = \int_{-\infty}^{\infty}[h(t)]dt$

For general forms of the function $f(t)$, we have to solve the forward problem in the full range, then apply $\mathcal{T}$ to reduce the size of the data. For example, Klibanov [19] suggests using generalised Fourier Harmonics. Whereas this is beneficial it still means the full cost of the forward solve. We thus turn our attention to those operators that can be computed efficiently. We will restrict the discussion to functions $t^n \exp(-at)$. This contains in the general form three special cases :

1. Fourier Transform: $n = 0, a$ imaginary

2. Laplace Transform: $n = 0, a$ real

3. Mellin Transform: n integer, $a = 0$

The questions to be addressed become the formation of the Objective function, and its derivatives. We will therefore examine these.

The objective function 3.2 becomes :

$$(7.10) \qquad \Psi^{\mathcal{M}} = \frac{1}{2} \sum_{j=1}^{S} \sum_{i=1}^{M_j} \left(\frac{y_{j,i}^{\mathcal{M}} - \mathcal{P}_{j,i}^{\mathcal{M}}[\mu, \kappa]}{\sigma_{j,i}^{\mathcal{M}}} \right)^2 = \sum_{j=1}^{S} \psi_j^{\mathcal{M}}$$

Furthermore, it is useful to consider more than one data type in the objective function :

$$(7.11) \qquad \Psi = \Psi^{\mathcal{M}_1} + \Psi^{\mathcal{M}_2} + \ldots \Psi^{\mathcal{M}_T}$$

which is represented in figure 1

The Jacobian now has a more complex form

$$(7.12) \qquad \begin{vmatrix} \vec{b}^{\mathcal{M}_1} \\ \vec{b}^{\mathcal{M}_2} \\ \cdot \\ \cdot \\ \cdot \\ \vec{b}^{\mathcal{M}_T} \end{vmatrix} = \begin{vmatrix} J_\mu^{\mathcal{M}_1} & J_\kappa^{\mathcal{M}_1} \\ J_\mu^{\mathcal{M}_2} & J_\kappa^{\mathcal{M}_2} \\ \cdot & \cdot \\ \cdot & \cdot \\ J_\mu^{\mathcal{M}_T} & J_\kappa^{\mathcal{M}_T} \end{vmatrix} \begin{vmatrix} \Delta\mu \\ \Delta\kappa \end{vmatrix}$$

Since $\mathcal{T}$ commutes with both $\mathcal{B}$ and $\frac{\partial}{\partial x_k}$ we have :

$$(7.13) \qquad \frac{\partial}{\partial x_k} \mathcal{M}_{\mathcal{T}}[\Phi] = \mathcal{M}_{\mathcal{T}}\left[\frac{\partial \Phi}{\partial x_k}\right]$$

thus the relations 6.10, 6.11 can be re-expressed

$$(7.14) \quad \Psi^{\mathcal{M}_T \prime} = -\mathcal{M}_{\mathcal{T}}[\vec{\mathcal{P}}'^T] R^{\mathcal{M}_T^{-1}} \mathcal{M}_{\mathcal{T}}[\vec{b}] = -J^{\mathcal{M}_T T} \mathcal{M}_{\mathcal{T}}[\vec{b}]$$

$$(7.15) \quad \Psi^{\mathcal{M}_T \prime\prime} = \mathcal{M}_{\mathcal{T}}[\vec{\mathcal{P}}'^T] R^{\mathcal{M}_T^{-2}} \mathcal{M}_{\mathcal{T}}[\vec{\mathcal{P}}'] - R^{\mathcal{M}_T^{-1}} \mathcal{M}_{\mathcal{T}}[\vec{\mathcal{P}}''] \mathcal{M}_{\mathcal{T}}[\vec{b}]$$

We will consider how to construct the submatrices $J_x^{\mathcal{M}}$ for each measurement operator $\mathcal{M}_{\mathcal{T}}$. The Fourier case was already dealt with in §6, and the Laplace domain can formally be considered as the Fourier domain with complex frequency $\omega = -\imath s$ [32], so we will consider only the Mellin problem in detail.

7.2.1. Mellin domain. As described in detail in [4] we make use of the relation

$$(7.16) \qquad \int_{-\infty}^{\infty} t h(t) dt = \imath \left. \frac{\partial \hat{H}(\omega)}{\partial \omega} \right|_{\omega=0}$$

where $\hat{H}(\omega)$ is the Fourier transform of $h(t)$. By differentiation of 5.1 with respect to ω we get :

$$(7.17) \qquad (\mathsf{K}(\kappa) + \mathsf{C}(\mu) + \alpha\mathsf{A} + \imath\omega\mathsf{B})\frac{\partial\vec{\Phi}(\omega)}{\partial\omega} = \frac{\partial\vec{Q}(\omega)}{\partial\omega} - \imath\mathsf{B}\vec{\Phi}(\omega)$$

As in the discussion following 6.4 we make the assumption that the source is constant in frequency, which for a δ-function source in time is reasonable. Therefore, in the following, the term $\frac{\partial\vec{Q}(\omega)}{\partial\omega} = 0$. Then for any vector $\vec{h}(t)$ we have formally that

$$(7.18) \qquad \begin{aligned} \mathcal{T}_{\mathrm{Me}(1)}[\vec{H}] &= \mathcal{T}_{\mathrm{Me}(1)}[\vec{h}(t)] = \int_{-\infty}^{\infty} t\vec{h}(t)\,dt \\ &= \imath\frac{\partial\vec{H}}{\partial\omega} = (\mathsf{K} + \mathsf{C} + \alpha\mathsf{A})^{-1}\mathsf{B}\vec{H} \end{aligned}$$

To find, for example, $\frac{\partial^2\vec{\Phi}}{\partial x_k \partial\omega}$ we differentiate 7.17 to give

$$(7.19) \qquad (\mathsf{K}(\kappa) + \mathsf{C}(\mu) + \alpha\mathsf{A} + \imath\omega\mathsf{B})\frac{\partial^2\vec{\Phi}(\omega)}{\partial x_k\partial\omega} + \mathsf{V}_k^{(x)}\frac{\partial\vec{\Phi}(\omega)}{\partial\omega} = -\imath\mathsf{B}\frac{\partial\vec{\Phi}(\omega)}{\partial x_k}$$

with $\frac{\partial\vec{\Phi}(\omega)}{\partial\omega}$ found from 7.17 and $\frac{\partial\vec{\Phi}(\omega)}{\partial x_k}$ from 6.4. Thus we require $n \cdot S \cdot (1 + N_{TOT})$ FEM solves and $n \cdot S \cdot N_{TOT}$ sparse matrix vector multiplications to construct the Jacobian for the Mellin Transform of order n.

To do the adjoint way, simply differentiate $J_{j,(x)_{i,k}}(\omega)$ in frequency :

$$(7.20) \qquad \frac{\partial}{\partial\omega}J_{j,(x)_{i,k}}(\omega) = \sigma_{j,i}^{-1}\left(\frac{\partial\vec{\Phi}_{m(i)}^{+\mathrm{T}}(\omega)}{\partial\omega}\mathsf{V}_k^{(x)}\vec{\Phi}_j(\omega) + \vec{\Phi}_{m(i)}^{+\mathrm{T}}(\omega)\mathsf{V}_k^{(x)}\frac{\partial\vec{\Phi}_j(\omega)}{\partial\omega}\right)$$

The cost is $n \cdot (S + M_{UNIQ})$ FEM solves, and $n \cdot M_{TOT} \cdot N_{TOT}$ sparse matrix vector multiplies. To get the gradient we only need to construct the weighted adjoint source, as before :

$$(7.21) \qquad \vec{v}_j^{T^+} = \sum_{i=1}^{M_j}\frac{b_{j,i}^T}{\sigma_{j,i}^T}\vec{Q}_{m(i)}^+$$

and solve

$$(7.22) \qquad (\mathsf{K} + \mathsf{C} + \alpha\mathsf{A})\vec{\eta}_j^{T^+} = \vec{v}_j^{T^+}$$

leading to :

$$(7.23) \qquad z_{(x)_k}^T = -\sum_{j=1}^{S}\mathcal{T}\left[\vec{\eta}_j^{T^+}\mathsf{V}_k^{(x)}\vec{\Phi}_j\right]$$

For example, the Mellin Transform of order 1 gives :

$$(7.24) \quad z_{(x)_k}^{T_{\text{Me}(1)}} = -\sum_{j=1}^{S} \left(\frac{\partial \bar{\eta}_j^{T^+}(\omega)}{\partial \omega} V_k^{(x)} \vec{\Phi}_j(\omega) + \bar{\eta}_j^{T^+}(\omega) V_k^{(x)} \frac{\partial \vec{\Phi}_j(\omega)}{\partial \omega} \right) \Bigg|_{\omega=0}$$

and in general, the Mellin Transform of order n gives rise to a gradient

$$(7.25) \quad z_{(x)_k}^{T_{\text{Me}(n)}} = -\sum_{j=1}^{S} \sum_{n'=1}^{n} \binom{n}{n'} \frac{\partial^{n'} \bar{\eta}_j^{T^+}(\omega)}{\partial \omega^{n'}} V_k^{(x)} \frac{\partial^{n-n'} \vec{\Phi}_j(\omega)}{\partial \omega^{n-n'}} \Bigg|_{\omega=0}$$

7.3. Use of normalized measurements. A useful choice of measurement operator are those that are *self-normalized*

$$(7.26) \qquad\qquad \bar{T} = \frac{T}{\mathcal{E}}$$

where $\mathcal{E}$ is simply the integrated intensity $\mathcal{E}[h(t)] = \int_{-\infty}^{\infty} h(t)dt$, which is the same as the DC intensity. Examples that have been suggested are :

time-gated intensity:

$$(7.27) \qquad\qquad \mathcal{M}_{\bar{E}(T)}[h(t)] = \frac{1}{\mathcal{E}[h(t)]} \int_0^T h(t)dt,$$

n-th temporal moment:

$$(7.28) \qquad\qquad \mathcal{M}_{\langle t^n \rangle}[h(t)] = \frac{1}{\mathcal{E}[h(t)]} \int_0^{\infty} t^n h(t)dt,$$

n-th central moment:

$$(7.29) \qquad\qquad \mathcal{M}_{c_n}[h(t)] = \frac{1}{\mathcal{E}[h(t)]} \int_0^{\infty} (t - \langle t \rangle)^n h(t)dt,$$

normalized Laplace transform:

$$(7.30) \qquad\qquad \mathcal{M}_{\bar{L}(s)}[h(t)] = \frac{1}{\mathcal{E}[h(t)]} \int_0^{\infty} e^{-st} h(t)dt \ .$$

Additional features could be chosen, such as the logarithmic slope of the temporal decay of Γ, the peak intensity, etc. However, we aim to chose features that satisfy two conditions : robustness of the experimental measurements with respect to systematic errors such as fluctuations of the source power, detector sensitivity, or fibre coupling losses, and secondly the ability of the forward model to generate the corresponding data efficiently. Both conditions are satisfied for measurement types (7.28), (7.29) and (7.30), since these are normalized and therefore not dependent on absolute intensity measurements, and they can be calculated directly by our forward model without the need of explicitly generating the temporal profile of Γ [4]. The time-gated intensity, $\mathcal{M}_{\bar{E}(T)}$, suffers from the need to explicitly compute the temporal function.

To find the problem derivative, consider :

$$(7.31) \qquad \bar{\mathcal{M}}[\Phi] = \frac{\mathcal{B}\mathcal{T}[\Phi]}{\mathcal{B}[\Phi]}$$

$$(7.32) \qquad \frac{\partial \bar{\mathcal{M}}[\Phi]}{\partial x_k} = \frac{\mathcal{B}\mathcal{T}\left[\frac{\partial \Phi}{\partial x_k}\right]}{\mathcal{B}[\Phi]} - \frac{\mathcal{B}\mathcal{T}[\Phi]}{\mathcal{B}[\Phi]^2}\mathcal{B}\left[\frac{\partial \Phi}{\partial x_k}\right]$$

Therefore, as discussed in [5], the Jacobian for $\bar{\mathcal{M}}$ can be constructed from

$$(7.33) \qquad J_{j_i}^{\bar{\mathcal{M}}} = \frac{1}{f_{j,i}^{\mathcal{E}}}\left(J_{j_i}^{\mathcal{M}} - f_{j,i}^{\bar{\mathcal{M}}} J_{j_i}^{\mathcal{E}}\right)$$

which may be constructed in full using either the direct or adjoint methods, for the algorithms of §6.3, or constructed a row at a time using the adjoint method for the algorithms of §6.2.

The gradient needs special consideration. From 7.33 we can write

$$(7.34) \qquad \mathbf{J}^{\bar{\mathcal{M}}^{\mathsf{T}}}\vec{\mathbf{b}}^{\bar{\mathcal{M}}} = \sum_j^S \sum_i^{M_j} J_{j_i}^{\bar{\mathcal{M}}} b_{j,i}^{\bar{\mathcal{M}}}$$

$$(7.35) \qquad = \sum_j^S \sum_i^{M_j} \frac{1}{f_{j,i}^{\mathcal{E}}}\left(J_{j_i}^{\mathcal{M}} - f_{j,i}^{\bar{\mathcal{M}}} J_{j_i}^{\mathcal{E}}\right) b_{j,i}^{\bar{\mathcal{M}}}$$

$$(7.36) \qquad = \sum_j^S \sum_i^{M_j} \frac{b_{j,i}^{\bar{\mathcal{M}}}}{\sigma_{j,i}^{\bar{\mathcal{M}}} f_{j,i}^{\mathcal{E}}}\left(\mathcal{T}\left[\vec{\Phi}_{m(i)}^{+^{\mathsf{T}}} \mathsf{V}_k^{(x)} \vec{\Phi}_j\right] - f_{j,i}^{\bar{\mathcal{M}}}\vec{\Phi}_{m(i)}^{+^{\mathsf{T}}} \mathsf{V}_k^{(x)} \vec{\Phi}_j\right)$$

We therefore need to construct two adjoint sources :

$$(7.37) \qquad \vec{v}_j^{\bar{\mathcal{M}}+}(0) = \sum_{i=1}^{M_j} \frac{b_{j,i}^{\bar{\mathcal{M}}} f_{j,i}^{\bar{\mathcal{M}}}}{\sigma_{j,i}^{\bar{\mathcal{M}}} f_{j,i}^{\mathcal{E}}} \vec{Q}_{m(i)}^+$$

$$(7.38) \qquad \vec{v}_j^{\bar{\mathcal{M}}+}(1) = \sum_{i=1}^{M_j} \frac{b_{j,i}^{\bar{\mathcal{M}}}}{\sigma_{j,i}^{\bar{\mathcal{M}}} f_{j,i}^{\mathcal{E}}} \vec{Q}_{m(i)}^+$$

From here we proceed as before, finding two solutions $\vec{\eta}_j^{\bar{\mathcal{M}}+}(0), \vec{\eta}_j^{\bar{\mathcal{M}}+}(1)$ and computing the gradient as :

$$(7.39) \qquad z_{(x)_k}^{\bar{\mathcal{M}}} = -\sum_j^S \left(\mathcal{T}\left[\vec{\eta}_j^{\bar{\mathcal{M}}+^{\mathsf{T}}}(1)\mathsf{V}_k^{(x)} \vec{\Phi}_j\right] - \vec{\eta}_j^{\bar{\mathcal{M}}+^{\mathsf{T}}}(0)\mathsf{V}_k^{(x)} \vec{\Phi}_j\right)$$

This method is applicable for measurement types (7.28) and (7.30). The nth central moment given by (7.29) can be constructed as a linear combination of the first n temporal moments and so the gradient requires n adjoint sources, constructed in a straightforward manner.

8. Discussion.

8.1. Summary of computational complexity. The basic cost of the various algorithms is as given in Table 1

TABLE 1

Comparison of complexity of algorithms.

Levenberg-Marquardt - direct :
  Non-linear step $C_{Setup} + S \cdot (1 + N_{TOT}) \cdot C_{FEM} + S \cdot N_{TOT} \cdot C_{SMV}$
  Linear step $C_{MI}(N_{TOT} \times N_{TOT})$
  Storage N_{TOT}^2
Levenberg-Marquardt - adjoint :
  Non-linear step $C_{Setup} + (S + M_{UNIQ}) \cdot C_{FEM} + M_{TOT} \cdot N_{TOT} \cdot C_{SVMV}$
  Linear step $C_{MI}(N_{TOT} \times N_{TOT})$
  Storage N_{TOT}^2
ART - explicit, adjoint :
  Non-linear step $C_{Setup} + (S + M_{UNIQ}) \cdot C_{FEM} + M_{TOT} \cdot N_{TOT} \cdot C_{SVMV}$
  Linear step $L \cdot M_{TOT} \cdot (C_{SP}(N_{TOT}) + C_{VT}(N_{TOT}))$
  Storage N_{TOT}^2
ART - implicit :
  Non-linear step $C_{Setup} + (S + M_{UNIQ}) \cdot C_{FEM}$
  Linear step $L \cdot M_{TOT} \cdot (N_{TOT} \cdot C_{SVMV} + C_{SP}(N_{TOT}) + C_{VT}(N_{TOT}))$
  Storage $(S + M_{UNIQ}) \cdot N_{TOT}$
Block-ART (variant 1) - implicit :
  Non-linear step $C_{Setup} + (S + M_{UNIQ}) \cdot C_{FEM}$
  Linear step $L \cdot M_{TOT} \cdot N_{TOT} \cdot C_{SVMV} +$
    $L \cdot \sum_j^S (C_{MV}(M_j \times N_{TOT}) + C_{MT}(M_j \times N_{TOT}))$
  Storage $(S + M_{UNIQ}) \cdot N_{TOT}$
Block-ART (variant 2) - implicit :
  Non-linear step *as variant 1*
  Linear step *as variant 1* $+ L \cdot \sum_j^S C_{MI}(M_j \times M_j)$
  Storage *as variant 1*
Block Steepest-Descent :
  Non-linear step $S \cdot (C_{Setup} + 2 \cdot C_{FEM} + C_{SVMV})$
  Linear step *Not Applicable*
  Storage $2N_{TOT}$
Conjugate-Gradients :
  Non-linear step $C_{Setup} + 2S \cdot C_{FEM} + S \cdot C_{SVMV}$
  Linear step $C_{LineSearch}$
  Storage $2N_{TOT}$

Here C_{Setup} is the cost of the assembly of the systems matrices K, C, A, and B, and C_{FEM} is the cost of the solver for these matrices for one right hand side. In most cases the C_{Setup} includes the cost of a Cholesky reduction, whence C_{FEM} is a Cholesky forward and backward substitution. C_{SMV} is the cost of a sparse matrix vector multiply $V_k^{(x)} \vec{\Phi}$ representing the multiplication of a nodal solution by a basis system matrix. C_{SVMV} is the cost of a sparse matrix quadratic form $\vec{\Phi}^T V_k^{(x)} \vec{\Phi}$ representing the inner product of a nodal solution and a nodal solution multiplied by a basis system matrix. $C_{MI}(n \times n)$ is the cost of a matrix inversion of a dense $n \times n$ matrix. $C_{SP}(n)$ is the cost of a scalar product of two vectors of length n, and $C_{VT}(n)$ is the cost of transposition of a vector of length n.

$C_{MV}(n \times m)$ and $C_{MT}(n \times m)$ are the equivalent costs for a $n \times m$ matrix. Finally, $C_{LineSearch}$ is the cost of searching and evaluating the objective function, used in Conjugate Gradients.

The above analysis applies in the DC case, and is the same in the Laplace Domain. In the fully-time domain case, the costs are uniformly scaled by T_{MAX}, and in the Mellin Domain by n, the order of the transform that is used. When using normalized measures, the costs are scaled by 2 in the normalized Laplace domain, $T_{MAX} + 1$ for normalized time-gating, and $n + 1$ for the normalized Mellin case (i.e. the temporal moments). In the Fourier domain, all vectors and matrices become complex which modifies the numerical methods that can be employed. In this case Cholesky reduction is not appropriate so C_{Setup} is reduced, but C_{FEM} is more costly, utilizing for example the biconjugate gradient method. Although this is not the only possibility, there are other situations wherein Cholesky reduction is not appropriate. Notably in the fully 3D problem where the system matrix bandwidth is very large, a sparse representation and partial-Cholesky preconditioned conjugate gradients is the method of choice.

8.2. Choice of basis for (μ, κ). The cost of the matrix-vector multiplications $V_k^{(x)} \vec{\Phi}$ depends on the sparsity of the system basis matrices, which in turn depends on the compactness of the support of $\{v^{(\kappa)}(\vec{r}), v^{(\mu)}(\vec{r})\}$. Choices that have been proposed are piecewise continuous on elements [5], piecewise continuous on an independent pixel grid [34], piecewise linear on a nodal basis (i.e. $v^{(x)}(\vec{r}) = u(\vec{r})$) [27], and piecewise linear on a second mesh, with lower resolution [28]. Each of these cases is compact making the adjoint methods preferable in most applications. However it is worth noting that other choices might cause a disadvantage to the adjoint method. An obvious example is a Fourier basis which might be a natural choice for a spectral expansion approach to the inverse problem, reconstructing lower frequencies first, before focussing on higher frequencies. Since this basis is not compact the direct method of Algorithm 1 combined with the Levenberg-Marquardt algorithm may still be appropriate. A wavelet basis, offering a heirarchical scheme with individual basis functions that are orthonormal and with compact support is an interesting possibility that has so far not been developed.

8.3. Regularization. A full discussion of regularization schemes is outside the scope of this paper, but some comments are in order. Regularization consists of the modification of the objective function to a form such as

(8.1) $$\Psi + \tau \Upsilon(x)$$

where $\Upsilon(x)$ is a function of the solution only, and represents *a priori* information, and τ is a *hyper-parameter* whose value is often hard to determine. Under these circumstances the Levenberg-Marquardt step 6.14 is modified to

$$(8.2) \qquad \Delta \vec{x} = \left(J^T J + \tau \Upsilon'' + \lambda I \right)^{-1} \left(J^T \vec{b} + \tau \Upsilon' \right)$$

and the gradient in conjugate-Gradient methods is modified to

$$(8.3) \qquad \vec{z} + \tau \Upsilon'$$

These modification are independent of whichever data type(s) are being used, and so add constant computational effort to the inverse problem.

A question that is more difficult to address is the application of regularization schemes to purely linear algorithms such as ART. Here the assumption is usually that the iterative nature of ART, combined with the relaxation parameter, is sufficient regularization, but strictly, it is not possible to derive a modification to ART that guarantees minimisation of any given modified objective functions such as 8.1. This remains an open question we will address in future publications.

9. Conclusions. Optical Tomography presents many challenges of interest to theoreticians and experimentalists alike. Many different approaches are being researched, but we feel that numerical methods based on optimising the parameters of a forward model are the only viable option for general complex geometries and inhomogeneous absorption and scattering distributions. Often it is thought that such methods pose an insurmountable computational burden, especially when the fully three-dimensional and time-dependent problem is addressed. In this paper we have presented a summary of Finite Element Method model-based iterative image reconstruction algorithms, and showed how these may be implemented in an efficient manner that we believe can remove this perceived barrier.

In our opinion the key to solving the problem lies in the use of appropriate operators to extract the optimal information from the potentially vast data sets, and in the use of adjoint methods for the computation of the problem derivatives. We concentrated in this paper on purely temporal operators, that could be computed efficiently using simple integral transforms. It may be that other operators exist, whose computational cost may be slightly more, yet which lead to faster convergence of the inverse problem. This is an on-going area of research in our group. Similarly, more efficient numerical methods are being developed, and the union of developments in these topics will, we feel, eventually lead to a more widespread recognition of their appropriateness.

Acknowledgements. We would like to thank Ken Hanson of Los Alamos National Laboratory, Frank Natterer of Münster University, and Ranadhir Roy of University College London for many useful discussions concerning adjoint methods in optimisation.

REFERENCES

[1] R.R. ALFANO AND J.G. FUJIMOTO, eds., *OSA TOPS on Advances in Optical Imaging and Photon Migration*, vol. 2, OSA, 1996.

[2] S.R. ARRIDGE, *Photon measurement density functions. Part 1: Analytical forms*, Appl. Opt., 34 (1995), pp. 7395–7409.

[3] S.R. ARRIDGE AND J.C. HEBDEN, *Optical imaging in medicine: II. Modelling and reconstruction*, Phys. Med. Biol., 42 (1997), pp. 841–853.

[4] S.R. ARRIDGE AND M. SCHWEIGER, *Direct calculation of the moments of the distribution of photon time of flight in tissue with a finite-element method*, Appl. Opt., 34 (1995), pp. 2683–2687.

[5] ——, *Photon measurement density functions. Part 2: Finite element calculations*, Appl. Opt., 34 (1995), pp. 8026–8037.

[6] S.R. ARRIDGE, M. SCHWEIGER, AND D.T. DELPY, *Iterative reconstruction of near-infrared absorption images*, in Inverse Problems in Scattering and Imaging, M.A. Fiddy, ed., vol. 1767, Proc. SPIE, 1992, pp. 372–383.

[7] S.R. ARRIDGE, M. SCHWEIGER, M. HIRAOKA, AND D.T. DELPY, *A finite element approach for modeling photon transport in tissue*, Med. Phys., 20 (1993), pp. 299–309.

[8] S.R. ARRIDGE, P. VAN DER ZEE, D.T. DELPY, AND M. COPE, *Reconstruction methods for infra-red absorption imaging*, in Time-Resolved Spectroscopy and Imaging of Tissues, B. Chance and A. Katzir, eds., vol. 1431, Proc. SPIE, 1991, pp. 204–215.

[9] M.S. BAZARAA, H. D. SHERALI, AND C.M. SHETTY, *Nonlinear Programming: Theory and Algorithms*, Wiley, second ed., 1993.

[10] B. CHANCE, M. MARIS, J. SORGE, AND M.Z. ZHANG, *A phase modulation system for dual wavelength difference spectroscopy of haemoglobin deoxygenation in tissue*, in Time-Resolved Laser Spectroscopy in Biochemistry II, J.R. Lakowicz, ed., vol. 1204, Proc. SPIE, 1990, pp. 481–491.

[11] S.B. COLAK, G.W. HOOFT, D.G. PAPAIOANNOU, AND M.B. VAN DER MARK, *3D backprojection tomography for medical optical imaging*, in Alfano and Fujimoto [1], pp. 294–298.

[12] D.T. DELPY, M. COPE, P. VAN DER ZEE, S.R. ARRIDGE, S. WRAY, AND J. WYATT, *Estimation of optical pathlength through tissue from direct time of flight measurement*, Phys. Med. Biol., 33 (1988), pp. 1433–1442.

[13] A.D. EDWARDS, J.S. WYATT, C.E. RICHARDSON, D.T. DELPY, M. COPE, AND E.O.R. REYNOLDS, *Cotside measurement of cerebral blood flow in ill newborn infants by near infrared spectroscopy*, Lancet, ii (1988), pp. 770–771.

[14] J.C. HEBDEN, S.R. ARRIDGE, AND D.T. DELPY, *Optical imaging in medicine: I. Experimental techniques*, Phys. Med. Biol., 42 (1997), pp. 825–840.

[15] J.C. HEBDEN, R.A. KRUGER, AND K.S. WONG, *Time resolved imaging through a highly scattering medium*, Appl. Opt., 30 (1991), pp. 788–794.

[16] D. ISAACSON, *Distinguishability of conductivities by electric current computed tomography*, IEEE Med. Im., 5 (1986), pp. 91–95.

[17] V. ISAKOV, *Inverse Problems in Partial Differential Equations*, Springer, New York, 1998.

[18] H. JIANG, K.D. PAULSEN, U.L. OSTERBERG, B.W. POGUE, AND M.S. PATTERSON, *Optical image reconstruction using frequency-domain data: Simulations and experiments*, J. Opt. Soc. Am. A, 13 (1995), pp. 253–266.

[19] M.V. KLIBANOV, T.R. LUCAS, AND R.M. FRANK, *A fast and accurate imaging algorithm in optical diffusion tomography*, Inverse Problems, 13 (1997), pp. 1341–1361.

[20] D.W. MARQUARDT, *An algorithm for least-squares estimation of nonlinear parameters*, J. SIAM, 11 (1963), pp. 431–441.

[21] R. MODEL, M. ORLT, M. WALZEL, AND R. HÜNLICH, *Reconstruction algorithm for near-infrared imaging in turbid media by means of time-domain data*, Appl. Opt., 14 (1997), pp. 313–324.

[22] J.D. MOULTON, *Diffusion modelling of picosecond laser pulse propagation in turbid media*, M. Eng. thesis, McMaster University, Hamilton, Ontario, 1990.

[23] K. MUELLER, R. YAGEL, AND F. CORNHILL, *The weighted-distance scheme: A globally optimizing projection ordering method for ART*, IEEE Med. Im., 16 (1997), pp. 223–230.

[24] F. NATTERER, *The Mathematics of Computerised Tomography*, Wiley, New York, 1986.

[25] F. NATTERER AND F. WÜBBELING, *A propagation-backpropagation method for ultrasound tomography*, Inverse Problems, 11 (1995), pp. 1225–1232.

[26] M.A. O'LEARY, D.A. BOAS, B. CHANCE, AND A.G. YODH, *Experimental images of heterogeneous turbid media by frequency-domain diffusing-photon tomography*, Opt. Lett., 20 (1995), pp. 426–428.

[27] K.D. PAULSEN AND H. JIANG, *Spatially-varying optical property reconstruction using a finite element diffusion equation approximation*, Med. Phys., 22 (1995), pp. 691–701.

[28] K.D. PAULSEN, P.M. MEANEY, M.J. MOSKOWITZ, AND J.J.M. SULLIVAN, *A dual mesh scheme for finite element based reconstruction algorithms*, IEEE Med. Im., 14 (1995), pp. 504–514.

[29] PH. I. TRANS. ROYAL SOC. B, *Near-infrared spectroscopy and imaging of living systems*, vol. 352, 1997.

[30] B.W. POGUE, M.S. PATTERSON, H. JIANG, AND K.D. PAULSEN, *Initial assessment of a simple system for frequency domain diffuse optical tomography*, Phys. Med. Biol., 40 (1995), pp. 1709–1729.

[31] S.S. SAQUIB, K.M. HANSON, AND G.S. CUNNINGHAM, *Model-based image reconstruction from time-resolved diffusion data*, in Medical Imaging: Image Processing, K.M. Hanson, ed., Proc. SPIE, 3034 (1997), pp. 369–380.

[32] M. SCHWEIGER AND S.R. ARRIDGE, *Direct calculation of the Laplace transform of the distribution of photon time of flight in tissue with a finite-element method*, Appl. Opt., 36 (1997), pp. 9042–9049.

[33] ———, *The finite element model for the propagation of light in scattering media: Frequency domain case*, Med. Phys., 24 (1997), pp. 895–902.

[34] M. SCHWEIGER, S.R. ARRIDGE, AND D.T. DELPY, *Application of the finite-element method for the forward and inverse models in optical tomography*, J. Math. Imag. Vision, 3 (1993), pp. 263–283.

[35] M. SCHWEIGER, S.R. ARRIDGE, M. HIRAOKA, AND D.T. DELPY, *The finite element model for the propagation of light in scattering media: Boundary and source conditions*, Med. Phys., 22 (1995), pp. 1779–1792.

[36] M. TAMURA, *Multichannel near-infrared optical imaging of human brain activity*, in Advances in Optical Imaging and Photon Migration, vol. 2, Proc. OSA, Proc. OSA, 1996, pp. 8–10.

[37] S.A. WALKER, S. FANTINI, AND E. GRATTON, *Back-projection reconstructions of cylindrical inhomogeneities from frequency domain optical measurements in turbid media*, in Alfano and Fujimoto [1], pp. 137–141.

[38] J.S. WYATT, M. COPE, D.T. DELPY, C.E. RICHARDSON, A.D. EDWARDS, S.C. WRAY, AND E.O.R. REYNOLDS, *Quantitation of cerebral blood volume in newborn infants by near infrared spectroscopy*, J. Appl. Physiol., 68 (1990), pp. 1086–1091.

SCATTERED RADIATION IN NUCLEAR MEDICINE:
A CASE STUDY ON THE BOLTZMANN
TRANSPORT EQUATION

HARRISON H. BARRETT†‡*, BRANDON GALLAS*†, ERIC CLARKSON‡,
AND ANNE CLOUGH§

Abstract. The objective in nuclear medicine is to deduce properties of a gamma-ray source inside a patient's body from measurements of the gamma rays that escape the body. The gamma rays are attenuated and scattered in complicated ways before they emerge from the body, and it is essential to have a good theoretical description of these processes before attempting the inverse problem. In this paper we make use of the Boltzmann transport equation to describe the variation of gamma-ray flux with position, direction and energy. After a tutorial derivation of the Boltzmann equation, we discuss various analytic solutions, showing the relation to the attenuated x-ray transform. Numerical methods are also discussed, and the relation of the gamma-ray distribution to measured data is elucidated. Then some approaches to the inverse-source problem are described.

1. Introduction. This paper deals with gamma-ray imaging, and in particular with a clinical modality known as *nuclear medicine*. Often the goal in nuclear medicine is to estimate or reconstruct the three-dimensional (3D) distribution of the gamma-ray source. In the mathematical literature this would be called an *inverse-source* problem, and in a clinical context it is called single-photon emission computed tomography or *SPECT*.

An important maxim, honored more in the breach than the observance, is: If you want to solve an inverse problem, concentrate on the forward problem. That is, with real data as opposed to simplistic simulations, it is crucial to have an accurate theory describing the data produced by a specified source; only then is it feasible to attempt to deduce properties of the source from the data.

The data, of course, are random, but gamma-ray sources rigorously obey Poisson statistics, so all we have to compute are the mean values of each detector output. The primary mathematical tool for this purpose is the *Boltzmann transport equation*, which relates the source to a distribution function that specifies the distribution of gamma rays as a function of position, direction and energy. From the distribution function the mean data can be computed.

Section 2 of this paper explains the clinical context of the SPECT problem and the physical processes that limit its solution. Then Sec. 3 presents a tutorial derivation of the Boltzmann equation in a form applicable to nuclear medicine.

---

*Dept. of Radiology, University of Arizona, Tucson, AZ 85724.
†Program in Applied Mathematics, University of Arizona, Tucson, AZ 85721.
‡Optical Sciences Center, University of Arizona, Tucson, AZ 85721-5067.
§Dept. of Mathematics, Marquette University, Milwaukee, WI 53233.

Section 4 discusses some analytical methods for solution of the Boltzmann equation, showing the relationship to two important integral transforms, the *x-ray transform* and the *attenuated x-ray transform*. As we shall see, the mean data are not directly given by these transforms, but they are nevertheless key building blocks in an accurate formulation of the forward problem.

In practice, however, the analytical expressions derived in Sec. 4 involve many nested integrals, and various approximations or simplifications are required to obtain numerical answers. Some important numerical methods are discussed in Sec. 5. Finally, in Sec. 6, we turn to inverse problems, beginning with a general discussion of the nature of inverse problems and what information about the object we can hope to determine. Then we apply these general considerations specifically to nuclear medicine problems. We show there how the mathematics of the forward problem developed in Secs. 4 and 5 can be used to good purpose in inverse problems.

2. Background. Nuclear medicine is an important medical-imaging modality in which pharmaceuticals labelled with radioactive isotopes are used to trace physiological functions or to detect pathologies [1,2]. The earliest example is the use of iodine isotopes to study the function of the thyroid gland. The thyroid uses iodine to synthesize hormones, and radioactive iodine behaves chemically just as stable iodine does. Isotopes such as ^{123}I and ^{131}I emit gamma rays that can penetrate through tissue and escape the body, and an external gamma-ray imaging system can form an image showing which areas of the gland have taken up the tracer.

As another example, thallium ions behave chemically like potassium and are therefore pumped across the membranes of functioning muscle cells. Radioactive thallium is a valuable tracer for studying the perfusion of the heart muscle and indicating the presence of a heart attack.

Other radioactive tracers concentrate in rapidly growing tumors or sites of inflammation. Indeed, there is scarcely an organ or function within the body that cannot be studied with the techniques of nuclear medicine.

The isotopes used in nuclear medicine emit gamma rays with energies in the range 80–500 keV (kilo-electron-Volts); the most common isotope is technetium (^{99m}Tc), which emits 140 keV gamma rays. At this energy, gamma rays have a mean free path in soft tissue of about 7 cm. If they must traverse a thickness L cm of tissue before escaping the body, they have a probability of $\exp(-L/7)$ of doing so without undergoing any interactions. Even if L is 10–20 cm, therefore, a reasonable fraction of the gamma-ray photons escape without interaction and contribute to the image.

Those photons that do interact in the body, however, are not simply absorbed. Instead, the dominant interaction process is Compton scattering, where gamma rays are scattered by electrons in the tissue. The scattered photons also have a reasonable chance of escaping the body and contributing to the image, but because of the random change of direction they convey relatively little useful information about the distribution of the tracer. The result is a broad, hazy background that obscures the details of interest in the gamma-ray image.

The usual approach to dealing with scattered radiation in nuclear medicine is to try to reject scattered photons. The basis for doing so is that the photon transfers part of its energy to an electron in a Compton interaction. If we can measure the photon energy, we can therefore distinguish scattered from unscattered photons. However, there are technological difficulties in the measurement, as we shall now illustrate with a numerical example.

By conservation of energy and momentum, it can be shown [3] that the scattered photon has an energy E given by

$$(1) \qquad \frac{1}{E} = \frac{1}{E_0} + \frac{1}{mc^2}(1 - \cos\theta_s),$$

where θ_s is the scattering angle, E_0 is the energy of the incident photon, m is the rest mass of the electron and c is the speed of light. Thus mc^2 is the rest-mass energy of the electron, numerically equal to 511 keV. For example, 45^0 scattering of a 140 keV photon gives a scattered photon of energy 129.6 keV, 90^0 scattering gives 109.9 keV and 180^0 scattering gives 90.4 keV.

The imaging detectors used in nuclear medicine can estimate the energy as well as the position of the gamma rays that strike them, but the energy resolution is relatively poor [4]. If an unscattered photon with $E_0 = 140$ keV is incident on the detector, then the estimated energy $\hat{E}$ is approximately a Gaussian random variable with standard deviation around 7 keV. In practice, photons are rejected if $\hat{E}$ is less than some preset threshold E_{th}. To avoid rejecting unscattered photons, E_{th} is often set to 2 standard deviations below E_0, or about 126 keV. As shown in the last paragraph, however, a 45^0 scattered photon has a true energy of 129.6 keV, and even a 90^0 scattered photon can, with reasonable probability, have an estimated energy in excess of this threshold. Hence, the threshold is a very crude means of rejecting scatter, unless E_{th} is set so high that many unscattered photons are also rejected.

3. The Boltzmann equation.

3.1. The distribution function. The *distribution function* is denoted $w(\mathbf{r}, \hat{n}, E, t)$, where $\mathbf{r} = (x, y, z)$ is a position vector in a 3D space, $\hat{n}$ is a unit vector in that space and E is the energy of a gamma ray photon. The distribution function is defined such that $w(\mathbf{r}, \hat{n}, E, t)dV\,d\Omega dE$ is the mean number of photons contained in volume dV centered on point $\mathbf{r}$, travelling in solid angle $d\Omega$ about direction $\hat{n}$, and having energies between E and $E + dE$ at time t. Unless the arguments are needed for clarity, we shall write $w(\mathbf{r}, \hat{n}, E, t)$ as w. Units of w are m^{-3} $(\text{ster})^{-1}$ $(\text{keV})^{-1}$ if energies are expressed in keV.

The Boltzmann equation is an integro-differential equation for the time derivative $\partial w / \partial t$. This derivative has contributions from the physical processes of absorption, emission, scattering and propagation of radiation, so we have

$$(2) \qquad \frac{\partial w}{\partial t} = \left[\frac{\partial w}{\partial t}\right]_{abs} + \left[\frac{\partial w}{\partial t}\right]_{em} + \left[\frac{\partial w}{\partial t}\right]_{sc} + \left[\frac{\partial w}{\partial t}\right]_{prop},$$

where the subscripts have the obvious meanings. We shall now examine each of these terms in succession.

3.2. Absorption. Consider a small volume ΔV containing ΔN identical atoms in a medium where the distribution function is w, and consider only photons travelling in a small solid angle $\Delta \Omega$ around direction $\hat{n}$ and having energies in a narrow range $(E, E + \Delta E)$. Then the number of photons per second per unit area crossing a plane normal to $\hat{n}$, called the *photon irradiance* I_0, is given by $I_0 = cw\Delta E \Delta \Omega$. We assume that the atoms absorb radiation independently of each other and that the number of photons absorbed is linearly proportional to the number incident.

With these assumptions, the total number of photons absorbed by this group of atoms in time Δt can be written as $\sigma_{abs} I_0 \Delta N \Delta t$, where the proportionality constant σ_{abs}, which has units of area, is called the *absorption cross section*. The photon irradiance is related to w by $I_0 = cw\Delta E \Delta \Omega$, so the change in w is given by

$$(3) \qquad \Delta w = -\frac{c\Delta N \sigma_{abs} w \Delta E \Delta \Omega \Delta t}{\Delta E \Delta V \Delta \Omega} = -\frac{\Delta N c \sigma_{abs} w \Delta t}{\Delta V}.$$

Hence,

$$(4) \qquad \left[\frac{\partial w}{\partial t}\right]_{abs} = -c\mu_{abs} w,$$

where μ_{abs} is the *absorption coefficient*, defined by

$$(5) \qquad \mu_{abs} = n_{abs}\sigma_{abs},$$

and $n_{abs} = \Delta N / \Delta V$ is the number of absorbing particles per unit volume. Since n_{abs} has units of reciprocal volume and σ_{abs} is an area, μ_{abs} has units of reciprocal length. The interpretation of this length will be seen in Sec. 4.

If there are several different kinds of absorbing atoms interspersed in the same volume, then we define

$$(6) \qquad \mu_{abs} = \sum_j n_{abs,j}\sigma_{abs,j},$$

where subscript j identifies a particular species of atom. With this definition, (4) still holds if each species absorbs radiation independently of the others, a valid assumption for x-ray or gamma-ray absorption.

Note that μ_{abs} can depend on the energy of the photon, position in the medium and time, in general.

3.3. Emission. The emission of photons by a radioactive source is described by a function $f(\mathbf{r}, t)$ called the *photon emission density*, defined such that $f\Delta V$ is the mean number of photons per second emitted in volume ΔV. Units of f are therefore photons/(sec·m^3) or equivalently sec^{-1}m^{-3}. The notation is meant to suggest that f is the quantity we wish to *find* in an inverse problem.

The emission density is closely related to the *activity* of the source, which is the total number of radioactive decays per second. SI units of activity are *becquerels*, abbreviated Bq, and $1\ Bq = 1$ sec^{-1}. The number of decays per second per unit volume of the source is its *specific activity*. If every radioactive decay resulted in exactly one photon, specific activity would be identical to f, but that is not usually the case. For our example of ^{99m}Tc, f is approximately 0.9 times the specific activity.

Though $f(\mathbf{r}, t)$ describes the spatial and temporal distribution of the source, it is often useful to include a specification of its angular and spectral properties as well. We therefore define another source density $S(\mathbf{r}, \hat{\mathbf{n}}, E, t)$ such that $S\Delta V\Delta\Omega\Delta E$ is the mean number of photons per second emitted in volume ΔV into solid angle $\Delta\Omega$ and having energy in $(E, E + \Delta E)$. With this definition, S is directly $[\partial w/\partial t]_{em}$. The units of S are m^{-3}(ster)$^{-1}$(keV)$^{-1}$(sec)$^{-1}$, which is the same as w except for the (sec)$^{-1}$.

Since all radioactive sources emit isotropically, S is independent of $\hat{\mathbf{n}}$. Moreover, for many isotopes used in nuclear medicine, most or all of the gamma rays have a single energy E_0. In this case, it follows immediately from the definitions of S and f that

$$(7) \qquad \left[\frac{\partial w}{\partial t}\right]_{em} = S = \frac{f}{4\pi}\delta(E - E_0),$$

where $\delta(\cdot)$ denotes the Dirac delta function.

3.4. Scattering. Scattering has two effects on the distribution function. Consider photons in a small volume element ΔV travelling in a small solid angle $\Delta\Omega$ and having energies in a narrow range $(E, E + \Delta E)$. The mean number of photons in this group is $w\Delta\Omega\Delta E\Delta V$. Scattering processes occurring in the volume element can either increase or decrease the number of photons in this group. The decrease comes about because photons in the group can change direction, energy or both as a result of scattering. On the other hand, photons not in the group under consideration can scatter *into* the angular range $\Delta\Omega$ and the energy band ΔE.

Scattering out of the group is described by exactly the same mathematics as in the absorption case; as far as removal from the group is concerned, there is no distinction between absorption and scattering. Thus we can write at once, by analogy to (4),

$$(8) \qquad \left[\frac{\partial w}{\partial t}\right]_{out} = -c\mu_{sc}w,$$

where μ_{sc}, called the *scattering coefficient* or *linear attenuation coefficient for scattering*, is defined by

$$(9) \qquad \mu_{sc} = n_{sc}\sigma_{sc},$$

and n_{sc} is the number of scatterers per unit volume. As in the absorption term, μ_{sc} can depend on position, time and photon energy.

Scattering into the group under consideration is more complicated. It has to involve integrals over energy and direction since photons of any energy or direction can, in principle, scatter into the group. On the other hand, no integral over position or time is needed since the scattering processes occur at a definite location and definite time. Thus we are looking for an integral transform that connects $w(\mathbf{r}, \hat{\mathbf{n}}, E, t)$ to $w(\mathbf{r}, \hat{\mathbf{n}}', E', t)$ for all other $\hat{\mathbf{n}}'$ and E'. This transform is linear in w since the photons do not interact with each other, just with the scatterers in the medium.

The general form of the term that describes scattering into the group of interest is thus

$$(10) \qquad \left[\frac{\partial}{\partial t}w(\mathbf{r}, \hat{\mathbf{n}}, E, t)\right]_{in} = \int_{4\pi} d\Omega' \int_E^\infty dE' K(\hat{\mathbf{n}}, E; \hat{\mathbf{n}}', E'|\mathbf{r}, t)w(\mathbf{r}, \hat{\mathbf{n}}', E', t).$$

The lower limit on the E' integral is E since there is always an energy loss in Compton scattering. The kernel $K(\hat{\mathbf{n}}, E; \hat{\mathbf{n}}', E'|\mathbf{r}, t)$ can depend on both the initial and final energy and direction, and it can also depend on position $\mathbf{r}$ in the medium and, for time-varying media, on the time t.

To simplify the notation, we denote the operator defined by (10) as $\mathcal{K}$ and write

$$(11) \qquad \left[\frac{\partial w}{\partial t}\right]_{scat} = \left[\frac{\partial w}{\partial t}\right]_{in} + \left[\frac{\partial w}{\partial t}\right]_{out} = \mathcal{K}w - c\mu_{sc}w.$$

To get an explicit form for the kernel of $\mathcal{K}$, we make use of the *differential scattering cross section* $\partial\sigma_{sc}/\partial\Omega$, defined for a single scatterer as the ratio of scattered photon intensity (scattered photon flux per unit solid angle) to the incident photon irradiance (incident photon flux per unit area). If the scatterers (electrons in our problem) act independently, the scattering kernel must be proportional to the density of scatterers, n_{sc}, times $\partial\sigma_{sc}/\partial\Omega$. Moreover, since it is reasonable to assume that the scatterers are randomly oriented, the dependence on $\hat{\mathbf{n}}$ and $\hat{\mathbf{n}}'$ is through the scalar product $\hat{\mathbf{n}} \cdot \hat{\mathbf{n}}'$, which is the same as $\cos\theta_s$ in (1).

The dependence of the kernel on energy must take account of the conservation rule (1) and must therefore involve a delta function. Since (1) shows that the initial energy E' is determined if the final energy E and $\cos\theta_s$ are specified, we have the choice of including a delta function of the form $\delta[E - \gamma_1(E', \cos\theta_s)]$, $\delta[E' - \gamma_2(E, \cos\theta_s)]$ or $\delta[\cos\theta_s - \gamma_3(E, E')]$, where the functions $\gamma_j(\cdot)$ are determined by solving (1). These delta functions all impose the constraint (1), but they differ from each other by Jacobians. As we shall demonstrate in a moment, for the first delta function listed, the kernel has the structure,

$$(12) \quad K(\hat{n}, E; \hat{n}', E'|\mathbf{r}, t) = cn_{sc}\frac{\partial\sigma_{sc}}{\partial\Omega}\delta\left[E - \left(\frac{1}{E'} + \frac{1}{mc^2}(1 - \cos\theta_s)\right)^{-1}\right].$$

By the usual rule for transforming delta functions, an equivalent form is

$$(13) \quad K(\hat{n}, E; \hat{n}', E'|\mathbf{r}, t) = cn_{sc}\frac{mc^2}{E^2}\frac{\partial\sigma_{sc}}{\partial\Omega}\delta\left[\cos\theta_s - 1 + mc^2\left(\frac{1}{E} - \frac{1}{E'}\right)\right].$$

To show that (12) is correct, consider a collimated beam of monoenergetic photons for which

$$(14) \qquad w(\mathbf{r}, \hat{n}', E', t) = A\delta(\hat{n}' - \hat{n}_0)\delta(E' - E_0),$$

where A is a constant and $\delta(\hat{n}' - \hat{n}_0)$ is an angular delta function, defined such that

$$(15) \qquad \int_{4\pi} d\Omega'\, \delta(\hat{n}' - \hat{n}_0)f(\hat{n}') = f(\hat{n}_0),$$

where $f(\hat{n})$ is a good function as defined by Lighthill [5]. The sifting property (15) can be extended to situations where $f(\hat{n})$ is a generalized function rather than a good function by noting that a generalized function can be approximated arbitrarily closely by a sequence of good functions [5]. With these considerations, (12) and (14) yield

$$(16) \quad [Kw](\mathbf{r}, \hat{n}, E, t) = cAn_{sc}\frac{\partial\sigma_{sc}}{\partial\Omega}\delta\left[E - \left(\frac{1}{E_0} + \frac{1}{mc^2}(1 - \cos\theta_{s0})\right)^{-1}\right],$$

where $\cos\theta_{s0} = \hat{n}\cdot\hat{n}_0$. If $\partial\sigma_{sc}/\partial\Omega$ depends on energy, it must be evaluated at energy E_0 as a result of the second delta function in (14).

The delta function in (16) shows that the photons have the correct energy, depending on the scatter angle θ_{s0} from the original beam direction. The right-hand side of (16) is also consistent with the definitions of $\partial\sigma_{sc}/\partial\Omega$ and w. From the definition of w, the incident photon irradiance is cA, so $cA\partial\sigma_{sc}/\partial\Omega$ is the scattered intensity per scatterer. The total scattered intensity can also be obtained by integrating Kw over a volume ΔV and over all energies, so it is given by $cAN_{sc}\partial\sigma_{sc}/\partial\Omega$, where $N_{sc} = n_{sc}\Delta V$ is the total number of scatterers in ΔV. Since this result is just what we would have obtained directly from the definition of $\partial\sigma_{sc}/\partial\Omega$ without going through the integrals, it verifies that the form (12) is correct.

3.5. Propagation. It is straightforward to compute the propagation term in the Boltzmann equation. In a short time interval Δt, photons at point $\mathbf{r}$ travelling in direction $\hat{\mathbf{n}}$ move to point $\mathbf{r} + c\Delta t \hat{\mathbf{n}}$, so

$$(17) \qquad w(\mathbf{r}, \hat{\mathbf{n}}, E, t) = w(\mathbf{r} + c\Delta t \hat{\mathbf{n}}, \hat{\mathbf{n}}, E, t + \Delta t).$$

Expanding the right-hand side in a Taylor series, retaining only the terms linear in the small quantities, and equating the result to the left-hand side yields

$$(18) \qquad \begin{aligned} w(\mathbf{r}, \hat{\mathbf{n}}, E, t) =\ & w(\mathbf{r}, \hat{\mathbf{n}}, E, t) + c\Delta t \hat{\mathbf{n}} \cdot \nabla w(\mathbf{r}, \hat{\mathbf{n}}, E, t) \\ & + \Delta t \left[\frac{\partial}{\partial t} w(\mathbf{r}, \hat{\mathbf{n}}, E, t) \right]_{prop}. \end{aligned}$$

The time derivative in the last term is the desired $[\partial w/\partial t]_{prop}$, so we have

$$(19) \qquad \left[\frac{\partial w}{\partial t} \right]_{prop} = -c\hat{\mathbf{n}} \cdot \nabla w.$$

3.6. The complete equation. Combining (4), (7), (14) and (19), we get the final form for the Boltzmann equation with a monoenergetic radiation source:

$$(20) \qquad \frac{\partial w}{\partial t} = -c\mu_{tot}w + \frac{f}{4\pi}\delta(E - E_0) + \mathcal{K}w - c\hat{\mathbf{n}} \cdot \nabla w,$$

where μ_{tot} is the total absorption coefficient, given in general by

$$(21) \qquad \mu_{tot} = \mu_{sc} + \mu_{abs} = n_{sc}\sigma_{sc} + n_{abs}\sigma_{abs}.$$

For soft tissues at the energies used in nuclear medicine, the absorption cross section σ_{abs} is negligible, so $\mu_{tot} \simeq n_{sc}\sigma_{sc}$. Since soft tissue contains mainly elements of low atomic number, the electron binding energies are small compared to the gamma-ray energies, so all electrons can participate approximately equally in the Compton scattering. Thus n_{sc} is the same as the electron density, denoted n_{el}.

4. Steady-state solutions of the Boltzmann equation. Because the transit time of photons across the body is only a few nanoseconds, the distribution function reaches its steady-state value very quickly, and we are usually interested in solutions where $\partial w/\partial t = 0$. When it is necessary to include the arguments, we shall denote a steady-state solution by $w(\mathbf{r}, \hat{\mathbf{n}}, E)$, without a time argument. In this section we examine several such solutions, using different terms in the Boltzmann equation to gain insight.

4.1. Emission and propagation only. If we consider only the emission and propagation of radiation, the steady-state Boltzmann equation for an isotropic, monoenergetic source is

$$(22) \qquad c\hat{\mathbf{n}} \cdot \nabla w = \frac{f}{4\pi} \delta(E - E_0).$$

To solve this equation, we can choose a Cartesian coordinate system such that the z axis is parallel to $\hat{\mathbf{n}}$. Then $\hat{\mathbf{n}} \cdot \nabla = \partial/\partial z$, and (22) reduces to an ordinary differential equation in z; it is easily integrated to yield

$$(23) \qquad w(x, y, z, \hat{\mathbf{n}}, E) = \frac{\delta(E - E_0)}{4\pi c} \int_{-\infty}^{z} dz' f(x, y, z').$$

No constant of integration is needed if there is no radiation source other than f.

Defining $\ell = z - z'$ and reverting to a general vector notation, we have

$$(24) \qquad w(\mathbf{r}, \hat{\mathbf{n}}, E) = \frac{\delta(E - E_0)}{4\pi c} \int_0^\infty d\ell\, f(\mathbf{r} - \hat{\mathbf{n}}\ell) = \frac{1}{c} \int_0^\infty d\ell\, S(\mathbf{r} - \hat{\mathbf{n}}\ell).$$

The interpretation of this equation is that $w(\mathbf{r}, \hat{\mathbf{n}}, E)$ is obtained by integrating the source function along a line in direction $\hat{\mathbf{n}}$ passing through the point $\mathbf{r}$. Only photons originating along this line can contribute to the distribution function with the specified $\mathbf{r}$ and $\hat{\mathbf{n}}$. Only positive values of ℓ are needed since photons originating at a point $\mathbf{r} + \hat{\mathbf{n}}|\ell|$ propagate away from the observation point $\mathbf{r}$.

Equation (24) defines an integral transform known as the *x-ray transform*. We shall denote it with the operator $\mathcal{L}$ and write (24) symbolically as

$$(25) \qquad w = \mathcal{L}S.$$

Note that we are including the factor $1/c$ in the definition of $\mathcal{L}$, so the operator is dimensionally a time.

4.2. Emission, propagation and absorption. With the absorption term, the steady-state Boltzmann equation is still easy to solve (with a suitable integrating factor), and we obtain [*cf.* (23)]

$$(26) \qquad \begin{aligned} w(x, y, z, \hat{\mathbf{n}}, E) = {}& \frac{\delta(E - E_0)}{4\pi c} \int_{-\infty}^{z} dz' f(x, y, z') \\ & \cdot \exp\left[-\int_{z'}^{z} dz'' \mu_{abs}(x, y, z'') \right]. \end{aligned}$$

In a vector notation analogous to (24), we can also write

$$(27) \qquad \begin{aligned} w(\mathbf{r}, \hat{n}, E) = {}& \frac{\delta(E - E_0)}{4\pi c} \int_0^\infty d\ell\, f(\mathbf{r} - \hat{\mathbf{n}}\ell) \\ & \cdot \exp\left[-\int_0^\ell d\ell' \mu_{abs}(\mathbf{r} - \hat{\mathbf{n}}\ell') \right]. \end{aligned}$$

Now $w(\mathbf{r}, \hat{\mathbf{n}}, E)$ is still found by integrating the source function along a line in direction $\hat{\mathbf{n}}$ passing through the point $\mathbf{r}$, but more distant points along this line contribute less because of the exponential attenuation factor.

We can now see the interpretation of μ_{abs}; if it is independent of $\mathbf{r}$, then radiation traversing a distance ℓ is attenuated by a factor of $\exp(-\mu_{abs}\ell)$, so μ_{abs} is the reciprocal of the distance required for attenuation by $1/e$. If μ_{abs} depends on position, however, the attenuation is determined by a line integral of μ_{abs}.

The integral in (27) defines the *attenuated x-ray transform* $\mathcal{L}_\mu$, and we write

$$(28) \qquad\qquad w = \mathcal{L}_\mu S.$$

4.3. Full equation. With the scattering term, the steady-state Boltzmann equation cannot be solved by simple integration, but an integral along $\hat{\mathbf{n}}$ is nevertheless useful. It transforms (20) to

$$
(29) \quad
\begin{aligned}
w(\mathbf{r}, \hat{\mathbf{n}}, E) = {}& \frac{\delta(E - E_0)}{4\pi c} \int_0^\infty d\ell\, f(\mathbf{r} - \hat{\mathbf{n}}\ell) \exp\left[-\int_0^\ell d\ell'\, \mu_{tot}(\mathbf{r} - \hat{\mathbf{n}}\ell')\right] \\
& + \frac{1}{c}\int_0^\infty d\ell\, [\mathcal{K}w](\mathbf{r} - \hat{\mathbf{n}}\ell) \exp[-\int_0^\ell d\ell'\, \mu_{tot}(\mathbf{r} - \hat{\mathbf{n}}\ell')].
\end{aligned}
$$

This equation is not a solution since the unknown w still appears in the second integral. With the operators defined above, however, (29) can be written as

$$(30) \qquad\qquad w = \mathcal{L}_\mu S + \mathcal{L}_\mu \mathcal{K} w,$$

or

$$(31) \qquad\qquad [\mathbf{I} - \mathcal{L}_\mu \mathcal{K}]\, w = \mathcal{L}_\mu S,$$

where $\mathbf{I}$ is the identity operator.

A formal solution is given by the Neumann series,

$$(32) \quad w = [\mathbf{I} - \mathcal{L}_\mu \mathcal{K}]^{-1} \mathcal{L}_\mu S = \mathcal{L}_\mu S + \mathcal{L}_\mu \mathcal{K} \mathcal{L}_\mu S + \mathcal{L}_\mu \mathcal{K} \mathcal{L}_\mu \mathcal{K} \mathcal{L}_\mu S + \dots.$$

It can be argued on physical grounds that the inverse must exist and hence that the series will converge.

The series in (32) can be interpreted as the attenuated x-ray transform of an effective source distribution S_{eff}, i.e.

$$(33) \qquad\qquad w = \mathcal{L}_\mu S_{eff},$$

where

$$(34) \qquad\qquad S_{eff} = S + \mathcal{K}\mathcal{L}_\mu S + \mathcal{K}\mathcal{L}_\mu \mathcal{K}\mathcal{L}_\mu S + \dots.$$

Successive terms in this series represent successively more scattering; photons that have scattered j times contribute the term $[\mathcal{K}\mathcal{L}_\mu]^j S$ to S_{eff}.

Unlike the actual source S, the effective source is not monoenergetic; the scattering processes result in an energy loss, so S_{eff} covers a broad spectrum extending up to E_0.

4.4. The imaging instrument. Before considering inverse problems, we need to determine how a measured image is related to the continuous, six-dimensional phase-space distribution function. Digital imaging systems acquire a finite set of measurements $\{g_m, m = 1, .., M\}$, which can be thought of as components of an $M \times 1$ column vector $\mathbf{g}$. These components are random variables in the sense that repeated measurements on the same source distribution S will yield different values. For a given source, the probability law on each g_m is Poisson, and two different components g_m and g_k ($m \neq$ k) are statistically independent (Barrett and Swindell, 1996). The mean of g_m is denoted by $\bar{g}_m$ and the variance of g_m is also given by $\bar{g}_m$. Equivalently, we can write

$$(35) \qquad g_m = \bar{g}_m + \epsilon_m,$$

where ϵ_m is a zero-mean random variable with variance $\bar{g}_m$ and $< \epsilon_m \epsilon_k > = \bar{g}_m \delta_{mk}$.

To get an expression for $\bar{g}_m$, we make only the minimal assumption that the imaging system responds linearly (on average) to the radiation incident on it. We can then set up a reference plane P somewhere between the radiation source and the detector. If we know the distribution function on this plane we can compute the mean output of each detector element as a linear functional of w. By the Riesz representation theorem, this functional must have the form,

$$(36) \qquad \bar{g}_m = \int_P d^2 r_p \int_0^\infty dE \int_{2\pi} d\Omega \, d_m(\mathbf{r}_p, \hat{\mathbf{n}}, E) \, w(\mathbf{r}_p, \hat{\mathbf{n}}, E),$$

where $\mathbf{r}_p$ is a vector specifying position on the plane P. Note that the integral over solid angle covers only 2π ster since photons on the plane but directed away from the detector do not contribute to the output.

The function $d_m(\mathbf{r}_p, \hat{\mathbf{n}}, E)$ in (36) is called the *detector response function* since it specifies how the mth detector element responds to photons at point $\mathbf{r}_p$ travelling in direction $\hat{\mathbf{n}}$ and having energy E. The detector system must necessarily include imaging elements such as collimators or pinholes so that information about the source distribution can be captured in the image, and the effect of these elements is contained in d_m. For example, if there is a cylindrical lead tube in front of the mth detector element and the plane P is immediately adjacent to this tube, then $d_m(\mathbf{r}_p, \hat{\mathbf{n}}, E)$ is zero unless $\mathbf{r}_p$ is within the opening of the tube and $\hat{\mathbf{n}}$ is within the narrow range of solid angles defined by the tube. The dependence on E is determined by the energy resolution of the detector and any thresholding or energy windowing that might be employed. If multiple energy windows are used, a separate value of the index m can be assigned to each.

4.5. The imaging equation. In operator form, (36) can be written as

$$\bar{\mathbf{g}} = \mathcal{M}w, \tag{37}$$

where $\mathcal{M}$ is the measurement operator defined by (36). If we substitute (32) into (37), we have

$$\bar{\mathbf{g}} = \mathcal{M} \left[\mathbf{I} - \mathcal{L}_\mu \mathcal{K} \right]^{-1} \mathcal{L}_\mu S = \mathcal{M} \mathcal{L}_\mu S + \mathcal{M} \mathcal{L}_\mu \mathcal{K} \mathcal{L}_\mu S + \ldots . \tag{38}$$

In imaging applications, we are interested in the relation between the spatial distribution of the source, $f(\mathbf{r})$, and the mean image $\bar{\mathbf{g}}$. Since the operator $\mathcal{M}[\mathbf{I} - \mathcal{L}_\mu \mathcal{K}]^{-1}\mathcal{L}_\mu$ in (38) is linear and S is proportional to f by (7), it must be possible to find a single operator $\mathcal{H}$ such that

$$\bar{\mathbf{g}} = \mathcal{H}f. \tag{39}$$

As an integral, (39) is equivalent to

$$\bar{g}_m = \int_\infty d^3r \, f(\mathbf{r}) h_m(\mathbf{r}) = \sum_{j=0}^{\infty} \int_\infty d^3r \, f(\mathbf{r}) h_m^{(j)}(\mathbf{r}), \tag{40}$$

where the index j indicates the various terms in (38). The subscript ∞ on the integral sign indicates an integral over the infinite 3D space, though in practice $f(\mathbf{r})$ will vanish outside some finite region of support.

The kernel $h_m^{(j)}(\mathbf{r})$ can be regarded as a *point response function*. If the source distribution is a point, so that $f(\mathbf{r}) = \delta(\mathbf{r} - \mathbf{r}_0)$, then $h_m^{(j)}(\mathbf{r}_0)$ is the contribution of photons that have scattered j times to the mth component of the mean image.

5. Explicit forms for the kernels. To understand the structure of the kernels $h_m^{(j)}(\mathbf{r})$, consider first the no-scattering term, $j = 0$. The relevant operator is $\mathcal{M}\mathcal{L}_\mu$, and the corresponding integral is given, from (26) and (36), by

$$\int_\infty d^3r \, f(\mathbf{r}) h_m^{(0)}(\mathbf{r}) = \int_P d^2r_p \int_0^\infty dE \int_{2\pi} d\Omega \, d_m(\mathbf{r}_p, \hat{\mathbf{n}}, E_0) \frac{\delta(E - E_0)}{4\pi c}$$

$$\times \int_0^\infty d\ell \, f(\mathbf{r}_p - \hat{\mathbf{n}}\ell) \exp\left[-\int_0^\ell d\ell' \, \mu_{tot}(\mathbf{r}_p - \hat{\mathbf{n}}\ell') \right]. \tag{41}$$

To identify the kernel, we define $\mathbf{r} = \mathbf{r}_p - \hat{\mathbf{n}}\ell$ and recognize that $\ell^2 d\ell d\Omega = d^3r$ and $\ell = |\mathbf{r}_p - \mathbf{r}|$, so

$$\int_\infty d^3r \, f(\mathbf{r}) h_m^{(0)}(\mathbf{r}) = \frac{1}{4\pi c} \int_\infty d^3r \, f(\mathbf{r})$$

$$\times \int_P d^2r_p \frac{d_m(\mathbf{r}_p, \hat{n}, E_0)}{|\mathbf{r}_p - \mathbf{r}|^2} \exp\left[-\int_0^{|\mathbf{r}_p - \mathbf{r}|} d\ell' \, \mu_{tot}\left(\mathbf{r}_p - \frac{\mathbf{r}_p - \mathbf{r}}{|\mathbf{r}_p - \mathbf{r}|}\ell'\right) \right], \tag{42}$$

where we have used the delta function to perform the E integral. By inspection,

$$h_m^{(0)}(\mathbf{r}) =$$

(43)
$$\frac{1}{4\pi c} \int_P d^2 r_p \frac{d_m(\mathbf{r}_p, \hat{\mathbf{n}}, E_0)}{|\mathbf{r}_p - \mathbf{r}|^2} \exp\left[-\int_0^{|\mathbf{r}_p - \mathbf{r}|} d\ell' \, \mu_{tot} \mathbf{r}_p - (\frac{\mathbf{r}_p - \mathbf{r}}{|\mathbf{r}_p - \mathbf{r}|})\ell'\right].$$

The factor $1/|\mathbf{r}_p - \mathbf{r}|^2$ arose from the change of variables $\mathbf{r} = \mathbf{r}_p - \hat{\mathbf{n}}\ell$, but it has an important physical interpretation; at the source point $\mathbf{r}$, an area element $d^2 r_p$ subtends a solid angle $\cos\theta_{np} d^2 r_p/|\mathbf{r}_p - \mathbf{r}|^2$, where θ_{np} is the angle between $\hat{\mathbf{n}}$ and the normal to plane P. The cosine is hidden in $d_m(\mathbf{r}_p, \hat{\mathbf{n}}, E_0)$, but the inverse-square factor appears explicitly.

The single-scatter term ($j = 1$) in (37) has the same structure as the no-scatter term except that $\mathcal{ML}_\mu$ operates on $\mathcal{KL}_\mu S$ rather than on S directly. We can therefore compute the kernel $h_m^{(1)}(\mathbf{r})$ in two steps, considering first the operator $\mathcal{KL}_\mu$ and then $\mathcal{ML}_\mu$. From (7), (10) and (27), the first of these operators has the form,

(44)
$$[\mathcal{KL}_\mu S](\mathbf{r}, \hat{\mathbf{n}}, E) = \int_\infty d^3 r' K(\hat{\mathbf{n}}, E; \frac{\mathbf{r} - \mathbf{r}'}{|\mathbf{r} - \mathbf{r}'|}, E_0|\mathbf{r})$$
$$\cdot \frac{f(\mathbf{r}')}{4\pi c|\mathbf{r} - \mathbf{r}'|^2} \exp[-\int_0^{|\mathbf{r} - \mathbf{r}'|} d\ell' \, \mu_{tot} \mathbf{r} - \frac{\mathbf{r} - \mathbf{r}'}{|\mathbf{r} - \mathbf{r}'|}\ell'],$$

where we have used the delta function to good end and made a change of variables, $\mathbf{r}' = \mathbf{r} - \hat{\mathbf{n}}'\ell'$.

The interpretation of (44) is straightforward. Photons of energy E_0 originate at $\mathbf{r}'$ and travel to $\mathbf{r}$, diminishing in number per unit area because of the inverse-square factor and the attenuation factor. At $\mathbf{r}$, they scatter into direction $\hat{\mathbf{n}}$ with energy E.

Now we can apply the operator $\mathcal{ML}_\mu$ to propagate these photons to the plane P where they are measured by the imaging system. The operator $\mathcal{L}_\mu$ has the effect of replacing $\mathbf{r}$ with $\mathbf{r}_p - \hat{\mathbf{n}}\ell$ everywhere, inserting another exponential factor and a $1/c$, and integrating over ℓ. The measurement operator $\mathcal{M}$ is implemented by multiplication by d_m and integration over the plane P, energy E and solid angle Ω. The overall kernel is thus

(45)
$$h_m^{(1)}(\mathbf{r}') = \int_P d^2 r_p \int_0^\infty dE \int_{2\pi} d\Omega \, d_m(\mathbf{r}_p, \hat{\mathbf{n}}, E)$$
$$\cdot \int_0^\infty d\ell \, K(\hat{\mathbf{n}}, E; \frac{\mathbf{r}_p - \hat{\mathbf{n}}\ell - \mathbf{r}'}{|\mathbf{r}_p - \hat{\mathbf{n}}\ell - \mathbf{r}'|}, E_0| \mathbf{r}_p - \hat{\mathbf{n}}\ell) \cdot \frac{1}{4\pi c^2|\mathbf{r}_p - \hat{\mathbf{n}}\ell - \mathbf{r}'|^2}$$
$$\cdot \exp[-\int_0^{|\mathbf{r}_p - \hat{\mathbf{n}}\ell - \mathbf{r}'|} d\ell' \, \mu_{tot} \mathbf{r}_p - \hat{\mathbf{n}}\ell - \frac{\mathbf{r}_p - \hat{\mathbf{n}}\ell - \mathbf{r}'}{|\mathbf{r}_p - \hat{\mathbf{n}}\ell - \mathbf{r}'|}\ell']$$
$$\cdot \exp[-\int_0^\ell d\ell' \, \mu_{tot} \mathbf{r}_p - \hat{\mathbf{n}}\ell'].$$

When integrated against f, this complicated expression gives the contribution to $\bar{g}_m$ of photons that have scattered exactly once. If patience suffices, higher-order terms can be computed similarly.

6. Numerical approaches to forward problems.

6.1. Discretization. A forward problem is to compute the effect produced by a specified source. In the previous section, we identified several such problems, depending on whether the effect was the distribution function w, the mean image $\bar{g}$, the point response functions $h_m(\mathbf{r})$ or the partial kernels $h_m^{(j)}(\mathbf{r})$. A finer taxonomy must be used when discussing numerical methods since we must then decide how to represent functions by discrete sets of numbers.

A discrete approximation of an arbitrary spatial function $f(\mathbf{r})$ is

$$(46) \qquad f_a(\mathbf{r}) = \sum_n \alpha_n \phi_n(\mathbf{r}), n = 1, ..., N,$$

where the subscript a denotes *approximate*. We shall refer to $\phi_n(\mathbf{r})$ as an *expansion function*, avoiding the common term *basis function* since the finite set $\{\phi_n(\mathbf{r})\}$ does not form a basis for the infinite-dimensional space of functions $f(\mathbf{r})$.

The most common expansion functions are volume elements or *voxels*. Typically, the voxels are $\Delta \times \Delta \times \Delta$ cubes arrayed on a regular cubic lattice, with the lattice spacing equal to Δ so that the voxels are contiguous; $\phi_n(\mathbf{r})$ is chosen to be a constant when $\mathbf{r}$ lies in the nth voxel and zero otherwise.

Many other sets of expansion functions are possible, however, and for each set there are many ways of choosing the coefficients $\{\alpha_n\}$. It might be appealing to compute α_n as a scalar product of $\phi_n(\mathbf{r})$ and $f(\mathbf{r})$, especially if the set $\{\phi_n\}$ is orthonormal, but computationally it is often easier to use simple point sampling, $\alpha_n = f(\mathbf{r}_n)$, where $\mathbf{r}_n$ is the center point of a voxel or other expansion function.

The distribution function is a function of six variables, x, y, z, E, θ and ϕ, where the latter two are the polar coordinates of $\hat{\mathbf{n}}$. Numerical computation of this function requires that all six variables be replaced by discrete indices, and again there is considerable arbitrariness in how we do so. The simplest approach is to use voxels for the spatial variables and similar rectangular bins in both angle and energy, but again many other choices are possible. For example, we shall see below that spherical harmonics are very useful for the angular dependence of w.

No matter how we choose the discretization, $w(\mathbf{r}, \hat{\mathbf{n}}, E)$ is represented by a set of numbers w_{ijklmn}, where i, j and k range over the spatial variables, l and m range over the angles and n specifies the energy. For notational simplicity we can compress the six indices into a single vector index $\mathbf{n}$, a 6D vector with integer components $(ijklmn)$. The sheer size of the set $\{w_\mathbf{n}\}$

should not be overlooked; if each variable is discretized to 100 values, there are 10^{12} elements in the set.

One forward problem is thus to compute or estimate the numbers $\{w_n\}$ for a specified source distribution. A more manageable forward problem is to estimate the mean image data produced by a given source distribution. No *ad hoc* discretization is needed in this case since real-world digital imaging systems acquire only a discrete set of measurements. The imaging system inherently maps a function of continuous variables to a discrete vector; we refer to such mappings as *continuous-to-discrete* or CD mappings. Specifically, both the mapping from w to $\bar{g}$ and the one from f to $\bar{g}$ are CD mappings, without any approximations arising from discretization.

Ideally, therefore, we would use continuous, analytic descriptions for the object, but this limits us to toy problems such as a uniform spherical source immersed in a spherical scattering medium. For more realistic problems, discrete representations are also needed for the source distribution f and the electron density n_{el}. (As noted in Sec. 3, the absorption cross section σ_{abs} is negligible for soft tissues at the energies used in nuclear medicine, so both μ_{tot} and the scatter kernel are fully determined by n_{el}.)

One way to test whether a particular object discretization is an adequate representation of the actual continuous object is just to choose a finer grid. If voxels of size $\frac{1}{2}\Delta \times \frac{1}{2}\Delta \times \frac{1}{2}\Delta$ yield essentially the same mean image as ones of size $\Delta \times \Delta \times \Delta$, then we can assume that the latter are adequate. Unfortunately, this test is rarely applied since the smaller voxels require 8 times the computation, so the adequacy of a particular choice of voxel size is usually a matter of faith and intuition.

6.2. Simplifications and approximations. A useful simplification is that n_{el} is independent of spatial position within the boundaries of the object; this assumption would be reasonable for imaging the abdomen since all soft tissues have about the same attenuation and scattering coefficients. For brain imaging the skull is substantially denser and more attenuating than the brain itself, but it is fairly thin and contributes relatively little to the line integrals of the attenuation coefficient, so treating the skull as soft tissue may result in only a small error.

A related approximation, also valid for soft tissues at nuclear-medicine energies, is that the differential scattering cross section is independent of photon energy. With this assumption and the one of the previous paragraph, μ_{sc} is a constant and the scatter kernel depends only on $\cos\theta_s$. All of the exponential attenuation factors are functions only of the distance between two voxels, so they can either be computed as needed or precomputed and stored in a lookup table. Similarly, the $\cos\theta_s$ factors, either computed on the fly or precomputed and stored, are easily incorporated into summations over angles.

In many cases, however, it is not a good approximation to treat the scattering medium as homogeneous. In cardiac imaging, for example, some of the photons pass through the lungs, which have only about one-third the electron density of soft tissue. Accurate computation of the mean image in these cases requires additional information about the spatial distribution of n_{el} in the specific patient, which can be obtained from an auxiliary computed-tomography image. The exponential attenuation factors can then be computed from this scan. Quantities related to the imaging system and not patient specific can be precomputed and stored, and it is feasible to compute $h^{(j)}(\mathbf{r})$ for j up to 2 and possibly higher [6].

Another common approximation is to consider just unscattered and singly scattered photons and ignore multiple scatter. This approach is likely to be valid for systems that reject photons with an estimated energy below a relatively high threshold. Multiply scattered photons have had multiple chances to lose energy and are therefore likely to be rejected, so it does not matter whether we model them accurately.

Some authors [7][8] have taken advantage of the fact that multiple scatter results in a broad, featureless spatial distribution; they have then proposed various empirical corrections that can be added to distributions computed with unscattered and singly scattered photons. The validity of these corrections is assessed with Monte Carlo computations for typical patient distributions of source activity and electron density.

6.3. Monte Carlo. The most common way of solving the Boltzmann equation is Monte Carlo simulation. Developed by Ulam and Metropolis in the Manhattan Project, this method uses a computer to simulate the trajectories of photons (or other particles) while keeping track of their positions, energies and directions. As the number of simulated trajectories approaches infinity, a histogram of the values of x, y, z, θ, ϕ and E approaches $\{w_n\}$. As noted above, however, the set $\{w_n\}$ may have of order 10^{12} elements, and it would require 10^{14} computations in order to have an average of 100 points per element.

For this reason, Monte Carlo methods are seldom used to estimate the full distribution function. A more tractable problem is estimating the mean image data. A single 2D image in nuclear medicine might be a 64×64 or 128×128 array, so the number of elements is of order 10^4 instead of 10^{12}. Even if we compute multiple images, for example at different projection angles in tomographic systems or for different energy windows, the problem remains feasible with modern computers.

An apparent drawback to estimating the image data by simulating photon trajectories is that most photons will make no contribution at all to the data set. (The collimators used in nuclear medicine typically accept only about 1 in every 10,000 photons.) It is possible, however, to run the trajectories backward starting from a point on the detector, so all simulated photons necessarily reach the detector. This *forced-detection* scheme and other *variance-reduction* methods can substantially reduce the sampling error in estimation of the mean data. For a thorough review of these tricks, see [9][10].

Monte Carlo calculations are inherently well suited to parallel computation. If N processors are available to compute trajectories and each has sufficient memory to store the source and scattering distributions as well as the outputs, then no message passing between processors is needed and the computation time for a fixed number of trajectories will be reduced by a factor of N.

6.4. Trickle-down theory. Numerical solution of the Boltzmann equation for the nuclear medicine problem is facilitated by the fact that Compton scatter always results in an energy loss. Other applications of Boltzmann equations, for example for describing electron transport in semiconductors, do not enjoy this advantage.

To show the trickle-down of photon energy explicitly, we use discrete energy bins and define

$$(47) \qquad w_k(\mathbf{r}, \hat{\mathbf{n}}) = w(\mathbf{r}, \hat{\mathbf{n}}, E_0 - k\Delta E), k = 0, ..., k_{max},$$

where $\Delta E \equiv E_0/k_{max}$ is the width of the energy bin.

The steady-state Boltzmann equation now takes the form

$$(48) \qquad -c\mu_{tot}w_k + s_k + \Delta E \sum_{j=0}^{k-1} \mathcal{K}_{kj}w_j - c\hat{\mathbf{n}} \cdot \nabla w_k = 0,$$

where $s_k(\mathbf{r})$ is the source distribution for the kth energy bin, given for a monoenergetic source by

$$(49) \qquad s_k(\mathbf{r}) = \frac{f(\mathbf{r})}{4\pi\Delta E}\delta_{k0},$$

and $\mathcal{K}_{kj}$ is the angular part of the operator $\mathcal{K}$ with its kernel sampled at $E = E_0 - k\Delta E$ and $E' = E_0 - j\Delta E$, i.e.

$$(50) \quad [\mathcal{K}_{kj}w_j](\mathbf{r}, \hat{\mathbf{n}}) = \int_{4\pi} d\Omega' \, K(\hat{\mathbf{n}}, E_0 - k\Delta E; \hat{\mathbf{n}}', E_0 - j\Delta E|\mathbf{r})w_j(\mathbf{r}, \hat{\mathbf{n}}').$$

Note that the kernel still depends on position $\mathbf{r}$ and that there is still an implicit delta function in it.

The trickle-down aspect of (48) is contained in the summation limits, which arise from the integration limits in (10); with discrete energy bins as we have defined them, the requirement $E' > E$ translates to $j < k$.

We have already obtained the solution to (48) for $k = 0$, which is the only bin with a real source in it. For this bin, the scattering term makes no contribution, and (28) shows that

$$(51) \qquad w_0 = \mathcal{L}_\mu s_0.$$

Next look at $k = 1$, where there is no true source term but the summation over j in (48) contains the single term $\mathcal{K}_{10}w_0$, which acts as an effective source of photons of energy $E_0 - \Delta E$. We then have

$$(52) \qquad w_1 = \mathcal{L}_\mu\mathcal{K}_{10}w_0 = \mathcal{L}_\mu\mathcal{K}_{10}\mathcal{L}_\mu s_0.$$

In the next bin, $k = 2$, we have two effective source terms since photons of energy $E_0 - 2\Delta E$ can be generated by scattering photons of energy E_0 or $E_0 - \Delta E$. Thus,

$$(53) \qquad w_2 = \mathcal{L}_\mu \left[\mathcal{K}_{20} w_0 + \mathcal{K}_{21} w_1 \right].$$

In general, we can compute w_k by the iteration rule,

$$(54) \qquad w_k = \mathcal{L}_\mu \sum_{j=0}^{k-1} \mathcal{K}_{kj} w_j.$$

Note that this series is fundamentally different from the Neumann series used in (32), where each term represents the contribution at all energies from a particular order of scattering. In (54), by contrast, each term gives the contribution at a particular energy from all orders of scatter.

To relate the set $\{w_k\}$ to the measured data, we can use (36) with the energy integral replaced by a sum to write

$$(55) \qquad \begin{aligned} \bar{g}_m &= \Delta E \sum_{k=0}^\infty \int_P d^2 r_p \int_{2\pi} d\Omega \, d_m(\mathbf{r}_p, \hat{\mathbf{n}}, E_k) w_k(\mathbf{r}_p, \hat{\mathbf{n}}) \\ &\equiv \sum_{k=0}^\infty [\mathcal{M}_k w_k]_m, \end{aligned}$$

where the meaning of the operator $\mathcal{M}_k$ can be discerned from the integral form. In many cases the imaging instrument will reject low-energy photons, so it may be a good approximation to truncate (55) at a relatively small value of k.

A simplification of (55) is frequently possible. Many gamma-ray cameras can be modeled well by assuming that their response as a function of position and angle is independent of the photon energy, so that

$$(56) \qquad d_m(\mathbf{r}_p, \hat{\mathbf{n}}, E) = d_m(\mathbf{r}_p, \hat{\mathbf{n}}) P_{acc}(E),$$

where $P_{acc}(E)$ is the probability that a photon of energy E is accepted into the recorded data. With this assumption, (55) becomes

$$(57) \qquad \begin{aligned} \bar{g}_m &= \Delta E \int_P d^2 r_p \int_{2\pi} d\Omega \, d_m(\mathbf{r}_p, \hat{\mathbf{n}}) \left[\sum_{k=0}^\infty P_{acc}(E_k) w_k(\mathbf{r}_p, \hat{\mathbf{n}}) \right] \\ &\equiv [\mathcal{M} w_{eff}]_m, \end{aligned}$$

where the effective distribution function is the actual one weighted with $P_{acc}(E)$, and $\mathcal{M}$ is now independent of energy.

6.5. Spherical harmonics. We still need to reduce the size of the discrete representation of w. Each w_k is a function of $\mathbf{r}$ and $\hat{\mathbf{n}}$, hence 5 variables, and we cannot store the result if we discretize each variable in, say, 100 steps. Fortunately, scattered radiation tends to be relatively isotropic, so the angular variation of w_k for $k > 0$ can be specified with a few numbers.

One way to do so is to expand w_k as

$$(58) \qquad w_k(\mathbf{r}, \hat{\mathbf{n}}) = \sum_{l=0}^{\infty} \sum_{m=-l}^{l} w_{klm}(\mathbf{r}) Y_{lm}(\hat{\mathbf{n}}),$$

where the Y_{lm} are spherical harmonics, normalized to form a complete, orthonormal set on the unit sphere. As written, (58) is exact but requires an infinite number of coefficients; the savings comes in if we can truncate the sum over l at some small number l_{max}. An extreme assumption, $l_{max} = 1$, reduces the Boltzmann equation to the diffusion equation [4], but in nuclear medicine values of l_{max} around 4–8 appear to be more reasonable. For $l_{max} = 4$, the angular variation of each w_k is specified by 25 numbers at each position $\mathbf{r}$. If we use a $100 \times 100 \times 100$ matrix for the spatial dependence and 20 energy bins, the entire distribution function can be stored in about 5×10^8 locations, which will fit in RAM on many modern workstations. Additional savings would accrue by using coarser spatial bins for larger k (lower energy).

Another advantage of the spherical-harmonic representation is that it facilitates implementation of the scatter operator $\mathcal{K}_{kj}$. It can be shown that this operator is diagonal in the spherical harmonics, while the propagation term couples each $w_{klm}(\mathbf{r})$ to at most six others, all at the same $\mathbf{r}$. Details of this decomposition in the context of neutron or electron transport can be found in [11][12][13], and the application to nuclear medicine will be published separately.

7. Inverse problems.

7.1. Analytical inverses. Many theoretical treatments of inverse problems regard both the object and the data as functions of continuous variables and try to find an analytical inverse to go from data function to object function. Discretization is then imposed at the end for computational reasons. An excellent example where this approach is successful is in x-ray computed tomography; the analytical solution is the inverse Radon transform, and its numerical implementation is the widely used algorithm, filtered backprojection [4].

In nuclear medicine, however, analytical solutions have proven elusive. If the attenuation coefficient is assumed to be independent of position within convex body contours and scatter is ignored altogether, the forward problem is described by the *exponential Radon transform* [14], which is closely related to our operator $\mathcal{L}_\mu$ but with $\mu = $ constant. An ideal parallel-bore collimator rotated around the patient in very fine angular steps would measure this transform, and an analytic inverse in this case was given by Tretiak and Metz [15][16]. No corresponding inverse has been found for inhomogeneous attenuation or for any problem involving scatter. Even if the medium is homogeneous and scattered photons are completely rejected, however, the inverse exponential Radon transform must ignore many details of the imaging system such as the depth-dependent blur of the collimator. For these reasons, analytic inverses play little role in the current practice of nuclear medicine.

On the other hand, mathematical investigations of the attenuation problem are actively continuing, often with fascinating results. Pan and Metz [17] have recently found a whole family of exact reconstruction algorithms, some with better noise performance than the Tretiak and Metz algorithm. Natterer [18] has even studied the challenging inverse problem where both the source distribution and the distribution of the attenuating medium are unknown and have to be determined from the emission data. An important mathematical result of this investigation was a consistency condition that must be satisfied by attenuated SPECT data [19]. Clarkson [20] has recently shown that the optimal Pan and Metz solution is equivalent to imposing a consistency condition.

Far less theoretical effort has been expended on the scatter problem; it is hoped that this paper will spur additional research in this direction.

7.2. Estimation of source parameters. When analytic inverses are not available, we can formulate the reconstruction problem as parameter estimation. For example, we can represent the source by an expansion analogous to (46) and then attempt to estimate the coefficients in the expansion. Unfortunately, this approach usually runs into a little-recognized fundamental contradiction.

To illustrate the difficulty, consider a general linear parameter θ, a linear functional of the source distribution in the form,

$$(59) \qquad \theta = \int_\infty d^3r \; f(\mathbf{r})t(\mathbf{r}),$$

where $t(\mathbf{r})$ can be thought of as a template placed over the source. We shall assume that both $t(\mathbf{r})$ and $f(\mathbf{r})$ are square-integrable and hence vectors in an $\mathbb{L}^2$ Hilbert space, which we call *object space*. This Hilbert space can be partitioned into two orthogonal subspaces, which we refer to as *null space* and *measurement space*. Null space is the space of all square-integrable functions $f_{null}(\mathbf{r})$ such that $\mathcal{H}f_{null}(\mathbf{r}) = 0$, where $\mathcal{H}$ is the system operator introduced in Secs. 4.5 and 4.6. Measurement space is the orthogonal complement of null space. Since the rank of the operator $\mathcal{H}$ is less than or equal to the number of measurements M, the dimension of measurement space is at most M, and null space is therefore necessarily infinite.

Any vector in object space can be projected onto either measurement space or null space. The projection operator onto measurement space is $\mathcal{H}^+\mathcal{H}$, where $\mathcal{H}^+$ is the Moore-Penrose pseudoinverse of $\mathcal{H}$. It follows, then, that the projection operator onto null space is $\mathbf{I} - \mathcal{H}^+\mathcal{H}$, where $\mathbf{I}$ is the unit operator in object space. Thus we can write

(60) $f(\mathbf{r}) = \mathcal{H}^+\mathcal{H}f(\mathbf{r}) + [\mathbf{I} - \mathcal{H}^+\mathcal{H}]\,f(\mathbf{r}) \equiv f_{meas}(\mathbf{r}) + f_{null}(\mathbf{r}),$

and similarly for the template $t(\mathbf{r})$.

Since θ has the form of a scalar product (f, t) in object space, and null space and measurement space are orthogonal subspaces of object space, we see that

(61) $\theta = (f_{meas}, t_{meas}) + (f_{null}, t_{null}).$

Except for contrived situations, we can be virtually certain that $f_{null} \neq 0$; no imaging system can capture all details of a real object. On the other hand, we are free to choose the template, so we can demand that t_{null} is at least approximately zero, in which case the second scalar product in (61) is negligible.

We shall say that the parameter θ is *estimable* [21][22] if and only if $t_{null} = 0$. If this condition is not satisfied, it is difficult to say anything about the quality of any estimate of θ. For example, we cannot use the usual definitions of bias or mean-square error since these quantities specify deviations from some true value of the parameter. But if the second term in (61) cannot be guaranteed to be zero, different true values of θ can result from different objects f that differ from each other by null functions and hence produce exactly the same mean data. There is no way to say which f and hence which θ is true. That is, we have no way of computing a bias or mean-square error for an estimate of θ unless

(62) $[\mathbf{I} - \mathcal{H}^+\mathcal{H}]\,t(\mathbf{r}) = 0.$

This condition is satisfied if $t(\mathbf{r})$ can be written as a linear combination of the point response functions $h_m(\mathbf{r})$ defined in (40).

One way around this impasse is to take a Bayesian perspective and assert that $f(\mathbf{r})$ is a random process and that we know *a priori* the probability density function for it. For example, we may assert that the system never views any object that contains null functions (and we can guarantee that this assertion is true by doing simulations where the simulated object, in fact, never contains null functions). Alternatively, we can use the assumed prior to compute an (assumed) average bias.

If all we know *a priori* is that $f(\mathbf{r})$ is a non-negative function of specified support, the best we can do if (62) is not satisfied is to compute an upper bound [23][24] on the bias of an estimate of θ.

7.3. Are the expansion coefficients estimable? Now let us apply the general considerations of Sec. 6.2 specifically to the coefficients in an expansion like (46). To avoid introducing any assumptions about the object, we write,

(63) $f(\mathbf{r}) = \sum_{n=0}^{N} \alpha_n \phi_n(\mathbf{r}) + \Delta f(\mathbf{r}),$

where $\Delta f(\mathbf{r})$ is defined simply as the true distribution $f(\mathbf{r})$ minus its approximate representation $f_a(\mathbf{r})$. The output of the mth detector for this general object can be expressed as

$$(64) \qquad g_m = \sum_{n=0}^{N} H_{mn}\alpha_n + \eta_m + \epsilon_m,$$

where

$$(65) \qquad H_{mn} \equiv \int_\infty d^3r\, h_m(\mathbf{r})\phi_n(\mathbf{r}),$$

and

$$(66) \qquad \eta_m = \int_\infty d^3r\, \Delta f(\mathbf{r})h_m(\mathbf{r}).$$

In matrix-vector form, we can write

$$(67) \qquad \mathbf{g} = \mathbf{H}\boldsymbol{\alpha} + \boldsymbol{\eta} + \boldsymbol{\epsilon},$$

where $\mathbf{H}$ is an $M \times N$ matrix and $\boldsymbol{\epsilon}$ and $\boldsymbol{\eta}$ are $M \times 1$ error vectors, the former arising from random (Poisson) noise and the latter from modeling error.

The major difference between $\boldsymbol{\epsilon}$ and $\boldsymbol{\eta}$ is that we can say very little about the properties of $\boldsymbol{\eta}$. Even if we regard $f(\mathbf{r})$ as a random process, it is difficult to specify the statistics of $\boldsymbol{\eta}$. It would surely be unrealistic, for example, to say that $\boldsymbol{\eta}$ is a zero-mean or that its components are uncorrelated.

We encountered a similar situation in Sec. 5.1 when we discussed discretization for the forward problem. There the suggested solution was just to choose a very fine grid so that $\Delta f(\mathbf{r})$ would consist of very fine details not likely to be imaged well by the system. Unfortunately, this remedy runs afoul of the estimability condition for an inverse problem if the expansion functions are voxels. Precisely because the system does not image fine details, a small voxel function must have a substantial component in null space, and there would be no way of defining a bias or mean-square error in an estimate of the associated coefficients α_n.

This dilemma can be restated in familiar terms by thinking about the number of measurements and the number of unknowns. If the number of measurements is fixed at M but we are free to choose the size of the discretization grid, we can choose between an underdetermined problem ($M < N$) and an overdetermined one ($M > N$). If we choose a large N, we can make the modeling error negligible but then we have more unknown parameters than we can uniquely estimate from the data. That is, even the discrete matrix operator $\mathbf{H}$ has a null space. On the other hand, if we use a coarse discretization, we can uniquely estimate the unknown parameters but we have no idea what the error in the data is.

One way, in principle, to resolve this dilemma is to use the singular functions of the CD operator $\mathcal{H}$ as the expansion functions in the object representation. If we denote these singular functions by $u_n(\mathbf{r}), n = 1,...,$ ∞, then an arbitrary object can be written, without approximation, as

$$(68) \qquad f(\mathbf{r}) = \sum_{n=1}^{\infty} \alpha_n u_n(\mathbf{r}).$$

If we denote the rank of $\mathcal{H}$ as R and order the singular functions by decreasing singular value, we can also write

$$(69) \qquad f(\mathbf{r}) = \sum_{n=1}^{R} \alpha_n u_n(\mathbf{r}) + \sum_{n=R+1}^{\infty} \alpha_n u_n(\mathbf{r}).$$

The first sum here is f_{meas} and the second is f_{null}. Since $\mathcal{H}\mathbf{u}_n = 0$ for $n > R$, we now have

$$(70) \qquad g_m = \sum_{n=0}^{R} H_{mn}\alpha_n + \epsilon_m,$$

where now H_{mn} is computed from (65) with ϕ_n replaced with u_n. Since $R \leq M$, the problem is fully determined, there is no modeling error and the statistics of the noise are easy to characterize. The problem is that we had to first solve the singular-value problem for $\mathcal{H}$, which we cannot do in very many cases.

Another way around the dilemma is to use the point response functions $h_m(\mathbf{r})$ as the expansion functions. Since these functions, sometimes called natural pixels [25], span measurement space, there is again no modeling error and the matrix $\mathbf{H}$ is square.

In summary, the expansion coefficients are estimable if the expansion functions are either the singular functions of $\mathcal{H}$ corresponding to nonzero singular values or the natural pixels. Ordinary pixel or voxel expansions, however, seldom have estimable coefficients.

7.4. Determining the parameters in underdetermined problems. One way to proceed from this point is to choose a relatively fine grid such that it is reasonable to assume that $\eta \simeq 0$ but for which N is greater than the rank R of $\mathcal{H}$, so the problem is underdetermined. Under this model, the coefficients are not estimable, but many reconstruction algorithms nevertheless seek an estimate $\hat{\alpha}$ in the form,

$$(71) \qquad \hat{\alpha} = \underset{\alpha}{\operatorname{argmin}} \left\{ \Psi_1(\mathbf{g}, \mathbf{H}\alpha) + \lambda\Psi_2(\alpha) \right\},$$

where $\Psi_1(\mathbf{g}, \mathbf{H}\alpha)$ is a scalar-valued functional designed to specify the distance, in some sense, between $\mathbf{g}$ and $\mathbf{H}\alpha$, while $\Psi_2(\alpha)$ is another scalar-valued functional intended to regularize or smooth the reconstructed image. For example, Ψ_1 might be a least-squares or weighted least-squares functional, or it might be the negative of the logarithm of the statistical likelihood of α given $\mathbf{g}$. In all practical cases, Ψ_1 will be minimized if $\mathbf{H}\alpha = \mathbf{g}$, so Ψ_1 is referred to as the *data-agreement functional*. The regularizing functional Ψ_2 could, for example, be the norm of the vector α or of its spatial gradient, or it could be the negative of the entropy of α. The parameter λ controls the tradeoff between the two terms in the minimization, hence the tradeoff between noise and image sharpness.

There are many algorithms for finding the minimum required by (71), and some of them allow incorporation of additional information, such as the requirement that the reconstructed image be nonnegative or that it have a specified support. Specific algorithms, however, are beyond the scope of this paper.

So long as $\Psi_1 + \lambda\Psi_2$ is strictly convex, the vector $\hat{\alpha}$ determined by (71) will be unique even though $\mathbf{H}$ may have a null space. It does not follow that $\hat{\alpha}$ is the best estimate in any sense, except the tautological one that it minimizes $\Psi_1 + \lambda\Psi_2$. In particular, $\hat{\alpha}$ is strongly influenced by the expansion functions chosen, by the functionals Ψ_1 and Ψ_2, by the free parameter λ, and of course by the actual system operator $\mathcal{H}$.

Nevertheless, if the discretization is fine enough that η is negligible, we can analyze the statistical properties of $\hat{\alpha}$. For a given object $f(\mathbf{r})$ and expansion functions $\phi_n(\mathbf{r})$, we can compute the mean vector and covariance matrix of $\hat{\alpha}$, and we can get an excellent approximation to the full N-dimensional probability density function, even when nonlinear minimization algorithms are used [26][27].

From these statistical properties and assumptions about the intended task of the imaging system, objective measures of image quality can be computed [28–31]. For example, the task may be to detect a tumor or other abnormality in an image, or it may be to distinguish a malignant from a benign lesion. With such tasks, we do not need to determine a bias or mean-square error, so the issue of estimability disappears. Rather than be concerned with how well the reconstructed image resembles the original object, which we cannot in principle determine, we specify the success of the imaging system and reconstruction algorithm by how well we can perform the given task.

7.5. Inverse-source problems with the Neumann series for scatter. We now assume that a fine discretization grid has been chosen, so the modeling error η is negligible, and that we want to estimate the coefficient vector α by some variant of the minimization principle (71). The remaining question is how to incorporate the detailed scatter modeling developed in Secs. 4 and 5. In this section we address this question using the

Neumann series (40) for scatter; a similar treatment could be based on the trickle-down series (54).

Substitution of (63) into (40) yields

$$(72) \qquad \bar{g}_m = \sum_{j=0}^{\infty} \sum_{n=0}^{N} H_{mn}^{(j)} \alpha_n + \sum_{j=0}^{\infty} \int_{\infty} d^3r \, \Delta f(\mathbf{r}) h_m^{(j)}(\mathbf{r}),$$

where

$$(73) \qquad H_{mn}^{(j)} \equiv \int_{\infty} d^3r \, h_m^{(j)}(\mathbf{r}) \phi_n(\mathbf{r}).$$

The matrix $\mathbf{H}^{(0)}$ corresponding to unscattered radiation is sparse, since there are relatively few paths by which unscattered gamma rays can contribute to a particular detector output. Moreover, it is this matrix that conveys information about fine spatial detail in the object.

If we simply ignore the matrices with $j > 0$, the reconstruction will be an image with background haze and blur from the scattered photons. There are three basic ways to overcome this problem: (a) perform a pre-construction correction of the data; (b) perform a post-reconstruction correction of the image; (c) do the problem right in the first place. The latter approach is simply to use the full $\mathbf{H}$ and not ignore the higher terms, but it incurs a computational penalty since $\mathbf{H}^{(j)}$ is not sparse for $j > 0$.

The pre-reconstruction correction method seeks to replace the observed data vector $\mathbf{g}$ with a corrected vector $\mathbf{g}'$ such that a solution that satisfies $\mathbf{H}^{(0)} \hat{\alpha} = \mathbf{g}'$ will also approximately satisfy $\mathbf{H} \hat{\alpha} = \mathbf{g}$. This condition would be exactly satisfied if

$$(74) \qquad \mathbf{g}' = \mathbf{g} - \Delta\mathbf{H}\hat{\alpha},$$

where

$$(75) \qquad \Delta\mathbf{H} = \sum_{j=1}^{\infty} \mathbf{H}^{(j)},$$

but of course (74) is of little use since it requires that we already have the solution to $\mathbf{H}\hat{\alpha} = \mathbf{g}$.

In practice, most pre-reconstruction correction methods amount to assuming some typical object, described by a coefficient vector α_p (where p stands for prototype) and defining the corrected data by

$$(76) \qquad \mathbf{g}' = \mathbf{g} - \Delta\mathbf{H}\alpha_p.$$

This method is quite computationally efficient, and it can work well for objects that resemble the prototype.

Post-reconstruction correction begins with an estimate obtained from
(71) with $\mathbf{H}$ replaced by $\mathbf{H}^{(0)}$. If we assume that λ is relatively small, the
minimization principle will yield an estimate $\hat{\alpha}^{(0)}$ such that

$$(77) \qquad\qquad \mathbf{H}^{(0)}\hat{\alpha}^{(0)} \simeq \mathbf{g}.$$

What we want, however, is a solution that satisfies

$$(78) \qquad\qquad \mathbf{H}\alpha \simeq \mathbf{g}.$$

If we write

$$(79) \qquad\qquad \alpha = \alpha^{(0)} - \Delta\alpha,$$

and neglect the term $\Delta\mathbf{H}\Delta\alpha$, then we require that

$$(80) \qquad \mathbf{H}\alpha = \mathbf{H}^{(0)}\alpha^{(0)} - \mathbf{H}^{(0)}\Delta\alpha + \Delta\mathbf{H}\alpha^{(0)} = \mathbf{g}.$$

With (77), we must therefore solve the problem,

$$(81) \qquad\qquad \mathbf{H}^{(0)}\Delta\alpha = \Delta\mathbf{H}\alpha^{(0)}.$$

The right-hand side is known from the initial solution $\alpha^{(0)}$ and knowledge of
the scatter matrices, so we can find $\Delta\alpha$ just by applying the minimization
principle again with $\mathbf{H}^{(0)}$. If it should turn out that $\Delta\mathbf{H}\Delta\alpha$ cannot be
neglected, the solution can be iterated.

**7.6. Post-reconstruction correction with the trickle-down
model.** Another form of post-reconstruction correction can be derived
from the trickle-down model. If the spatial and angular dependence of
the detector response is independent of the photon energy, the mean data
are given by (57). With (49), (52) and (53), we can write

$$
\begin{aligned}
(82) \qquad \bar{\mathbf{g}} = \frac{P_{acc}(E_0)}{4\pi\Delta E}\mathcal{M}\mathcal{L}_\mu \Bigg[& 1 + \frac{P_{acc}(E_1)}{P_{acc}(E_0)}\mathcal{K}_{10}\mathcal{L}_\mu \\
& + \frac{P_{acc}(E_2)}{P_{acc}(E_0)}(\mathcal{K}_{20}\mathcal{L}_\mu + \mathcal{K}_{21}\mathcal{L}_\mu\mathcal{K}_{10}\mathcal{L}_\mu) + ... \Bigg] f(\mathbf{r}).
\end{aligned}
$$

When we use the discrete representation (63) for $f(\mathbf{r})$ and ignore the modeling error, (82) takes the form

$$(83) \qquad\qquad \bar{\mathbf{g}} = \mathbf{H}_0(\mathbf{I} + \mathbf{K})\alpha,$$

where the $M \times N$ matrix $\mathbf{H}_0$ now includes the factor $P_{acc}(E_0)$, $\mathbf{I}$ is the
$N \times N$ unit matrix and $\mathbf{K}$, an $N \times N$ matrix obtained by summing the
trickle-down series and discretizing, accounts for all scatter effects in the
data.

Now suppose that we have obtained an initial solution $\hat{\alpha}_0$ which satisfies

$$(84) \qquad\qquad \mathbf{g} \simeq \mathbf{H}_0\hat{\alpha}_0.$$

The desired solution to

$$(85) \qquad\qquad \mathbf{g} \simeq \mathbf{H}_0(\mathbf{I} + \mathbf{K})\hat{\alpha},$$

can be found by a Neumann expansion,

$$(86) \qquad\qquad \hat{\alpha} = \left[\mathbf{I} - \mathbf{K} + \mathbf{K}^2 + ...\right]\hat{\alpha}_0.$$

This approach was demonstrated in 1985 by Clough [32] with a two-dimensional object and a simplified scatter model. With the computer power that is now available, it is feasible to extend the approach to 3D objects and to use the full Boltzmann equation and trickle-down series to compute $\mathbf{K}$.

It is, however, worth restating the assumptions needed to get to (86). We had to use a discrete object representation and assume that the modeling error was negligible (even if that meant that the coefficients were not strictly estimable), we had to assume the detector was linear and that the spatial and angular dependence of the detector response was independent of the photon energy, and we had to assume that the primary goal of the reconstruction was to produce an estimate of the object coefficients such that $\mathbf{H}\hat{\alpha} \simeq \mathbf{g}$. In practice, some regularization or smoothing will be needed to control noise; it is our view that the choice of the regularization parameter and the quality of the resulting image must then be judged by objective, task-based metrics.

8. Summary and conclusions. We have shown that the Boltzmann transport equation provides a comprehensive framework for describing the flux of gamma rays produced by a radioactive tracer in a patient's body. The Boltzmann equation includes the effects of photon generation, absorption, propagation and scattering. When combined with a general linear model for an image detector, the Boltzmann equation yields an accurate description of the imaging system as a continuous-to-discrete mapping. When a discrete object model is assumed, this theory leads to a matrix description which forms the starting point for inverse-source problems. Included in the imaging matrix is a series of terms describing the scattering process. This series can be formulated in two distinct ways: either each term represents a specific number of scattering events, or it represents the contribution from all scattering events to a specific photon energy.

From a classical estimation-theoretic viewpoint, it makes little sense to attempt to estimate the coefficients in the discrete object expansion. If many terms are used in the expansion, the object model will lead to an accurate description of the data, but the coefficients will not be estimable.

Because of the inevitable null functions in the system, infinitely many different true objects will produce the same data, so there is no way of saying how close any estimated object comes to the truth. On the other hand, if only a coarse object description is used (large voxels, for example), it may be possible to estimate the coefficients and define estimation errors, but then modeling errors produce a large and unknown effective noise source in the data.

Most reconstruction algorithms ignore this fundamental contradiction, adopt some discrete object representation and then determine the coefficients by minimizing a suitable functional. We sketched above some ways of carrying out this procedure using the Boltzmann-based theory of the forward problem. If the object description is accurate enough that modeling errors can be neglected, we can compute the mean, variance and covariance of the estimates of the coefficients, and in many cases even their full probability density function, for any specified object. From this statistical information we can compute objective, task-based figures of merit for image quality.

We also note the tremendous impact that high-performance computing has on the problems posed in this paper. In Sec. 5 we emphasized the large amount of memory required to store even the most efficient representations of the distribution function. Moreover, the highest possible number of floating-point operations per second are needed for executing the computations required in implementing the forward models or solving any inverse problem. And, as demanding as these requirements are, the computations needed for task-based assessment of image quality are even more so, since they inherently involve large training sets or ensembles of objects. Image quality is fundamentally defined in statistical terms, so adequate sampling of the object statistics is essential.

Acknowledgements. The authors have benefitted from helpful discussions with Tim White, Craig Abbey, Jack Denny, Kyle Myers and Robert Wagner.

This work was supported by the National Institutes of Health (NIH) through grants no. PO1 CA23417 and RO1 CA52643, but the views presented here do not represent an official position of NIH.

REFERENCES

[1] P.J. Early and D.B. Sodee, *Principles and Practice of Nuclear Medicine*, Mosby-Year Book, Inc., Toronto, second edition, 1995.

[2] M.N. Maisey, K.E. Britton, and D.L. Gilday, editors, *Clinical Nuclear Medicine*, J.B. Lippincott Co., Philadelphia, PA, USA, 2nd edition, 1991.

[3] K. Krane, *Modern Physics*, John Wiley & Sons, Inc., Toronto, 1983.

[4] H.H. Barrett and W. Swindell, *Radiological Imaging: Theory of Image Formation, Detection and Processing, Vols. I and II*, Academic Press, New York, 1981; paperback edition, 1996.

[5] M.J. Lighthill, *An Introduction to Fourier Analysis and Generalized Functions*, Cambridge University Press, Cambridge, England, 1958.

[6] R.G. Wells, A. Celler, and R. Harrop, "Analytical calculation of photon distributions in SPECT projections", IEEE Trans. Nuc. Sci., 1997.

[7] C. Lowry and M. Cooper, "The problem of Compton scattering in emission tomography: a measurement of its spatial distribution", Phys. Med. Biol. 32:1187, 1987.

[8] P. Msaki, B. Axelsson, C.M. Dahl, and S.A. Larsson. "Generalized scatter correction method in SPECT using point scatter distribution functions", J. Nuc. Med. 28(12):1861–1869, 1987.

[9] D.R. Haynor, R.L. Harrison, T.K. Lewellen, A.N. Bice, C.P. Anson, S.B. Gillispie, R.S. Miyaoka, K.R. Pollard, and J.B. Zhu, "Improving the efficiency of emission tomography simulations using variance reduction techniques", IEEE Trans. Nuc. Sci. 37(2):749–753, 1990.

[10] M. Ljungberg, *Development and evaluation of attenuation and scatter correction techniques for SPECT using the Monte Carlo method*, Ph.D. Thesis, University of Lund, Sweden, 1990.

[11] A.M. Weinberg and E.P. Wigner, *The Physical Theory of Neutron Chain Reactors*, U. Chicago Press, 1958.

[12] N. Ben Abdallah and P. Degond, "On a hierarchy of macroscopic models for semiconductors", J. Math. Physics 37(7):3306–3333, 1996.

[13] C.P. Carpenter and F.W. Metzger, "General spherical harmonics formulation of plasma Boltzmann equation", J. Math. Phys. 2(5):694–701, 1961.

[14] F. Natterer, *The Mathematics of Computed Tomography*, Wiley, New York, 1986.

[15] O. Tretiak and C.E. Metz, "The exponential Radon transform, SIAM J. Appl. Math. 39:341–354, 1980.

[16] A. Clough and H.H. Barrett, "Attenuated Radon and Abel transforms", J. Opt. Soc. Am. 73:1590–1595 1983.

[17] X. Pan and C.E. Metz, "Non-iterative methods and their noise characteristics in 2D SPECT reconstruction, to be published.

[18] F. Natterer, "Determination of tissue attenuation in emission tomography of optically dense media", Inverse Problems 9:731–736, 1991.

[19] F. Natterer and H. Herzog, "Attenuation correction in positron emission tomography, Mathematical methods in the Applied Sciences 15:321–330, 1992.

[20] E. Clarkson, "Consistency conditions and reconstruction methods for the exponential Radon transform", submitted to Inverse Problems, 1998.

[21] A. Albert, *Regression and the Moore-Penrose Pseudoinverse*, Academic Press, New York, 1972.

[22] H.H. Barrett, "Objective assessment of image quality: effects of quantum noise and object variability", J. Opt. Soc. Am. A 7:1266–1278 1990.

[23] E. Clarkson and H.H. Barrett, "A bound on null functions for digital imaging systems with positivity constraints", Opt. Lett. 22:814–815, 1997.

[24] Eric Clarkson and H.H. Barrett, "Bounds on null functions of linear digital imaging systems", accepted for publication in J. Opt. Soc. Am. A.

[25] M.H. Buonocore, W.R. Brody and A. Macovski, IEEE Trans. Biomed. Eng. BME-28:69–78, 1981.

[26] H.H. Barrett, D.W. Wilson and B.M.W. Tsui, "Noise properties of the EM algorithm: I Theory", Phys. Med. Biol. 39:833–846 1994.

[27] J.A. Fessler, "Mean and variance of implicitly defined biased estimators (such as penalized maximum likelihood): Applications to tomography", IEEE Trans. in Image Proc. 5(3):493–506, 1996.

[28] H.H. Barrett, J.L. Denny, R.F. Wagner, and K.J. Myers, "Objective assessment of image quality. II Fisher information, Fourier crosstalk, and figures of merit for task performance", J. Opt. Soc. Am. A 12(5):834–852, 1995.

[29] H.H. Barrett, Jie Yao, Jannick Rolland, and Kyle J. Myers, "Model observers for assessment of image quality", Proc. Nat. Acad. Sci. 90:9758–9765 1993.

[30] H.H. Barrett, T.A. Gooley, K.A. Girodias, J.P. Rolland, T.A. White, and J. Yao, "Linear discriminants and image quality, Image and Vision Computing", 10(6):451–460, 1992.
[31] C.K. Abbey and H.H. Barrett, "Practical issues and methodology in assessment of image quality using model observers", Proc. SPIE, vol. 3032, 1997.
[32] A.V. Clough, "A mathematical model of single-photon emission computed tomography", University of Arizona Ph. D. dissertation 1985.

MATHEMATICAL ASPECTS OF RADIATION THERAPY TREATMENT PLANNING: CONTINUOUS INVERSION VERSUS FULL DISCRETIZATION AND OPTIMIZATION VERSUS FEASIBILITY

YAIR CENSOR*

Abstract. A mathematical formulation of the radiation therapy problem consists of a pair of forward and inverse problems. The inverse problem is to determine external radiation beams, along with their locations, profiles, and intensities, that will provide a given dose distribution within the irradiated object. We discuss the inverse problem in its fully discretized formulation.

1. Introduction. This paper deals with radiation *teletherapy* where beams of penetrating radiation are directed at the lesion (tumor) from an external source. The other radiation delivery mode which involves direct implantation of radioactive sources inside the lesion, called *brachytherapy*, is not included in our discussion. Chapter 11 of the book by Censor and Zenios [9] and Brahme's special issue [6] and references therein, as well as the tutorial review of Altschuler, Censor, and Powlis [2], can be used as introductory material to this area.

Based on understanding of the physics and biology of the situation, there are two principal aspects of radiation teletherapy that call for mathematical modelling. The first is the calculation of the *radiation dose* which is a measure of the actual energy absorbed per unit mass everywhere in the irradiated tissue. In dose calculation, termed *dosimetry*, the relevant physical and biological characteristics of the irradiated object and the relevant information about the radiation source (geometry, physical nature, intensity, etc.) serve as input data. The result (output) of the calculation is a *dose function* whose values are the dose absorbed as a function of location inside the irradiated body.

The second aspect is the *mathematical inverse problem* of the first. In addition to all physical and biological parameters of the irradiated object we assume here that the relevant information about the capabilities and specifications of the available *treatment machine* (i.e., radiation source) is given. Based on medical diagnosis, knowledge, and experience, the physician prescribes a *desired dose function* to the case. The output of this problem should be a *radiation intensity function* whose values are the radiation intensity at the source as a function of source location, that would result in a dose function which is identical to the desired one. To be of practical value, this resulting radiation intensity function must be implementable, in a clinically acceptable form, on the available treatment machine.

*Department of Mathematics, University of Haifa, Mt. Carmel, Haifa 31905, Israel. E-mail: yair@mathcs2.haifa.ac.il

In what follows we discuss, from a mathematician's point of view, two main modelling dilemmas: (i) continuous inversion versus full discretization, and (ii) optimization versus feasibility.

Much of current *radiation therapy treatment planning* (RTTP) is still done in two dimensions where only a single plane through the center of the target is considered. RTTP is also still done mostly in a trial-and-error fashion by picking a machine setup that gives rise to a certain external radiation intensity field (function) and then using a forward-problem-solver software package to determine the resulting dose function, see Figure 1. If the discrepancy between this dose function and the prescribed dose function is unacceptable then some changes are made to the machine setup and the process is repeated until the physician and dosimetrist are satisfied with the resulting dose function. Only then actual patient treatment is performed.

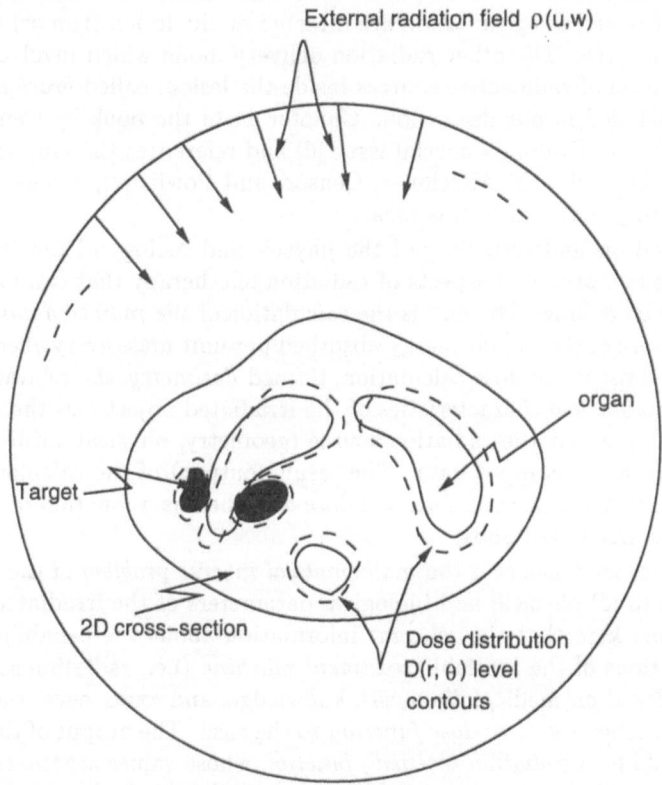

FIG. 1. *2D-RTTP, an external radiation field $\rho(u, w)$ results in a dose distribution $D(r, \theta)$.*

Such 2D-RTTP has achieved success due to accumulated experience and also because of the ever increasing quality and speed of forward-problem-solvers.

Automated solution of the inverse problem of RTTP should be useful in handling difficult planning cases, particularly in 3D–RTTP, see Figure 2. There, it would be much more difficult to reach an acceptable plan by trial-and-error because of the multitude of potential directions from which the 3D object can be irradiated. Nonetheless, even a 2D discussion, as given here, is enough to expose the nature of the dilemmas that we consider.

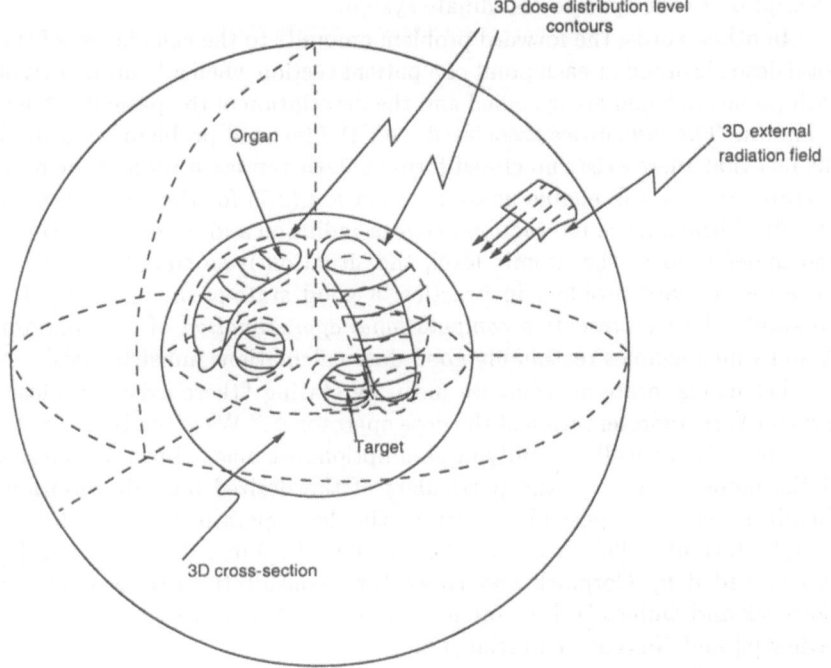

FIG. 2. *3D–RTTP, fully 3D cross section, external radiation field and dose distribution.*

In addition to the references given in the sequel we recommend also Mackie et al. [16], Raphael [18], Webb [19], and Xing and Chen [20].

2. Problem definition and the continuous model. Let $D(r, \theta)$ be a real-valued nonnegative function, of the polar coordinates r and θ, whose value is the dose absorbed at a point in the patient's planar cross-section coincident with the plane of the machine's gantry motion. This is the *dose function*, or dose distribution. A *ray* is a directed line along which radiated energy travels away from the *source*, i.e., the *teletherapy source*. Rays are parametrized by variables u and w in some well-defined way and the real-valued nonnegative function $\rho(u, w)$ represents the *radiation intensity* along the ray (u, w) due to a point source on the gantry circle. The continuous forward problem of RTTP is the following. Assume that the cross section Ω of the patient and its radiation absorption characteristics are known. Given a radiation intensity function $\rho(u, w)$ for $0 \leq u < 2\pi$ and $-W \leq w \leq W$, find the dose function $D(r, \theta)$ for all $(r, \theta) \in \Omega$ from the formula

$$(2.1) \qquad\qquad D(r, \theta) = \Delta[\rho(u, w)](r, \theta),$$

where Δ is the *dose operator*. This operator relates the dose function to the radiation intensity function. See, e.g., [8] or [9, Chapter 11], for a description of the specific coordinate system.

In other words, the forward problem amounts to the calculation of the total dose absorbed at each point of a patient section when all parameters of each radiation beam are specified and the description of the patient section is known. The difficulties associated with the forward problem stem from the fact that there exists no closed-form analytic representation of the dose operator Δ that will enable us to use equation (2.1) for the calculation of $D(r, \theta)$. Although the interaction between radiation and tissue is measured and understood at the atomic level, the situation is so complex that, to solve the forward problem in practice, a good state-of-the-art computer program, which represents a *computational approximation* of the operator Δ and which enables reasonably good dose calculations, must be used.

Let us elaborate on what we mean by stating "there exists no close-form analytic representation of the dose operator Δ." We actually mean the following: If drastically simplifying assumptions are made about the physics of the model as well as the particulars of the desired dose distribution, then it is sometimes possible to express the dose operator in a closed-form analytic formula. This has been done first by Brahme, Roos and Lax [4] and extended by Cormack and co-workers, consult the review paper of Cormack and Quinto [12] for further references. See also Brahme's recent review [5] and Goitein's editorial [13].

In current practice of RTTP, when dose calculations are performed to verify the dose that will result from a proposed treatment plan, the goal is to obtain results that are as accurate as possible. To achieve this, various empirical data, which are often condensed in look-up tables, are incorporated into the forward calculation. Thus, the true forward calculation, or true dose operator, is not represented by a closed-form analytic relation between the radiation intensity function $\rho(u, w)$ and the dose function $D(r, \theta)$, but by a software package that calculates $D(r, \theta)$ from $\rho(u, w)$. Thus, what we really mean by saying that there is no closed-form analytic expression for Δ is that we choose to adhere to the software representation rather than compromise by allowing simplifying assumptions that might lead to a closed-form analytic mathematical formula.

The *inverse problem* of radiation therapy is the treatment planning problem:

Given a description of the patient section, the dose prescribed for the target, and the maximum permissible doses to the target, critical organs, and other tissues, calculate the external configuration and relative intensities of radiation sources (i.e., the radiation field) that will deliver the specified radiation doses (or some acceptable approximation thereof).

Assuming that the cross section Ω of the patient and its radiation absorption characteristics are known, and given a prescribed dose function $D(r, \theta)$, the problem is to find a radiation intensity function $\rho(u, w)$ such that equation (2.1) holds, or $\rho(u, w) = \Delta^{-1}[D(r, \theta)]$ where Δ^{-1} is the inverse operator of Δ. This is the inversion problem that we want to solve, in a computationally tractable way, although no closed-form analytic mathematical representation is available for the dose operator Δ. The dose at (r, θ) is the sum of the dose contributions from the sources at all the different gantry angles. Thus

$$
(2.2) \qquad D(r, \theta) = \sum_{i=1}^{S} y_i D_i(r, \theta),
$$

where, for each $i = 1, 2, \ldots, S$, the value $D_i(r, \theta)$ is the dose deposited at point (r, θ) by a beam of unit intensity from the ith source, and y_i is the time the ith beam is kept on.

It will be assumed here that the dose $D_i(r, \theta)$ can be calculated accurately once the beam parameters and patient section information are specified. That is, we assume that we can solve the forward problem and calculate $D(r, \theta)$ accurately from (2.2). This assumption is confirmed by innumerable direct measurements in water and tissue-equivalent phantoms.

Whereas a dose distribution that solves the forward problem is always obtained for a specified radiation intensity field, the inverse problem may have no solution at all, since some prescribed dose functions may be unobtainable from any radiation field.

3. Discretization of the problem. In the approach presented here, we adhere to the computerized calculation of the dose operator Δ. Full discretization of the problem at the outset is used to circumvent the difficulties associated with the inversion of Δ. We also neglect the effect of scatter. The patient's cross section Ω is discretized into a grid of points represented by $\{(r_j, \theta_j) \mid j = 1, 2, \ldots, J\}$. Define $\Delta_j[\rho]$ by

$$
(3.1) \qquad \Delta_j[\rho] = [\Delta\rho](r_j, \theta_j)
$$

and call Δ_j a *dose functional*, for every $j = 1, 2, \ldots, J$. Acting on a radiation intensity function $\rho(u, w)$, the functional Δ_j provides $\Delta_j[\rho]$, which is the dose absorbed at the jth grid point of the patient's cross section Ω due to the radiation intensity field ρ.

In continuing the discretization process of the problem it is assumed that a set of I *basis radiation intensity fields* is fixed and that their nonnegative linear combinations can give adequate approximations to any radiation intensity field we wish to specify. This is done by discretizing the region $0 \leq u < 2\pi$, $-W \leq w \leq W$ in the (u, w)-plane into a grid of points given by $\{(u_i, w_i) \mid i = 1, 2, \ldots, I\}$. A radiation intensity function

$$(3.2) \qquad \sigma_i(u, w) = \begin{cases} 1, & \text{if } (u, w) = (u_i, w_i), \\ 0, & \text{otherwise,} \end{cases}$$

is a *unit intensity ray* and serves as a member of the set of basis intensity fields, $i = 1, 2, \ldots, I$. A desired radiation intensity function ρ that solves the inverse problem is approximated by

$$(3.3) \qquad \widehat{\rho}(u, w) = \sum_{i=1}^{I} x_i \sigma_i(u, w),$$

where x_i is the intensity of the ith ray, and it is required that $x_i \geq 0$, for all $i = 1, 2, \ldots, I$. Once the grid points are fixed, any radiation intensity function $\widehat{\rho}$ that can be presented as a nonnegative linear combination of the rays is uniquely determined by the coefficients x_i, $1 \leq i \leq I$. The vector $x = (x_i)$, in the I-dimensional Euclidean space $\mathbb{R}^I$, is referred to as the *radiation vector* or *basic solution*.

Further, assume that the dose functionals Δ_j are linear and continuous. This assumption cannot be mathematically verified due to the absence of an analytic representation of Δ or Δ_j, but it is a reasonable assumption based on the empirical knowledge of Δ_j. Using linearity and continuity of all Δ_j's, we can write $\Delta_j[\rho] \simeq \Delta_j[\widehat{\rho}] = \sum_{i=1}^{I} x_i \Delta_j[\sigma_i]$. For $j = 1, 2, \ldots, J$, and $i = 1, 2, \ldots, I$, denote by

$$(3.4) \qquad a_{ij} = \Delta_j[\sigma_i]$$

the dose deposited at the jth point (r_j, θ_j) in the patient's cross section Ω due to a unit intensity ray $\sigma_i(u, w)$. The *fully discretized inverse problem* of RTTP then becomes to find a radiation vector $x \in \mathbb{R}^I$ such that

$$(3.5) \qquad A^T x = b, \quad x \geq 0,$$

where $A = (a_{ij})$ is the $I \times J$ matrix with elements as in (3.4) and $b = (b_j) \in \mathbb{R}^J$ is the discretized desired dose vector.

This fully discretized model calls for the quantities a_{ij} which can be precalculated with any state-of-the-art forward-problem-solver. If the latter is beam-driven the apportionment of beam dose per unit intensity among all rays, into which the beam has been discretized, is necessary, see Censor, Altshuler and Powlis [8], Powlis et al. [17]. Numerous iterative techniques are available for the solution of (3.5), both in the consistent case, see, e.g., the recent review of Bauschke and Borwein [3], and the inconsistent case, e.g., Combettes [11], Byrne and Censor [7].

The tendency to make the discretization finer results in very large values of I and J. As long as the available treatment machines cannot deliver such finely discretized radiation intensity fields, we need an additional computational step after a solution vector x^* (or approximation thereof) of the system (3.5) has been obtained. This is a "consolidation" step in which a clinically acceptable machine setup, usually at few (up to 5–6) beam positions, is derived from the fully discretized solution vector x^*, see [17]. To sum up, the fully discretized model is not difficulties-free, but it offers a route of circumventing the inversion problem of the computational dose operator Δ without compromising on any of the heuristics and empiricism involved in it.

4. Optimization versus feasibility.

4.1. Feasibility. The *feasibility formulation* relaxes the equality (2.1). Let $\overline{D} = \overline{D}(r, \theta)$ and $\underline{D} = \underline{D}(r, \theta)$ be two dose functions whose values represent upper and lower bounds, respectively, on the permitted and required dose inside the patient's cross section. A radiation therapist defines $\overline{D}$ and $\underline{D}$ for each given case and will accept as a solution to the RTTP problem any radiation intensity function $\rho(u, w)$ that satisfies

$$(4.1) \qquad \underline{D}(r, \theta) \leq \Delta[\rho(u, w)](r, \theta) \leq \overline{D}(r, \theta), \quad \text{for all} \quad (r, \theta) \in \Omega.$$

In target regions (tumors) the lower bound $\underline{D}$ is usually the important factor because the dose there should exceed some given value. In critical organs and other healthy tissues $\underline{D}(r, \theta) = 0$, so that $\overline{D}(r, \theta)$ is the dose that cannot be exceeded. Any solution $\rho(u, w)$ that fulfills (4.1), for given $\overline{D}$ and $\underline{D}$, is a *feasible solution* to the RTTP problem. In order to discretize (4.1) we must specify the dose functions $\overline{D}$ and $\underline{D}$ at the grid points by giving, for all $j = 1, 2, \ldots, J$,

$$(4.2) \qquad \overline{D}(r_j, \theta_j) = \overline{D}_j, \qquad \underline{D}(r_j, \theta_j) = \underline{D}_j,$$

thus converting (4.1) into a finite system of *interval inequalities*

$$(4.3) \qquad \underline{D}_j \leq \Delta_j[\rho] \leq \overline{D}_j, \quad j = 1, 2, \ldots, J.$$

Denoting hereafter by $\overline{D}$ ($\underline{D}$) the J-dimensional column vector whose jth element is $\overline{D}_j$ ($\underline{D}_j$), the inverse problem of RTTP is restated as follows:

Given vectors $\overline{D}$ and $\underline{D}$ of permitted and required doses at J grid points in the patient's cross section Ω, find a radiation intensity distribution $\rho = \rho(u, w)$ such that (4.3) holds. The *fully discretized feasibility inverse problem* of RTTP then becomes the linear interval feasibility problem of finding a vector $x \in \mathbb{R}^I$ such that

$$(4.4) \qquad \underline{D}_j \leq \sum_{i=1}^{I} x_i d_{ij} \leq \overline{D}_j, \quad j = 1, 2, \ldots, J,$$

$$x_i \geq 0, \quad i = 1, 2, \ldots, I.$$

Let the set of pixels in the discretized patient cross section be denoted by $N = \{1, 2, \ldots, J\}$. Organs within the patient section are then defined as subsets of N. The subsets $B_k \subset N$, where $k = 1, 2, \ldots, K$ denote K critical organs to be spared from excessive radiation. Let the values b_k denote the corresponding upper bounds on the dose permitted in each critical organ. The subsets $T_q \subset N$, where $q = 1, 2, \ldots, Q$, denote Q target regions. Let the values t_q denote the corresponding prescribed lower bounds for the absorbed dose in each. All the B_k and T_q are pairwise disjoint. The set of pixels inside the patient section that are not in any B_k or T_q are called the *complement*, denoted as the subset $C \subset N$, and c is the upper bound for the total permitted dose there. It is assumed that the definition of all subsets B_k, T_k, and C and the prescription of all b_k, t_q, and c are given by the radiotherapist as input data for the discretized treatment planning problem.

Problem (4.4) then becomes the following system of linear inequalities, which we call the *basic model*:

$$(4.5) \qquad \sum_{i=1}^{I} d_{ij} x_i \leq b_k, \quad \text{for all} \quad j \in B_k, \ k = 1, 2, \ldots, K,$$

$$(4.6) \qquad t_q \leq \sum_{i=1}^{I} a_{ij} x_i, \quad \text{for all} \quad j \in T_q, \ q = 1, 2, \ldots, Q,$$

$$(4.7) \qquad \sum_{i=1}^{I} a_{ij} x_i \leq c, \quad \text{for all} \quad j \in C,$$

$$(4.8) \qquad x_i \geq 0, \quad \text{for all} \quad i = 1, 2, \ldots, I.$$

With b_k, t_q, and c given and the a_{ij}'s calculated from (3.4), the mathematical question represented by the basic model (4.5)–(4.8) is to find a nonnegative solution vector $x^* = (x_i^*)$ for a system of linear inequalities. The remarks about clinical acceptability of x^* from the end of the last section apply also here.

We first proposed this fully discretized feasibility inverse problem in Altschuler and Censor [1], see also [9, Section 11.7] for a brief review of other approaches and references.

4.2. Optimization. When it comes to discussing an optimization approach to RTTP we must distinguish between two different kinds of optimization problems depending on the space in which they are formulated. One possibility is to define an objective function $f : \mathbb{R}^I \to \mathbb{R}$, i.e., over the space of radiation intensity vectors x and use either the system (3.5) or the constraints (4.5)–(4.8) as the feasible set. For example, choosing $f(x) = \frac{1}{2}\|x\|^2$ ($\|\cdot\|$ stands for the Euclidean norm) and solving a minimization problem will lead to a minimum-norm solution vector x^*. I.e., a feasible vector closest to the origin so that the total radiation intensity is smallest possible in the Euclidean norm sense. A special-purpose iterative minimization method such as Hildreth's algorithm, see, e.g., [9], applies in this case.

Regardless of the specific choice of f, in this approach the interval-constrained optimization problem

$$\begin{aligned} &\min f(x) \\ \text{s.t. } &\alpha \le A^T x \le \beta, \\ &x \ge 0, \end{aligned}$$

(4.9)

is still aiming at solution of the fully discretized formulation of the inverse problem. A solution vector x^* will represent a radiation field that will deliver a dose which is both feasible (i.e., adheres to the upper and lower doses imposed by the physician) and is optimal in the sense of the objective function f. This approach of optimization in the space of radiation intensity vectors will be called henceforth *radiation intensity optimization.*

The second possibility for introducing an optimization problem in RTTP is to use (3.5) or (4.5)–(4.8) as constraints but choose an objective function $g : \mathbb{R}^J \to \mathbb{R}$ defined over the space of dose vectors. Such objective functions may be either *biological,* or *physical.* Biological objective functions represent knowledge (statistical or other) about various biological mechanisms that affect our ability to control the disease. An example is the conditional probability of having tumor control without severe injury, denoted in RTTP literature by P_+. Physical objective functions aggregate physical features which are important for tumor control and prevention of normal tissue complications, such as dose variance over target volume or peak dose to organs at risk. A thorough discussion of biological and physical objective functions can be found in Brahme [5]. Let us call this kind of optimization, over the space of dose vectors, *dose optimization.*

5. Discussion. The trade-off between the continuous model and full discretization has already been explained in Section 3. Brahme reaches also a conclusion in favor of full discretization and says [5, p. 216]: "... In either case it is very useful to transform the relevant integral equation into an algebraic form by discretizing the transport quantities along the coordinates of the free variables."

The question of feasibility versus optimization is not crucial if only radiation intensity optimization (as defined above) is considered. This is so because both the feasibility formulation and the optimization formulation (regardless of the particular choice of the objective function $f(x)$) occur in the same space (of radiation intensity vectors) and, thus, aim at a solution of the discretized inverse problem. Therefore, the difference between these two formulations is, from the mathematical point of view, only technical. Recently, Cho et al. [10] reported on the advantage of the feasibility approach over a global optimization model solved by simulated annealing. In contrast, the dose optimization (as defined above) approach leads to a problem of the form

$$\min g(A^T x)$$
$$(5.1) \qquad \text{s.t.} \quad \alpha \leq A^T x \leq \beta,$$
$$x \geq 0.$$

If a set of dose vectors $b^\ell \in \mathbb{R}^J$, for $\ell = 1, 2, \ldots, L$, each of which represents a deliverable treatment plan, are given, then the values of a biological or physical dose objective function $g(b^\ell)$ can be calculated for each and compared. Choosing the plan with lowest $g(b^\ell)$ in such circumstances means that we are merely doing a comparison (among rival plans) which are given (i.e., constructed in some way prior to the comparison).

In case when the composite function $g(A^T x)$ is simple enough (the approach of (5.1) can still be efficiently used for solving directly the (discretized) inverse problem in its full generality. Otherwise, the inversion problem has to be abandoned and the optimization can be performed with respect to only few parameters of the external radiation field. See, for example, Gustafsson [14] and Gustafsson, Lind and Brahme [15]. This is done while other important parameters are left out of the optimization problem and must be given as input to the process, see also the discussion in [9, Section 11.7].

The question whether to adhere to the mathematical inverse problem (and possibly confront a difficulty when translating a radiation intensity solution vector x^* into an implementable and clinically acceptable treatment plan) or to use biological or physical objective functions in the space of dose vectors (and thereby possibly compromise on the full generality of the inverse problem)—remains unsettled.

Acknowledgements. The author gratefully acknowledges useful discussions with Martin Altschuler, Gabor Herman, William Powlis, and the comments of Christopher Borgers on an earlier version of the paper. Part of the material presented here appears in [9, Chapter 11]. This work was partially supported by NIH grant HL–28438 at the Medical Image Processing Group (MIPG), Department of Radiology, University of Pennsylvania, Philadelphia, PA, USA, and by the Israeli Science Foundation founded by The Israel Academy of Sciences and Humanities. The work of Ms. Patricia Brick and her team at IMA in preparing this manuscript in Tex is greatly appreciated.

REFERENCES

[1] ALTSCHULER, M.D., AND CENSOR, Y., *Feasibility solutions in radiation therapy treatment planning.* In J.R. Cunningham, D. Ragan, and J. Van Dyk, editors, *Proceedings of the Eighth International conference on the Use of Computers in Radiation Therapy,* IEEE Computer Society Press, Silver Spring, MD, USA, 1984, pp. 220–224.

[2] ALTSCHULER, M.D., CENSOR, Y., AND POWLIS, W.D., *Feasibility and optimization methods in teletherapy planning,* in: *Advances in Radiation Oncology Physics* (J.A. Purdy, Ed.), Medical Physics Monograph No. 19, American Association of Physicists in Medicine, American Institute of Physics, Inc., New York, 1992, pp. 1022–1057.

[3] BAUSCHKE, H.H. AND BORWEIN, J.M., *On projection algorithms for solving convex feasibility problems,* SIAM Review, 38:367–426, 1996.

[4] BRAHME, A., ROOS, J–E., AND LAX, I., *Solution of an integral equation encountered in rotation therapy,* Physics in Medicine and Biology, 27:1221–1229, 1982.

[5] BRAHME, A., *Treatment optimization using physical and radiological objective functions,* In: A.R. Smith (Editor), *Medical Radiology: Radiation Therapy Physics,* Springer-Verlag, Berlin, Germany, 1995, pp. 209–246.

[6] BRAHME, A (Editor), *Optimization of the Three-Dimensional Dose Delivery and Tomotherapy,* Special Issue of : International Journal of Imaging Systems and Technology, Volume 6(1), Spring 1995.

[7] BYRNE, C. AND CENSOR, Y., *Proximity function minimization and the convex feasibility problem for jointly convex Bregman distances,* Technical Report, February 1997.

[8] CENSOR, Y. ALTSCHULER, M.D., AND POWLIS, W.D., *On the use of Cimmino's simultaneous projections method for computing a solution of the inverse problem in radiation therapy treatment planning,* Inverse Problems, 4:607–623, 1988.

[9] CENSOR, Y. AND ZENIOS, S.A., *Parallel Optimization: Theory, Algorithms, and Applications,* Oxford University Press, New York, 1997.
(See http://www.oup-usa.org/gcdocs/gc_019510062X.html on the internet.)

[10] CHO, P.S., LEE, S., MARKS II, R.J., REDSTONE, J.A, AND OH, S., *Comparison of algorithms for intensity modulated beam optimization: projections onto convex sets and simulated annealing,* In: Leavitt, D.D., and Starkschall, G. (Editors), *Proceedings of the XII International Conference on the Use of Computers in Radiation Therapy (May 27–30, 1997, Salt Lake City, Utah, USA),* Medical Physics Publishing, Madison, Wisconsin, 1997, pp. 310–312.

[11] COMBETTES, P.L., *Inconsistent signal feasibility problems: Least-squares solutions in a product space,* IEEE Transactions on Signal Processing, SP–42:2955–2966, 1994.

[12] CORMACK, A.M. AND QUINTO, E.T., *The mathematics and physics of radiation dose planning using X-rays,* Contemporary Mathematics, 113:41–55, 1990.

[13] GOITEIN, M., *The inverse problem,* International Journal Radiation Oncology Biology Physics, 18:489–491, 1990.

[14] GUSTAFSSON, A., *Development of a versatile algorithm for optimization of radiation therapy,* Ph.D Thesis, Department of Medical Radiation Physics, University of Stockholm, Sweden, 1996.

[15] GUSTAFSSON, A., LIND, B.K., AND BRAHME, A., *A generalized pencil beam algorithm for optimization of radiation therapy,* Medical Physics, 21:343–356, 1994.

[16] MACKIE, T.R., HOLMES, T.W., RECKWERDT, P.J., AND YANG, J., *Tomotherapy: Optimized planning and delivery of radiation therapy,* International Journal of Imaging Systems and Technology, 6:43–55, 1995.

[17] POWLIS, W.D., ALTSCHULER, M.D., CENSOR, Y., AND BUHLE, E.L. JR., *Semi-*

automatic radiotherapy treatment planning with a mathematical model to satisfy treatment goals, International Journal Radiation Oncology Biology Physics, 16:271–276, 1989.

[18] RAPHAEL, C., *Radiation therapy treatment planning: An L^2 approach*, Applied Mathematics and Computation, 52:251–277, 1992.

[19] WEBB, S., *Optimization of conformal radiotherapy dose distribution by simulated annealing*, Physics in Medicine and Biology, 34:1349–1370, 1989.

[20] XING, L. AND CHEN, G.T.Y., *Iterative methods for inverse treatment planning*, Physics in Medicine and Biology, 41:2107–2123, 1996.

EARLY RESULTS ON GENERAL VERTEX SETS AND TRUNCATED PROJECTIONS IN CONE-BEAM TOMOGRAPHY

ROLF CLACKDOYLE*, MICHEL DEFRISE†, AND FRÉDÉRIC NOO‡

Abstract. This paper summarizes a talk titled "Some recent results in cone-beam tomography" which was presented at the IMA workshop "Computational Radiology and Imaging: Therapy and Diagnostics" March 17-21, 1997.

A cone-beam projection of some object is a collection of ray-sums through the object, where all the rays converge on a single "vertex point" in space. Usually this vertex point is outside the object, and it is often assumed that from each vertex point, every non-zero ray-sum through the object is available. If some of these ray-sums are not available, the cone-beam projection is called a truncated projection.

Several algorithms are available to reconstruct the object from its cone-beam projections, under the assumptions that the vertex point travels along a suitable path in space and that no projection is truncated. There has been progress recently on algorithms that relax these assumptions. Effective algorithms for handling certain kinds of truncated data will be discussed, and an algorithm that is able to reconstruct from a discrete, unordered set of vertex points (but with no truncated projections) will be presented.

Images obtained from these algorithms will be presented for the cases of computer-simulated data, and for data taken from a large-area CT scanner.

1. Introduction. The development of cone-beam tomography as an imaging tool has been accelerating over the past five years, as software and hardware advances leapfrog to meet applications demands and generate new imaging configurations. Cone-beam scanners providing tomographic information for non-destructive examination and for medical diagnosis are becoming more prevalent. A major step forward was the development in 1984 of the FDK algorithm by L. A. Feldkamp *et al* [4], which provided a fast and robust method for image reconstruction when the cone-beam vertex point traverses a circle surrounding the object of interest. With the general availability of an effective and easily implementable algorithm, cone-beam scanners have been built using image-intensifiers to gather two-dimensional x-ray projections, and cone-beam and pinhole collimators have been used on conventional Single Photon Emission Computed Tomography (SPECT) medical scanners to image radiotracer distributions in medical imaging. Using small focal spot x-ray sources, microtomography systems have also appeared, and impressive images of small objects, even insects [10], have been reported.

The condition for tomographically complete data is usually attributed to Tuy [19], and requires that the cone-beam vertex point (the source position in x-ray systems, or the focal point of the cone-beam collimator in SPECT systems) traverse a path around the object which is intersected by

---

*University of Utah
†Free University of Brussels
‡University of Liège

any plane which cuts the object to be imaged. Although the usual circular path does not satisfy this condition, the missing information is, for most applications, relatively small, unless large cone-angles are used or detailed quantitative images are required. Tuy's 1983 paper did not provide an image reconstruction algorithm and the vertex paths satisfying the completeness condition were necessarily non-planar and not as convenient to implement as a circular path. Suitable algorithms started to appear a few years later, the most effective of which were based on the celebrated result of Grangeat [5, 6] who provided a formula linking cone-beam data to a certain region of the three-dimensional Radon transform of the object. This result plays a similar role to the well-known central-section projection theorem (or Fourier-slice theorem) in classical tomography. Rebinning algorithms appeared, analogous to the Fourier method in classical tomography, and in the early 1990's, general filtered-backprojection style algorithms were published [8, 3]. Some cone-beam scanners are now appearing which can collect sufficient (tomographically complete) data sets and require these new algorithms for image reconstruction. General iterative methods are also being used for cone-beam reconstruction from complete projection data sets as previously prohibitive reconstruction times are gradually offset by dramatic improvements in computer power.

High quality quantitative x-ray systems for medical diagnosis still await the development of inexpensive area detectors capable of high dynamic range measurements. Such detectors are starting to appear, but currently consist of just a few rows of detector pixels, and can hardly be considered true cone-beam systems. However, this trend is likely to continue, and the first medical cone-beam x-ray CT systems will probably have active detector areas of roughly 50 cm in width by 10 cm of height. This truncated detector geometry has already sparked considerable research effort on reconstruction algorithms.

In this paper, existing reconstruction techniques will be summarized, and some recent advances will be described. Only analytic (direct) reconstruction methods are considered; iterative methods generally do not require theoretical advances specific for cone-beam tomography. In section 2 the cone-beam problem is described mathematically, and the two main tools, Radon inversion and Grangeat's formula, are introduced. Two classes of reconstruction algorithms are presented in section 3, the Radon rebinning method and filtered-backprojection. Sections 4 and 5 describe some preliminary results on general vertex sets and truncated projection data, and in section 6, execution times for some of the algorithms are discussed.

2. Fundamentals.

2.1. The cone-beam problem. We will use f to represent the physical object under investigation, so $f : \mathbb{R}^3 \to \mathbb{R}$, and we assume that f lies in some sufficiently smooth class of functions to justify all the manipulations in this article.

We define the cone-beam operator $\mathcal{X}$, which takes f and forms cone-beam projections g, according to

$$(2.1) \qquad \mathcal{X}f = g(v, \alpha) = \int_0^\infty f(v + l\alpha)dl$$

where $\alpha \in S^2$. For a physical example, consider an x-ray source at position $v \in \mathbb{R}^3$, and an object with density $f(x)$ positioned such that each point $x \in \operatorname{supp} f$ lies between v and a large x-ray detector. The measurements on the detector can be converted to estimates of $g(v, \alpha)$ which are ray sums of the density of the object along line segments originating at v and traveling in the direction α; see figure 1. Except when stated otherwise (section 5), we assume that the detector is large enough to measure all non-zero values of $g(v, \alpha)$.

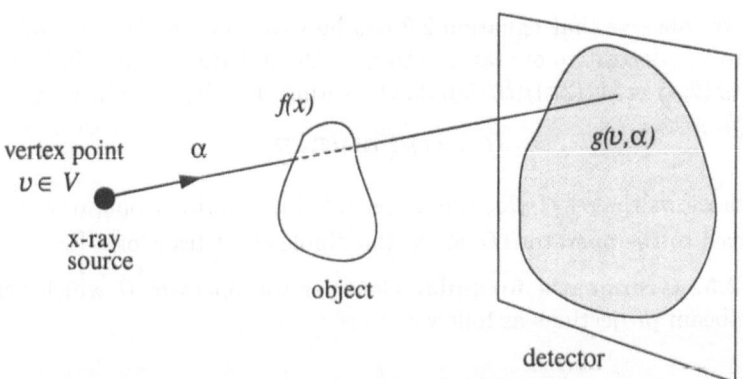

FIG. 1. *Illustration of the cone-beam operator $\mathcal{X}$. A line integral of the object density $f(x)$ is measured. Equation 2.1 gives the formula for $\mathcal{X}f = g$.*

The cone-beam problem can now be stated: find f, given g for $v \in V$ where the vertex set V satisfies Tuy's condition [19]. Tuy's condition was originally posed for the case where V is a curve in space, but it can equally well be phrased for arbitrary subsets V of $\mathbb{R}^3$ as follows. A vertex set V satisfies the tomographic completeness condition if for all planes P with $P \cap \operatorname{supp} f \neq \emptyset$, there exists some vertex $v \in V \cap P$. (Strictly speaking,

Tuy's condition requires that the intersection $V \cap P$ is not tangential to V.)

2.2. The three-dimensional Radon transform and inverse. The starting point for inversion of equation 2.1 is the Radon transform pair. We use $\mathcal{R}_3$ to represent the three-dimensional (3-d) Radon transform, so

$$(2.2) \qquad \mathcal{R}_3 f = r(\theta, s) = \int_{x \cdot \theta = s} f(x)\, dx = \int_{\mathbb{R}^3} f(x)\, \delta(x \cdot \theta - s)\, dx$$

where $\theta \in S^2$ and $s \in \mathbb{R}$.

Conceptually, $r(\theta, s)$ is the plane integral of f over the plane $P_{\theta,s}$ with unit normal vector θ and displacement s from the origin: $P_{\theta,s} = \{x : x \cdot \theta = s\}$. The formula for the inverse 3-d Radon transform is well-known (see e.g. [11]), and can be written as

$$(2.3) \qquad \mathcal{R}_3^{-1} r = f(x) = \frac{-1}{8\pi^2} \int_{S^2} \frac{\partial^2}{\partial s^2} r(\theta, s)|_{s = x \cdot \theta}\, d\theta$$

The duality in equations 2.2 and 2.3 lead us to introduce the formal adjoint $\mathcal{R}_3^*$ which (assuming, as everywhere in this paper, that we use L_2 norms on the various spaces) is easily verified to be

$$(2.4) \qquad \mathcal{R}_3^* r = \int_{S^2} r(\theta, s)|_{s = x \cdot \theta}\, d\theta$$

We observe that equation 2.3 can be written as $\mathcal{R}_3^{-1} r = -\frac{1}{2}\mathcal{R}_3^* \mathcal{D}_3^{\,2}\, r$, where the derivative operator acting on the 3-d Radon domain is defined as $\mathcal{D}_3 r(\theta, s) = (1/(2\pi))(\partial/\partial s) r(\theta, s)$. Noting that $\mathcal{D}_3^* = -\mathcal{D}_3$, we see that

$$(2.5) \qquad f = \tfrac{1}{2}(\mathcal{R}_3^* \mathcal{D}_3^*)(\mathcal{D}_3 \mathcal{R}_3) f$$

which shows that $\frac{1}{\sqrt{2}}\mathcal{D}_3 \mathcal{R}_3$ can be regarded as a unitary operator. Barrett referred to the operator $i\mathcal{D}_3 \mathcal{R}_3$ as the dipole-sheet transform [1].

2.3. Grangeat's formula. Consider the operator $\mathcal{G}$ which acts on cone-beam projections as follows:

$$(2.6) \qquad \mathcal{G}g = G(v, \theta) = \int_{S^2} g(v, \alpha)\, h(\alpha \cdot \theta)\, d\alpha$$

where h is some fixed generalized function satisfying $h(\lambda x) = h(x)/\lambda^2$ for $\lambda > 0$. This operator is important because it links cone-beam data g to the 3-d Radon transform:

$$(2.7) \qquad G(v, \theta) = (r(\theta, s) * h(-s))|_{s = v \cdot \theta}$$

where the symbol "$*$" refers to the convolution with respect to the variable s.

Equation 2.7 can be established formally as follows. After equation 2.1 is substituted into equation 2.6, a change to rectangular coordinates from spherical polar coordinates is carried out followed by some simple manipulations involving the δ-function.

$$
\begin{aligned}
G(v, \theta) &= \int_{S^2} \int_0^\infty f(v + l\alpha) \, h(\alpha \cdot \theta) \, dl \, d\alpha \\
&= \int_{S^2} \int_0^\infty f(v + l\alpha) \, h(l\alpha \cdot \theta) \, l^2 \, dl \, d\alpha \\
&= \int_{\mathbb{R}^3} f(v + x) \, h(x \cdot \theta) \, dx = \int_{\mathbb{R}^3} f(x) \, h(x \cdot \theta - v \cdot \theta) \, dx \\
&= \int_{\mathbb{R}^3} f(x) \int_{-\infty}^\infty \delta(x \cdot \theta - s) \, h(s - v \cdot \theta) \, ds \, dx \\
&= \int_{-\infty}^\infty \int_{\mathbb{R}^3} f(x) \, \delta(x \cdot \theta - s) \, dx \, h(s - v \cdot \theta) \, ds \\
&= (r(\theta, s) * h(-s))\,|_{s = v \cdot \theta}
\end{aligned}
$$

Using different notation and variables, this result has been derived by Tuy [19], Grangeat [5], Smith [18] and others, using either the classical tomographic "ramp filter" $h(x) = \int |\omega| \exp(2\pi i \omega x) d\omega$, or using the derivative kernel $h(x) = \int i\omega \exp(2\pi i \omega x) d\omega = (1/(2\pi)) \, \delta'(x)$. Grangeat's formula uses this derivative kernel, and from now on we exclusively consider $h(x) = (1/(2\pi)) \, \delta'(x)$ in equation 2.6, so the conclusion, equation 2.7, becomes $G(v, \theta) = -(1/(2\pi)) \, (\partial/\partial s) \, r(\theta, s)\,|_{s = v \cdot \theta}$, and in operator notation, $\mathcal{G}\mathcal{X} f (v, \theta) = -\mathcal{D}_3 \mathcal{R}_3 f (\theta, s = v \cdot \theta)$.

Note that $\mathcal{G}\mathcal{X} f$ is a cone-beam projection function with variables (v, θ), whereas $\mathcal{D}_3 \mathcal{R}_3 f$ is a function in the Radon domain with variables (θ, s). We express this connection using the operator $\mathcal{K}$ which acts on a function in the Radon domain and performs the appropriate conversion: $\mathcal{K}r (v, \theta) = r(\theta, v \cdot \theta)$. For a function r defined on planes, $\mathcal{K}r$ is a function of cone-beam projection variables. For each vertex point v and direction θ, the quantity $\mathcal{K}r (v, \theta)$ is associated with the plane $P_{\theta, s = v \cdot \theta}$ passing through the vertex point v and with normal vector θ, and $\mathcal{K}r (v, \theta)$ is given the same value that r assigns to this plane.

Finally, we can succinctly express the Grangeat result as

(2.8) $$\mathcal{G}\mathcal{X} = -\mathcal{K}\mathcal{D}_3 \mathcal{R}_3$$

3. Reconstruction algorithms.

3.1. The Radon method.
By analogy with the Fourier reconstruction method of classical tomography, there is a corresponding algorithm for cone-beam tomography which we call the Radon Method. Recall that Fourier reconstruction is based on the central-slice theorem which describes how a parallel projection can be modified (by taking its 1-d Fourier transform) to give a subset of coefficients of the 2-d Fourier transform of the

unknown object. The algorithm proceeds by processing each projection and interpolating onto a pre-defined rectangular grid the subset (in this case a "central slice") of the 2-d Fourier transform. Once all projections are processed, a 2-d inverse Fourier transform is performed to complete the image reconstruction.

In cone-beam tomography, the relevant space is the 3-d Radon domain, and the linking theorem is Grangeat's formula. Conceptually, each cone-beam projection is converted to the Radon domain, and the values interpolated onto a 3-d grid (where each grid point corresponds to a plane-a single point in the Radon domain). When all projections have been processed, the inverse transform is executed, in this case $(1/2)(\mathcal{D}_3 \mathcal{R}_3)^*$, to achieve the reconstruction of f.

In the Radon Method, just as in the Fourier method, the crucial step that most affects the quality of the reconstruction is the interpolation. Also, just as in the Fourier method, the projection data must satisfy the completeness condition in order that the entire Radon domain is measured.

In summary, the Radon Method consists of three steps:
1. For each projection
   (a) convert $g(v, \alpha)$ to $G(v, \theta)$
   (b) assign $G(v, \theta) \to r'(\theta, s)$ using $s = v \cdot \theta$
2. Normalize the array $r'(\theta, s)$ for multiple assignments $n(\theta, s)$
3. Inverse transform: $f = (1/2)(\mathcal{D}_3 \mathcal{R}_3)^* r'$

Several remarks are warranted. First, the Radon array $r'(\theta, s)$ requires samples suitable for performing the inverse transform of step 3. For most implementations, the s-variable would be sampled uniformly and the direction variable θ would be sampled using the usual polar and azimuthal angles. Second, the 3-d interpolation step has the geometric interpretation that each value of $G(v, \theta)$ is assigned to the plane $P_{\theta, s=v \cdot \theta}$ which contains the vertex point v and has normal vector θ. Interpolation is required to map sampled values of $G(v, \theta)$ to the samples of $r'(\theta, s)$. Third, Tuy's condition ensures that all Radon values $r'(\theta, s)$ will be available, since the plane $P_{\theta, s}$ must contain some $v \in V$, thus $v \cdot \theta = s$ and $G(v, \theta)$ supplies the value $r'(\theta, s)$. On the other hand, if more than one such $v \in V$ exist then $r'(\theta, s)$ will be available from multiple, say $n(\theta, s)$ cone-beam projections and $r'(\theta, s)$ must be normalized by the number of contributions, as indicated in step 2.

The various implementations of this algorithm differ mainly in how the interpolation step is performed. Just as in the Fourier reconstruction method, the accuracy of the interpolation largely determines the quality of the reconstruction. Different researchers have investigated "pushing" where an available $G(v, \theta)$ is assigned to multiple values of $r'(\theta, s)$ according to where (v, θ) falls in the predefined Radon grid. "Pulling" refers to interpolation performed where each Radon grid point (θ, s) is assigned a weighted sum of values $G(v, \theta)$. A more elaborate normalizing scheme is required when pushing than when pulling. In [20], type I and type II interpolations correspond to the pulling and pushing approaches.

In the reconstructions shown in figure 2, a more subtle pulling and pushing scheme devised by F. Noo was applied. For each pair of consecutive vertex samples along the curve, the subset of potential Radon values was first identified (pushing) then for each (θ, s) sample, in the selected subset of the Radon domain, the value of $r'(\theta, s)$ was obtained by (pulling) linear interpolation between the two consecutive vertices. The normalizing procedure only needed to keep track of how many times each (θ, s) sample was selected at the pushing stage.

This scheme, and virtually all published implementations of the Radon method, rely on the concept of a vertex *path*, where V consists of a curve (or collection of curves). Interpolation is only ever performed between successive vertex samples along the curve. However, in section 4 below, we discuss an application of the Radon method to handling vertices sampled from more general sets.

3.2. Cone-beam filtered backprojection. Following again the analogy with classical tomography, we now discuss so-called "Filtered Backprojection" (FBP) in the cone-beam context. The cone-beam backprojection operation is described by $\mathcal{X}^*$, the formal adjoint of $\mathcal{X}$, and is given by the formula

$$(3.1) \qquad \mathcal{X}^* g(x) = \int_V \frac{1}{|x-v|^2} g(v, \alpha)\big|_{\alpha = \frac{x-v}{|x-v|}} \, dv$$

For a FBP type algorithm, we are looking for an inversion formula of the form $f(x) = \mathcal{X}^* g^F(x)$ for some filtered projections g^F.

As we shall see below, the operation that must be applied to each projection $g(v, \cdot)$ to yield $g^F(v, \cdot)$ is linear, but not necessarily shift-invariant. Thus "filtering" is not the correct term, but is nevertheless used (sometimes as "shift-variant filtering") as a reminder that the projection by projection processing corresponds to classical 2-d FBP.

A fast derivation of the cone-beam FBP formula uses equations 2.5 and 2.8 as follows, where we assume the existence of $\mathcal{M}$ such that $\mathcal{K}^* \mathcal{M} \mathcal{K} = \mathcal{I}$ with $\mathcal{I}$ representing the identity operator.

$$\begin{aligned} f &= \tfrac{1}{2} (\mathcal{D}_3 \mathcal{R}_3)^* (\mathcal{K}^* \mathcal{M} \mathcal{K})(\mathcal{D}_3 \mathcal{R}_3) f \\ &= \tfrac{1}{2} (-\mathcal{K} \mathcal{D}_3 \mathcal{R}_3)^* \mathcal{M} (-\mathcal{K} \mathcal{D}_3 \mathcal{R}_3) f \\ &= \tfrac{1}{2} (\mathcal{G} \mathcal{X})^* \mathcal{M} (\mathcal{G} \mathcal{X}) f \\ &= \tfrac{1}{2} \mathcal{X}^* \mathcal{G}^* \mathcal{M} \mathcal{G} g \\ &= \tfrac{1}{2} \mathcal{X}^* g^F \end{aligned}$$

The filtering operation is thus given by $g^F = \tfrac{1}{2} \mathcal{G}^* \mathcal{M} \mathcal{G} g$. The operator $\mathcal{G}$ was defined in equation 2.6, and it is easily verified that $\mathcal{G}$ is self-adjoint, so $\mathcal{G}^* = \mathcal{G}$. All dependency on the particular vertex set V is buried in the

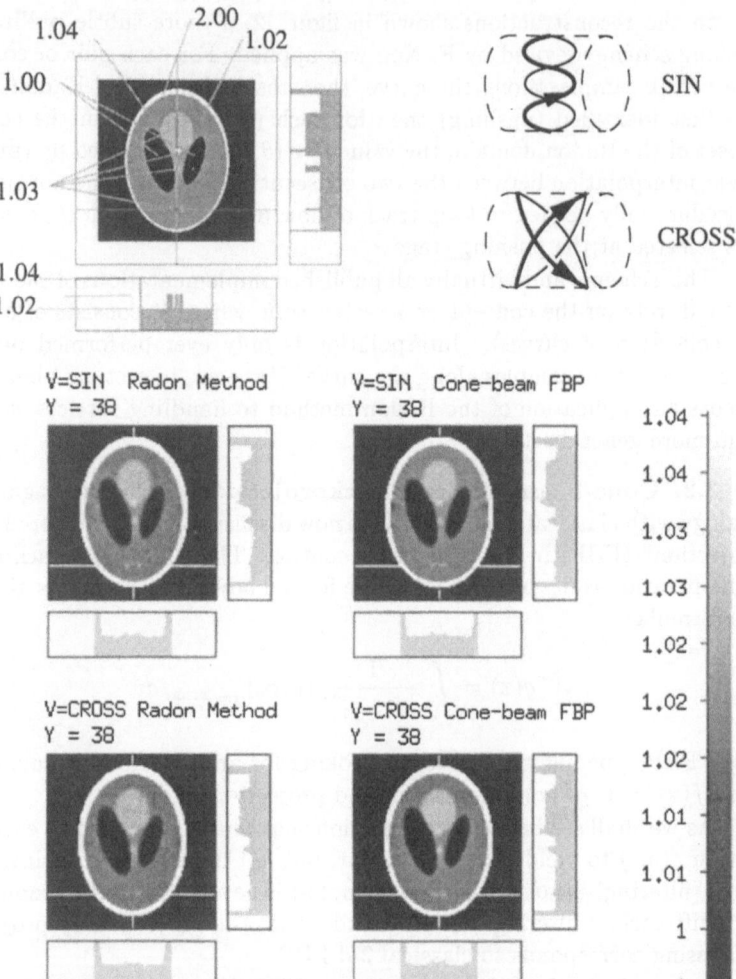

FIG. 2. *Computer simulations: examples of reconstructions using the Radon method and using cone-beam FBP.* Upper left: *a single slice through the object being imaged, which is the 3-d Shepp phantom consisting of 12 ellipsoids of small intensity differences.* Upper right: *sketch (not to scale) of 2 vertex paths from which cone-bean projections were simulated; there were 256 vertex samples on each path.* Left column: *slices through reconstructions using the Radon method.* Right column: *slices through reconstructions using the cone-beam FBP.*

operator $\mathcal{M}$. The expression for $\mathcal{K}^*$ is given by

$$(3.2) \qquad \mathcal{K}^* G(\theta, s) = \int_{x \in V \cap \theta^\perp} G(s\theta + x, \theta)\, d\mu(x)$$

where $\mu(x)$ is a suitable invariant measure on planes in $\mathbb{R}^3$, depending on the nature of the vertex set V.

We now consider V to be a piecewise smooth curve in space, parameterized by arclength with the parameter λ with domain Λ, a compact subset of $\mathbb{R}$. Thus we have $v(\lambda) \in V$, $v(\Lambda) = V$, and $|v'(\lambda)| = 1 \; \forall \lambda \in \Lambda$, so in equation 3.1 we can replace $\int_V (\cdot) dv$ with $\int_\Lambda (\cdot)|_{v=v(\lambda)} d\lambda$. For notational convenience the dependence of the vertex point v on the parameter λ is sometimes omitted and we write, with a slight abuse of notation, v_i for $v(\lambda_i)$ and v_i' for $v'(\lambda_i)$.

For such vertex paths, we evaluate $\mathcal{K}^*$ using inner products, $\langle \mathcal{K}r, G \rangle = \int_{S^2} \int_\Lambda \mathcal{K}r(v(\lambda), \theta) \, G(v(\lambda), \theta) \, d\lambda d\theta = \langle r, \mathcal{K}^* G \rangle$, to obtain

$$(3.3) \qquad \mathcal{K}^* G(\theta, s) = \sum_{i=1}^{n(\theta,s)} \frac{G(v_i, \theta)}{|v_i' \cdot \theta|}$$

where $n(\theta, s) = |P_{\theta,s} \cap V|$ is the number of non-tangential intersections of the plane $P_{\theta,s}$ with the vertex path V, and $v_1, v_2, \ldots, v_n$ are the vertices which satisfy the equations $v_i \cdot \theta = s$ and $|v_i' \cdot \theta| \neq 0$ (i.e. the vertices lying in the plane $P_{\theta,s}$, except any vertices at tangent points to the plane). To further simplify the notation, when we write the condition $v_i \cdot \theta = s$ below, we implicitly require $|v_i' \cdot \theta| \neq 0$ also.

By direct calculation we find solutions $\mathcal{M}$ of the equation $\mathcal{K}^* \mathcal{M} \mathcal{K} = \mathcal{I}$ of the multiplicative form

$$(3.4) \qquad MG(v, \theta) = G(v, \theta)|v' \cdot \theta| M(v, \theta)$$

where the function $M(v, \theta)$ must satisfy the normalization equation

$$(3.5) \qquad \sum_{\substack{i=1 \\ (v_i \cdot \theta = s)}}^{n(\theta,s)} M(v_i, \theta) = 1$$

for each (θ, s).

Thus when V is a vertex path we can write $g^F = \frac{1}{2} \mathcal{G}^* \mathcal{M} \mathcal{G} g$ where $\mathcal{M}$ is multiplication by $|v' \cdot \theta| M(v, \theta)$. There are many possibilities for the normalization function $M(v, \theta)$. One simple example is to take the constant $M(v, \theta) = 1/n(\theta, v \cdot \theta)$. Another example arises by considering using $(\mathcal{K}\mathcal{K}^*)^{-1}$ for $\mathcal{M}$. Since G is in the range of $\mathcal{K}$ (which is the same as the range of $\mathcal{G}\mathcal{X}$), we have $G(v_i, \theta) = G(v_j, \theta)$ whenever $v_i \cdot \theta = v_j \cdot \theta$. Consequently we find

$$(3.6) \qquad \mathcal{K}\mathcal{K}^* G(v, \theta) = \sum_{\substack{i=1 \\ (v_i \cdot \theta = v \cdot \theta)}}^{n(\theta, v \cdot \theta)} \frac{G(v_i, \theta)}{|v_i' \cdot \theta|} = G(v, \theta) \sum_{i=1}^{n(\theta, v \cdot \theta)} \frac{1}{|v_i' \cdot \theta|}$$

so we can let $\mathcal{M} = (\mathcal{K}\mathcal{K}^*)^{-1}$ be multiplication by $1/(\sum_{i=1}^{n}(|v' \cdot \theta|)^{-1})$. The corresponding normalization function $M(v,\theta)$ is given by $1/M(v,\theta) = |v' \cdot \theta| \sum_{i=1}^{n} (|v_i' \cdot \theta|)^{-1}$ which indeed satisfies the normalization equation 3.5. More discussion on selecting a normalization function $M(v,\theta)$ suitable for digital implementation can be found in [3].

For the vertex paths, Kudo and Saito [8] and Defrise and Clack [3] independently formulated the cone-beam FBP results in 1994, and gave details for implementation in a coordinate system appropriate for cone-beam projections measured on planar detectors. Figure 2 shows some images from reconstructions using the cone-beam FBP approach.

We remark that the derivations have all tacitly assumed that the domain of all functions of (θ, s) is restricted to just those planes $P_{\theta,s}$ which intersect the support of f. Since we also assume Tuy's condition is satisfied, we are assured that $n(\theta, s)$ is never zero. Without these two assumptions, the equation $\mathcal{K}^*\mathcal{M}\mathcal{K} = \mathcal{I}$ could never hold for bounded vertex sets V, and $\mathcal{K}^*$ would not be well-defined in equations 3.2 and 3.3 unless the right-hand-sides were interpreted as zero when the integration or summation arguments (respectively) are empty.

4. Vertex paths to vertex sets. In practice, all cone-beam imaging systems collect data for a finite vertex set $V = \{v_1, v_2, \ldots v_n\}$. Ideally, these vertices constitute a fine sampling of some vertex path $v(\Lambda)$ and any one of the algorithms described in section 3 could be applied. This sampling assumption is not adequate for all applications, for example if the positioning of the vertices is not naturally linked to a path, such as in the FASTSPECT imaging system [7] consisting of several dozen pinholes in fixed positions; or simply if the sampling is too coarse to maintain an adequate approximation to the intended path.

In this section, we describe some advances on approaches to handle a finite vertex set $\{v_1, v_2, \ldots v_n\}$ independently of the order of the vertices. More details on the research summarized in this section can be found in [13]. The first approach considers vertices pairwise and (conceptually) connects close vertex pairs until a path satisfying Tuy's condition is formed. Define

$$(4.1) \qquad \varepsilon_2 = \inf\{\varepsilon : \bigcup_{|v_i - v_j| \le \varepsilon} [v_i, v_j] \text{ satisfies Tuy's condition}\}.$$

The notation $[v_i, v_j]$ refers to the line segment connecting v_i to v_j. The value of ε_2 indicates the length of the longest line segment necessary for $V_2 = \bigcup\{[v_i, v_j] : |v_i - v_j| \le \varepsilon_2\}$ to satisfy Tuy condition. However, if $\varepsilon_2 = \infty$, then V_2 does not satisfy Tuy's condition because the vertices lie in the half-space on one side of some plane intersecting supp f. The vertex set is considered to be a sampling of this tomographically complete path, with vertices at the endpoints of each of its component segments.

In the implementation described in [13], a slightly different path is constructed, which avoids connecting too many segments just to accommodate a few poorly sampled directions θ normal to a set of parallel planes intersecting supp f. But whatever specific path is constructed from paired vertices, it would almost certainly be unsuitable for the FBP techniques described in section 3.2. These methods assume a well-discretized smooth curve and require a good approximation to the tangent vector at each vertex. However, the Radon Method described in section 3.1 is amenable to these kinds of jagged paths. We have implemented and experimented with this approach using several examples of vertex sets and simulated objects. Figure 3 shows some results and illustrates the difficulties that might be encountered when using the FBP algorithm.

Like the first approach, our second approach also converts the discrete vertex set, which can not possibly satisfy Tuy's condition, to a set that can. We define

$$(4.2) \qquad \varepsilon_1 = \inf\{\varepsilon : \bigcup_i (v_i + B_\varepsilon) \text{ satisfies Tuy's condition.}\}$$

where $B_\varepsilon = \{x : |x| \leq \varepsilon\}$ is the ball of radius ε. The corresponding vertex set is $V_1 = \bigcup_i (v_i + B_{\varepsilon_1})$ whereby each vertex point is replaced by a small ball. Note that V_1 is a 3-d set. Also, unlike the vertex-pair approach, a finite value of ε_1 is always achieved.

Again, in our implementation of this approach each direction was considered separately. For each $\theta \in S^2$, we defined $\varepsilon_\theta = \inf\{\varepsilon : \text{all planes}$ with normal θ intersecting supp f also intersect $\bigcup_i (v_i + \varepsilon_\theta[-\theta, \theta])\}$. Thus the vertex set was given by $\tilde{V}_1 = \bigcup_i (v_i + \tilde{B})$ using the subset $\tilde{B}$ of B_{ε_1} defined by $\tilde{B} = \bigcup\{\varepsilon_\theta[-\theta, \theta] : \theta \in S^2\}$, and $\tilde{V}_1$ also satisfies Tuy's condition. A variation of the Radon Method was used for reconstruction, whereby Radon values $r'(\theta, s)$ received contributions from vertices v_i only when $v_i + \varepsilon_\theta[-\theta, \theta]$ intersected the plane $P_{\theta,s}$. The contribution was given a weight which depended on ε_θ and the distance of the plane to the center of the line segment $v_i + \varepsilon_\theta[-\theta, \theta]$.

Any finite vertex set can be processed using this "single vertex" approach, unlike the vertex-pair method which requires $\varepsilon_2 < \infty$. In particular, the single-vertex algorithm can be applied to a set of uniformly-spaced vertices lying on a circle. Figure 4 shows a reconstruction of simulated cone-beam data for this circular vertex set, with a comparison of the FDK algorithm applied to the same data. The general single-vertex method removes the bias present in the FDK reconstruction, because of the implicit data interpolation which is inherent to the method.

As a demonstration of the power of the discrete vertex set approach, we show in figure 5 images reconstructed from data taken from a randomly placed set of vertices. It is unlikely that any vertex-path based approach could handle these data.

5. Region-of-interest imaging and truncation. We consider the interior problem in cone-beam tomography. The aim is to reconstruct some

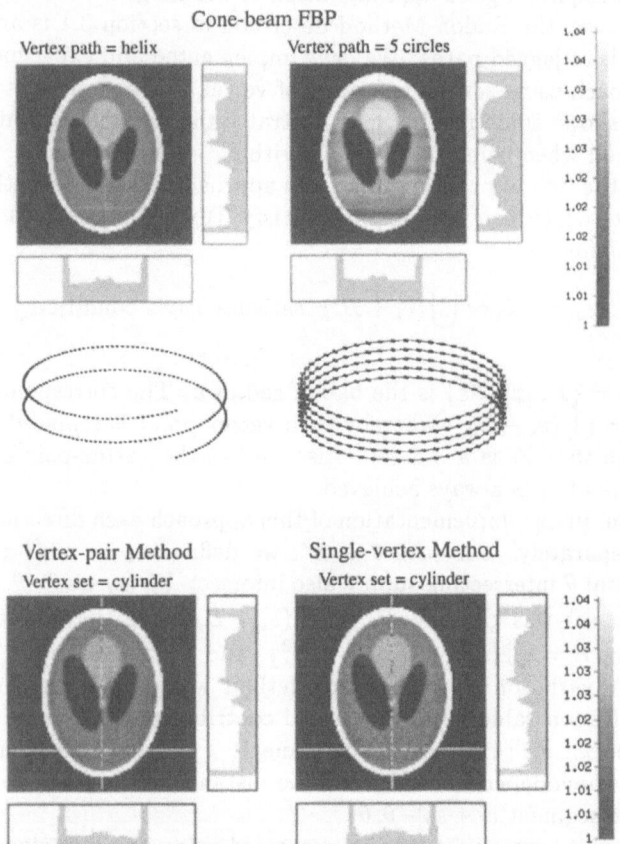

FIG. 3. *Handling general finite vertex sets. Upper left: cone-beam FBP algorithm applied to data gathered on a well-sampled helix. Upper right: cone-beam FBP algorithm applied to data uniformly sampled on the surface of cylinder (250 vertices). The algorithm attempts to reconstruct this data assuming a path consisting of 5 circles. Lower left: vertex-pair approach applied to the 250 vertices on the cylinder. Lower right: single-vertex approach applied to the 250 vertices on the cylinder. (Reprinted from [12, 13] with permission. ©1997 IEEE)*

region-of-interest of the object. For convenience, we assume that $0 \in \text{supp} f$ and the region-of-interest is (or is contained in) the ball B_R of radius R. The goal is to reconstruct $f(x)$ for $x \in B_R \subseteq \text{supp} f$. At each vertex position, we assume the detector is large enough to measure the cone-beam projection of B_R. If some portion of the cone-beam projection of

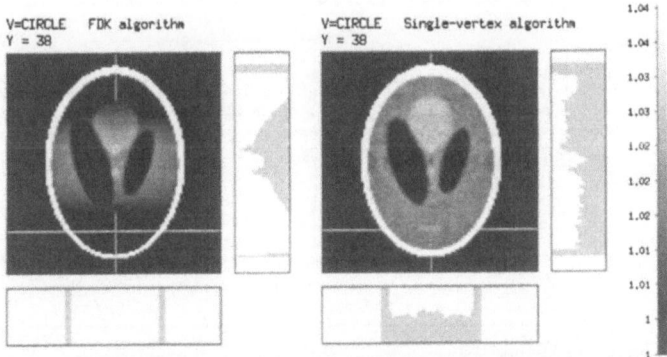

FIG. 4. *Comparison of the FDK algorithm with the single-vertex approach. Simulated projection data was from 256 vertex points sampled uniformly on a circle of radius 350 mm. The 3-d Shepp phantom has height 184 mm. Left: result using the FDK algorithm. Right: result using single-vertex approach. Note that the greyscale is highly compressed (as in all the figures) to reveal low contrast differences in the phantom. This compressed scale exaggerates the bias in the FDK reconstruction.*

supp $f \setminus B_R$ misses the detector, the projection is said to be truncated.

We remark that for reconstruction of the region-of-interest B_R, Tuy's condition on the vertex set V may be relaxed to apply only to B_R [19]: for every plane P satisfying $P \cap B_R \neq \emptyset$, there exists some $v \in V \cap P$. Provided there is no truncation, the theory carries through with this less restrictive sufficiency condition. For $x \in B_R$ (i.e. $|x| \leq R$) the step $f(x) = (1/2)\mathcal{R}_3^* \mathcal{D}_3^* r'$ only requires values $r'(\theta, s)$ for $|s| \leq R$. These values $r'(\theta, s)$ are obtained from $\mathcal{G}g(v, \theta)$ provided $|v \cdot \theta| = |s| \leq R$, i.e. provided all planes lying within a distance R of the origin contain some vertex point v. Thus $f(x)$ can be reconstructed for $x \in B_R$ provided V satisfies the relaxed sufficiency condition.

However, no such relaxed condition applies to the cone-beam projections. The evaluation of $G(v, \theta)$ ($G(v, \theta) = \mathcal{G}g(v, \theta)$ from equation 2.6) uses the projection values $g(v, \alpha)$ for all α. However, by analyzing the operator $\mathcal{G}$, it can be surmised that a certain amount of projection truncation is admissible provided the corresponding plane information $r'(\theta, s)$ is available from other projections. See [2] for more details.

A significant advance by H. Kudo was reported in [9]. We do not attempt to explain all the details of this method, but we will sketch the main ideas and show some simulation results.

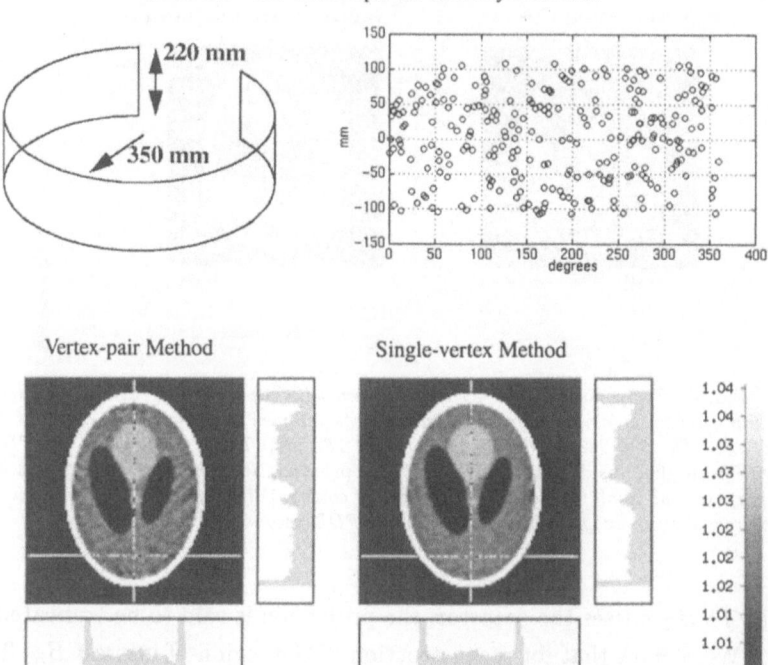

FIG. 5. *Reconstruction from a randomly placed set of vertices. Here 256 vertices were located on the wall of an imaginary cylinder of height 220 mm and radius 350 mm. Left: reconstruction using the vertex-pair approach. Right: reconstruction using the single-vertex approach. (Reprinted from [12, 13] with permission. ©1997 IEEE)*

The version of the interior problem addressed in [9] is called the long object problem. The support of f is taken to be a tall cylinder, and the region of interest is a disk of the same radius and orientation inside the cylinder. We define a set of axes $\{e_1, e_2, e_3\}$ centered in the cylinder with e_3 lying in the axial direction; see figure 6. For each vertex point, it is convenient to consider the detector as being oriented parallel to the e_3 axis.

The method is based on the cone-beam FBP algorithm, with the vertex path consisting of two subpaths V_1 and V_2. Although more general possibilities exist, for the purposes of exposition we fix V_1 to be a circle, and V_2 to be a line segment. The circle lies in the plane $e_3^{\perp}$; the line segment is parallel to e_3, meets the circle and the positive e_2 axis simulta-

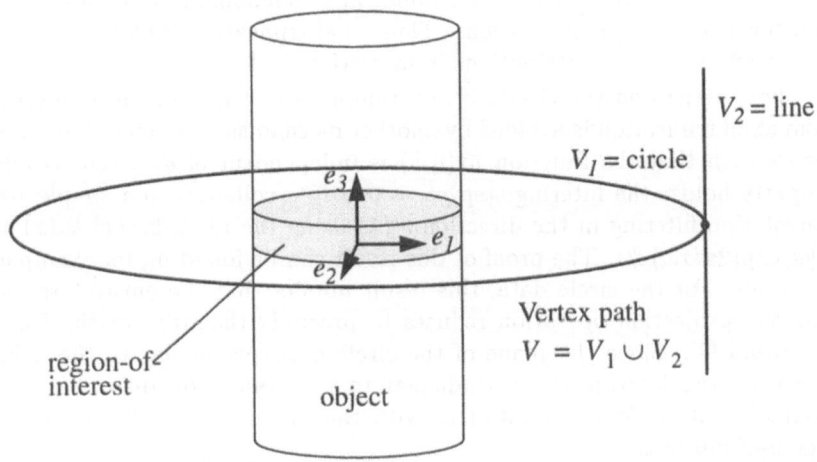

FIG. 6.   *The long object problem. The tall cylinder represents* supp *f, and the shaded disk is the region-of-interest to be reconstructed. The axes* $\{e_1, e_2, e_3\}$ *are centered in the cylinder with* e_3 *along the cylinder axis. For the explanation of Kudo's approach, we let the vertex path be the union of the circle and line oriented as shown.*

neously at its midpoint, and is long enough so that $V = V_1 \cup V_2$ satisfies Tuy's condition with respect to the region-of-interest.

A special choice of the function $M(v, \theta)$ is made in the filtering step $\frac{1}{2}\mathcal{G}^* \mathcal{M} \mathcal{G} g$ (recall from section 3.2 that there is some freedom in selecting the function M). For all vertices $v \in V_1$ on the circle, the function $M(v, \theta)$ is assigned the value $1/2$. For the remaining vertices, $v \in V_2$, the function $M(v, \theta)$ is defined so as to satisfy the normalization equation $\sum M(v_i, \theta) = 1$. Thus,

$$(5.1) \qquad M(v, \theta) = \begin{cases} 1/2 & \text{if } v \in V_1 \\ 1 & \text{if } v \in V_2 \text{ and } V_1 \cap P_{\theta, v \cdot \theta} = \emptyset \\ 0 & \text{otherwise} \end{cases}$$

With this choice of M, the contribution $r'(\theta, s)$ to the reconstruction is obtained either from vertices on the circle, or from a vertex on the line, but not both. If the plane $P_{\theta, s}$ intersects both the circle V_1 and the line V_2, the potential contribution from the vertex on the line is ignored.

One important consequence of this choice of $V = V_1 \cup V_2$ with this normalizing function $M(v, \theta)$ is that axial truncation of projections from vertex set V_2 does not affect these vertices' contribution to the reconstruction of the region-of-interest. The details can be found elsewhere, but a geometric explanation is possible. For each vertex $v \in V_2$, the contributing planes $P_{\theta, v \cdot \theta}$ have their normals θ nearly parallel to e_3, $|\theta \cdot e_3| \approx 1$, because more oblique planes are eliminated when they intersect the circle V_1. These contributing planes intersect the detector in lines that are nearly perpendicular to e_3, and it is the values in a neighborhood of these lines that form the plane contributions. Thus axial truncation (unless extreme) has no effect on the contributions from vertices in V_2.

For vertices on the circle, V_1, contamination of the region-of-interest from axial truncation is avoided by another mechanism. In general, for any vertex such that the function $M(v, \theta)$ is independent of θ, a remarkable property holds: the filtering step $g^F = \mathcal{G}^* \mathcal{M} \mathcal{G} \, g$ collapses to a simple 1-d convolution filtering in the direction $v'(\lambda)$ using the ramp kernel $h_r(x) = \int |\omega| \exp(2\pi i x \omega) \, d\omega$. The proof of this result can be found in, for example, [8, 3, 16]. For the circle data, this result implies that the entire filtering and backprojection operation reduces to precisely the steps of the FDK algorithm [4]. Since the plane of the circle is perpendicular to the axial direction, the filtering is perpendicular to the truncation direction, and axial truncation does not interfere with the processing of the projected region-of-interest.

For much more on this approach to handling axial truncation and the long object problem, see [9, 15, 16, 14].

We illustrate a reconstruction of the 3-d Shepp phantom using this method, and by naïvely applying the cone-beam FBP method. Figure 7 clearly shows the benefit of using a theoretically accurate algorithm to handle projection truncation.

6. Execution times. The computer time required to execute the various algorithms depends mainly on the specific parameters of the application. We give an example here of a fairly large reconstruction problem arising in an industrial application.

A large cone-beam x-ray CT scanner, called the DRCT scanner, has been assembled by staff at the Idaho National Engineering and Environmental Laboratory (INEEL), Idaho, U.S.A. The system consists of a large immobile detector of approximate dimensions 1.1 m × 1.3 m, with an x-ray source which can translate vertically at a fixed distance (about 2 m) from the detector. The object being scanned, usually a standard-sized drum (height 90 cm, diameter 55 cm) containing a variety of internal structure and objects, is positioned on a platform which has rotate and translate capability. More details on the scanner and its applications can be found in [17]. Typically about 200 projections are measured, for positions sampled on a variety of possible vertex paths. Some small amount of axial truncation of the projections occurs, depending on the height of the x-ray source.

In figure 8, some sections through the reconstruction of a drum consisting mainly of alternating slabs of graphite are displayed; and in figure 9, a region-of-interest reconstruction of a drum containing a range of smaller objects, usually containers, is shown. For these reconstructions, the vertex path consisted of a circle and three lines, and a slightly modified version of the algorithm by Kudo was applied.

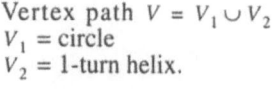 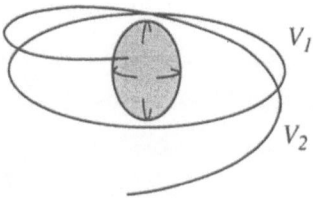

Vertex path $V = V_1 \cup V_2$
V_1 = circle
V_2 = 1-turn helix.

Cone-beam FBP

Untruncated projections   Truncated projections

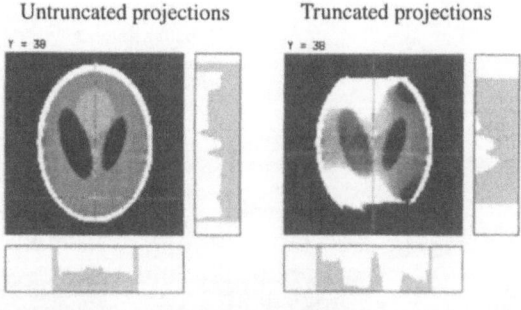

Kudo's Method

Untruncated projections   Truncated projections

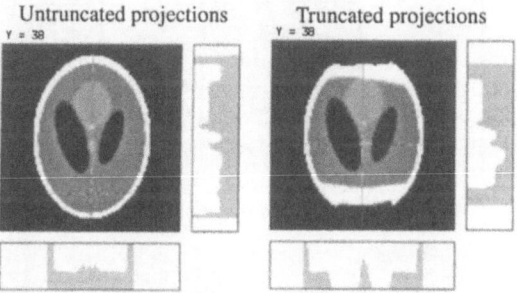

FIG. 7. *Example of Kudo's method of handling axial truncation. Projection data of the 3-d Shepp phantom were generated for a circle and 1-turn helix path. (The helical path plays the role of the line segment in the algorithm description.) Left side: reconstruction slices from untruncated data using the cone-beam FBP algorithm, and Kudo's method. Right side: same slices for same algorithms, except that the data were truncated axially by 33%. (Reprinted from [16] with permission. ©1998 IEEE)*

There were 237 vertex samples (180 on the circle, 19 on each line segment), and each projection consisted of about 700 x 1000 square pixels of approximate size $(1.5 \text{ mm})^2$. For a 3-d image representation in voxels of size $(1.7 \text{ mm})^3$, a total of 384 x 384 x 540 voxels were needed to cover

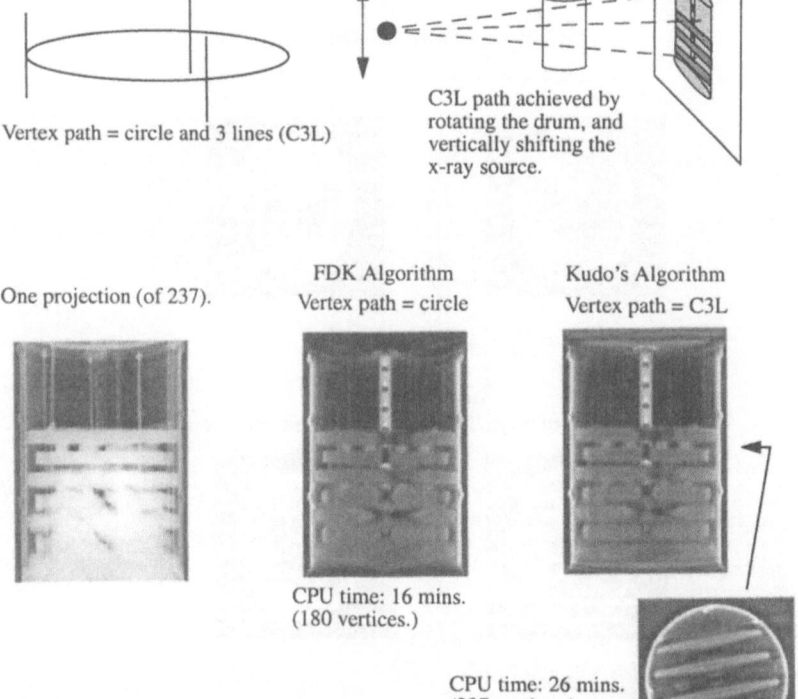

Vertex path = circle and 3 lines (C3L)

C3L path achieved by rotating the drum, and vertically shifting the x-ray source.

One projection (of 237).

FDK Algorithm
Vertex path = circle

Kudo's Algorithm
Vertex path = C3L

CPU time: 16 mins.
(180 vertices.)

CPU time: 26 mins.
(237 vertices.)

FIG. 8. *Reconstruction of a drum from the INEEL DRCT scanner. The vertex path was the union of a circle and 3 lines, oriented as shown upper right, and implemented by performing one complete rotation of the drum in 2 degree steps, followed by three linear scans with the drum fixed and the x-ray source translated vertically. Bottom left: one of the projection of the graphite-layers drum, showing mild axial truncation. Bottom middle: reconstruction using the FDK algorithm from the data taken on the circle only. Note the poor recovery of the lower graphite slabs. Bottom right: a vertical and transaxial slice through the reconstruction (processing all vertices on the path) using Kudo's approach. No corrections were made for beam-hardening or scatter before applying the tomographic reconstruction algorithms. (Data acquired on the DRCT scanner. T. J. Roney, INEEL, Idaho Falls, U.S.A.) (Reprinted from [15] with permission. ©1996 IEEE)*

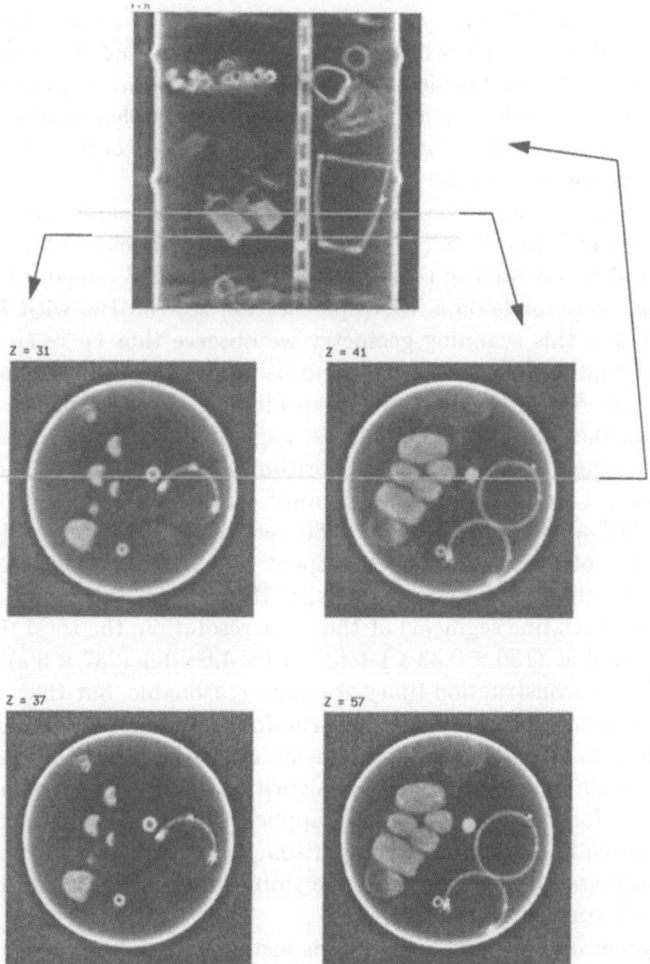

FIG. 9. *Reconstruction of a drum containing various internal features. The drum scanned on the DRCT scanner contained a vertical cylinder with high density disks, and several miscellaneous objects, many of which were containers, partially filled with sand. Top three images: a vertical and 2 transaxial slices are show with the relative positions indicated. The reconstruction array consisted of $128 \times 128 \times 160$ $(5$ $mm)^3$ voxels. Bottom two images: the same two transaxial slices but using $256 \times 256 \times 89$ voxels of size $(2.5$ $mm)^3$.*

the extent of the drum. These dimensions are larger than what is usually encountered in 3-d imaging; it represents a linear system of some 150 million equations in 75 million unknowns. In practice, the data are compressed (at the cost of some resolution) and a coarser representation of the object is used in order to achieve reconstructions in a reasonable time.

Using the Kudo method, the total reconstruction time is proportional to the number of vertices times a component which conveniently splits between the time required to perform the backprojection $f = \mathcal{X}^* g^F$ and the time for the filtering $g^F = \mathcal{G}^* \mathcal{M} \mathcal{G} g$ or $g^F = h_r * g$. Therefore we write $t_T = |V|(t_B + t_F)$, where t_T is the total time, and $|V|$ is the number of vertices. The backprojection time per vertex, t_B, is proportional to the number of voxels (size of the output) in our implementation; and the filtering time per vertex, t_F, depends on the number of detector pixels on each projection (size of the input).

In tables 1 and 2, we have recorded some measurements of t_F and t_B for various input sizes (number of projection pixels) and output sizes (number of voxels used to represent the reconstructed image). These measurements were made on a SUN Sparcstation 20, 75MHz, with 440 Mbyte of RAM. For this scanning geometry we observe that $t_F \approx t_B$ when the pixel size matched the voxel size (and using the full shift-variant filtering operation $\mathcal{G}^* \mathcal{M} \mathcal{G}$). However, the ramp-filtering operation is over ten times faster and the filtering time is almost negligible compared to backprojection time when using the FDK algorithm. For example, at the coarsest resolution $((4.3 \text{ mm})^2$ pixels and $(5 \text{ mm})^3$ voxels) indicated in tables 1 and 2, the FDK algorithm applied to 180 vertices on a circle requires about 16 minutes, of which 15 minutes is spent on backprojection. However, to process the circle and three lines (with 180 vertices on the circle and 19 vertices on each line segment) at the same resolution, the total time would be calculated as $(180 \times 0.33 \text{ s }) + (3 \times 19 \times 4.6 \text{ s }) + (237 \times 5 \text{ s}) =26$ minutes. These reconstruction times are quite reasonable, but they apply to a moderate sized problem. For reconstruction at the highest resolution, corresponding to the original measurements on the scanner, the times would be 4 hours 20 minutes for the FDK algorithm on 180 vertices, and 7 hours 30 minutes for the Kudo algorithm applied to the 237 vertices of the circle and three lines path. For comparison, the cone-beam FBP algorithm, which processes all vertices using the shift-variant filter, would require 13 hours for reconstruction.

In general, the Kudo algorithm is faster than the cone-beam FBP algorithm because the circular subpath V_1 can be filtered very quickly. Given that the FBP algorithm cannot properly handle axially-truncated data, it is especially practical to use vertex paths which satisfy the requirement for Kudo's approach. (These paths usually contain a circular or planar subpath perpendicular or nearly perpendicular to the truncation direction. See [9] for the "extended completeness condition" on vertex paths.) However, there are limitations to the Kudo approach. Most of the data from the V_2

TABLE 1
Filtering time per vertex: t_F

t_F	Projection size: 192 x 256 pixels (4.3 mm x 4.3 mm)	Projection size: 384 x 512 pixels (2.2 mm x 2.2 mm)	Projection size: 768 x 1024 pixels (1.1mm x 1.1mm)
$g^F = \mathcal{G}^*\mathcal{M}\mathcal{G}g$	4.6 seconds	20 seconds	120 seconds
$g^F = h_r*g$	0.33 seconds	1.8 seconds	8.6 seconds

TABLE 2
Backprojection time per vertex: t_B

t_B	No. of voxels: 128 x 128 x 180 $(5\ mm)^3$	No. of voxels: 256 x 256 x 360 $(2.5\ mm)^3$	No. of voxels: 320 x 320 x 450 $(2\ mm)^3$	No. of voxels: 384 x 384 x 540 $(1.7\ mm)^3$
$f = X^*g^F$	5 seconds	25 seconds	46 seconds	78 seconds

subpath is poorly utilized, because unlike the FBP approach, the choice of $M(v,\theta)$ (see equation 5.1) effectively discards any information not already measured by the V_1 subpath. Since much of the reconstruction time is spent extracting and processing a small amount of information from the V_2 subpath, it seems reasonable to measure fewer samples from it. One avenue of further research would be to determine what relative sampling rates of V_1 and V_2 would be appropriate when using Kudo's method.

7. Acknowledgements. Financial support for the work described in this article included a grant from the Whitaker Foundation and a contract with the INEEL. RC was supported for one month by the Institute for Mathematics and Its Applications during the preparation of the manuscript. FN and MD are supported by the Belgian National Fund for Scientific Research.

REFERENCES

[1] H. H. BARRETT, *Dipole-sheet transform*, J. Opt. Soc. Am., 72, 468-475, 1982.

[2] R. CLACK, M. DEFRISE, *Overview of reconstruction algorithms for exact cone-beam tomography*, SPIE Proceedings series, 2299, Mathematical Methods in Medical Imaging III, San Diego, CA, 1994

[3] M. DEFRISE, R. CLACK, *A cone-beam reconstruction algorithm using shift-variant filtering and cone-beam backprojection*, IEEE Trans. Med. Imag., 13, 186-195, 1994.

[4] L. A. FELDKAMP, L. C. DAVIS, J. W. KRESS, *Practical cone-beam algorithm*, J. Opt. Soc. Amer. A, A6, 612-619, 1984.

[5] P. GRANGEAT, *Analyse d'un système d'imagerie 3D par reconstruction à partir de radiographies X en géométrie conique*, Ph.D. Thesis, Ecole Nationale Supérieure des Télécommunications, France, 1987.

[6] P. GRANGEAT, *Mathematical framework of cone-beam 3D reconstruction via the first derivative of the Radon transform*, Mathematical methods in tomography, Herman, Louis and Natterer (eds.), Lecture Notes in Mathematics 1497, 66-97, Springer-Verlag, Berlin 1991.

[7] W. P. KLEIN, H. H. BARRETT, I. W. PANG, D. D. PATTON, M. M. ROGULSKI, J. D. SAIN, *FASTSPECT: Electrical and mechanical design of a high-resolution dynamic SPECT imager*, Conference Record of the 1995 Nuclear Science Symposium and Medical Imaging Conference, San Francisco, CA, 931-933, 1996.

[8] H. KUDO, T. SAITO, *Derivation and implementation of a cone-beam reconstruction algorithm for nonplanar orbits*, IEEE Trans. Med. Imag., 13, 196-211, 1994.

[9] H. KUDO, T. SAITO, *An extended completeness condition for exact cone-beam reconstruction and its application*, IEEE Conference Record of the 1994 Nuclear Science Symposium and Medical Imaging Conference, Norfolk, VA., 1710-1714, 1995.

[10] E. J. MORTON, S. WEBB, J. E. BATEMAN, L. J CLARKE, C. G. SHELTON, *Three-dimensional x-ray microtomography for medical and biological applications*, Phys. Med. Biol., 35, 805-820, 1990.

[11] F. NATTERER, *The Mathematics of Computerized Tomography*, ed. Wiley: New York, 1986.

[12] F. NOO, R. CLACK, M. DEFRISE, *Cone-beam Reconstruction from General Discrete Vertex Sets*, IEEE Conference Record of the 1996 Nuclear Science Symposium and Medical Imaging Conference, Anaheim, CA., 1496-1500, 1997.

[13] F. NOO, R. CLACK, M. DEFRISE, *Cone-beam Reconstruction from General Discrete Vertex Sets using Radon Rebinning Algorithms*, IEEE Trans. Nucl. Sci., 44, 1309-1316, 1997.

[14] F. NOO, R. CLACK, T. J. RONEY, T. A. WHITE, *Symmetrical vertex paths for exact reconstruction in cone-beam C.T.*, Phys. Med. Biol., 43, 797-810, 1998.

[15] F. NOO, M. DEFRISE, R. CLACK, T. J. RONEY, T. A. WHITE, S. G. GALBRAITH, *Stable and Efficient Shift-Variant Algorithm for Circle-plus-Lines Orbits in Cone-Beam C.T.*, Proceedings ICIP-96: 1996 International Conference on Image Processing, Lausanne, September 1996, P. Delogne (ed). IEEE, Ceuterick, Leuven, Belgium, 539-542, 1996.

[16] F. NOO, M. DEFRISE, R. CLACK, *Direct reconstruction of cone-beam data acquired with a vertex path containing a circle*, IEEE Trans. Imag. Proc. 7, 854-867, 1998.

[17] T. J. RONEY, S. G. GALBRAITH, T. A. WHITE, M. O'REILLY, R. CLACK, M. DEFRISE, F. NOO, *Feasibility and applications of cone-beam X-ray imaging for*

containerized waste, Proceedings for the 4th Nondestructive Assay and Nondestructive Examination Waste Characterization Conference (Salt Lake City, 1995), 295-324, 1995.

[18] B. D. SMITH, *Image reconstruction from cone-beam projections: Necessary and sufficient conditions and reconstruction methods*, IEEE Trans. Med. Imag., 4, 14-25, 1985.

[19] H. TUY, *An inversion formula for cone-beam reconstruction*, SIAM J. Appl. Math., 43, 546-522, 1983.

[20] Y. WENG, G. L. ZENG, G. T. GULLBERG, *A reconstruction algorithm for helical cone-beam SPECT*, IEEE Trans. Nuc. Sci., 40, 1092-1101, 1993.

MATHEMATICAL PROBLEMS IN MICROWAVE MEDICAL IMAGING

DAVID COLTON* AND PETER MONK*

Abstract. The change in tissue structure due to cancer causes the index of refraction of the tissue to change. Hence the question is posed if it is possible to detect the presence of cancer by microwave imagining. In this paper we survey recent results we have obtained on the use of microwaves to detect leukemia in the bone marrow of the leg that suggest that this is possible. Our emphasis in this paper is on the mathematical problems that arise in this investigation.

Key words. Microwaves, leukemia, inverse scattering.

AMS(MOS) subject classifications. 35R30, 78A70, 92C50

1. Introduction. In this paper we will report on some recent progress we have made on the possibility of using microwaves to image leukemia in the upper part of the lower leg. This portion of the body is a common location for leukemia to occur and difficulties arise in both its detection and the monitoring of its treatment by chemotherapy. The primary source of these difficulties is that the material displaced by the newly created cells is roughly the same density as the new cells and hence the overall change in density is small. This means that methods based on density differentials are of limited utility. Because of this fact, the common approach to bone marrow interrogation is currently through the use of needle aspiration, a procedure that is uncomfortable for the patient and cannot be repeated often. These considerations have motivated us (at the suggestion of Dr. Richard Albanese of the Biomechanics and Modeling Branch of Brooks Air Force Base) to investigate the possibility of using microwave imagining since the increase in capacity due to cell proliferation should cause the dielectric constant of the bone marrow to decrease and the conductivity to increase by several orders of magnitude in regions of such proliferation. This paper is devoted to a description of our recent investigations in this direction and in particular to the mathematical problems that have arisen in trying to develop a realistic and practical model for testing the feasibility of such an approach.

There are three main problems in trying to use microwaves to monitor cell proliferation in bone marrow. The first is that the living human body is dispersive with a poorly understood dispersion relation. This suggests that the use of continuous wave interrogation at a fixed frequency may be more appropriate than pulses to interrogate the bone marrow. The second problem is that at frequencies suitable for medical imaging the inverse scattering problem of determining the index of refraction from measured

---

*Department of Mathematical Sciences, University of Delaware, Newark, Delaware, 19716, U.S.A.

microwave scattering data is nonlinear. The third, and most serious, problem is that the inverse scattering problem just referred to is improperly posed in the sense that small variations in the measured data can cause large variations in the reconstructed model of the tissue.

In recent years there have been a number of methods developed to solve inverse scattering problems such as the one mentioned in the previous paragraph. One of the more successful of these methods is diffraction tomography which relies on the Rytov approximation and leads to a linear integral equation for the determination of the index of refraction n [12, 20]. The method of diffraction tomography is in general only suitable for n a slowly varying function of position. In particular, this method is appropriate for the detection of tumors in the breast but not for locating leukemia in the bone marrow. A second method for solving inverse scattering problems is the one of nonlinear optimization. This method reformulates the inverse problem as one in nonlinear optimization which is solved by some iterative method [4, 14]. We have previously used this method to study the problem of detecting leukemia in the leg [6, 7, 8, 9, 10]. Unfortunately, these methods are in general time consuming and, for complicated problems, can suffer from local minima. This becomes particularly true for three dimensional problems where now the optimization algorithm must incorporate boundary conditions across the interfaces of the different tissues being interrogated.

We have recently proposed a new approach to the inverse scattering problem that is different from diffraction tomography and nonlinear optimization and which we believe is very promising for a number of problems in medical imaging [6]. The approach we have in mind is based on the fact that a complete reconstruction of the scattering object is often far more than is needed. In particular, for the case of the detection and location of leukemia in the leg by microwaves, all that is needed is to determine if there is in fact cancer present and if so what is the extent of the proliferated tissue. The approach for doing this that we have recently developed [3], [6], [11] determines the *support* of anomalies in a background medium rather than the *values* of the index of refraction in the entire medium. Furthermore, the method is exact, i.e. no approximations are made on the index of refraction other than mild smoothness assumptions and there are no assumptions made on the size of the wave number. The resulting mathematical problem that needs to be solved is very simple (leading us to call our method the *simple method*): a linear integral equation with the right hand side a function of y needs to be solved to determine if y is in the support of the anomaly. The purpose of this paper is to describe our "simple method" for the case of detecting leukemia in the upper part of the lower leg which we will model as a two dimensional inverse scattering problem. Before embarking on our project, we hasten to emphasize that the problem we are considering in microwave medical imaging is only one of many in this field. For a further discussion of the use of microwaves in medicine we refer the reader to [1], [2] and [16].

2. The mathematical model. We will now formulate a simple, but still hopefully realistic, mathematical model of the medical imaging problem described above. In order to do this, we need to make a number of assumptions. The main assumptions are as follows:

1. The leg is immersed in water in order to make a better impedance match between the object under test with the host medium;
2. The portion of the leg imaged is viewed as a cross section of a cylinder with the permittivity ε and the conductivity σ varying only along a plane perpendicular to the cylinder;
3. The index of refraction for a healthy leg is known (e.g. computed at a previous time by the method used in [10]);
4. It is possible to ignore the fact that the human body is anisotropic and heterogeneous and assume instead that we are dealing with a piecewise homogeneous isotropic medium
5. The presence of arteries and veins is ignored.

Given the above assumptions, we now assume that the incident electric field is a time-harmonic line source of frequency ω positioned parallel to the leg and polarized parallel to the axis of the leg in the direction $\vec{k}$. Then the total electric field $E(x)e^{-i\omega t}\,\vec{k}, x \in R^2$, satisfies

$$(2.1) \qquad \triangle_2 E + k^2 n(x)E = 0 \ ; \ x \in R^2\backslash\{y\}$$

$$(2.2) \qquad E(x) = \frac{i}{4}H_0^{(1)}(kn_0|x - y|) + E^s(x)$$

$$(2.3) \qquad E^s \text{ is bounded in } R^2$$

where $x = y$, $|y| = a$, is the location of the source in the water region, $H_0^{(1)}$ is a Hankel function of the first kind of order zero,

$$(2.4) \qquad n(x) = \frac{1}{\varepsilon_0}\left(\varepsilon(x) + i\frac{\sigma(x)}{\omega}\right)$$

is the index of refraction, ε_0 is the permittivity of water at frequency ω, $n(x) = n_0^2$ for x in the water region and $k = \omega/c$ where c is the speed of light in water. Note that since the conductivity σ is positive in the water region we can replace the more familiar radiation condition [4] by the boundedness condition (2.3).

Now suppose the index of refraction n_h of the healthy leg is known (note that for notational ease we are calling the index of refraction n instead of n^2). We are interested in determining the support of $n - n_h$ from measurements of E on the circle $|x| = a$. More specifically, given

$E = E(x; y)$ on $|x| = a$ for sources at $y = y_m$, $1 \leq m \leq M$, $|y_m| = a$, we want to determine the support of $n(x) - n_h(x)$ for $|x| < a$. Note that, in the presence of noise, only a small number of Fourier coefficients of E on $|x| = a$ give significant information. This is due to the fact that, on $|x| = a$, $E^s = E^s(a, \theta)$ is infinitely differentiable with respect to θ and hence the Fourier coefficients $\{a_j\}$ decay faster than any power of j. This observation implies that the inverse scattering problem formulated above is improperly posed. Furthermore, since E depends on n, this inverse problem is also nonlinear.

The inverse scattering problem discussed above is to determine an anomaly in a piecewise constant background from measured near field data corresponding to point sources as incident fields. In the next section, for the sake of simplicity, we will present the mathematical ideas of how to do this for the special case of an anomaly in a constant background and plane wave incidence with the far field pattern as data. At the end of this discussion we will then describe the technical modifications needed for piecewise constant background, line sources as incident fields and near field data.

3. The simple method for a simple case. To illustrate the basic idea of the simple method as well as some of the mathematical problems associated with its derivation we now consider the scattering in R^2 of a time-harmonic plane wave by a bounded inhomogeneous medium D with smooth boundary ∂D having unit outward normal ν. We further assume that the background medium is homogeneous and nonabsorbing. In this case [4] the scattering problem (2.1-2.3) is replaced by

$$(3.5) \qquad \Delta_2 u + k^2 n(x) u = 0 \qquad \text{in} \qquad R^2$$

$$(3.6) \qquad u(x) = e^{ikx \cdot d} + u^s(x)$$

$$(3.7) \qquad \lim_{r \to \infty} \sqrt{r} \left(\frac{\partial u^s}{\partial r} - iku^s \right) = 0$$

where $r = |x|$, $d \in \Omega := \{x \in R^2, |x| = 1\}$ and we assume that the wave number k is positive and $m := 1 - n$ is piecewise continuously differentiable with compact support. In particular the anomaly in the background medium is defined by those values of x for which $m(x)$ is different from zero. Equations (3.5-3.7) imply that [4]

$$u^s(x) = \frac{e^{ikr}}{\sqrt{r}} u_\infty(\hat{x}; d) + O(r^{-3/2})$$

as r tends to infinity where $\hat{x} = x/|x|$ and u_∞ is the *far field pattern* of the scattered wave u^s. The inverse scattering problem we are now concerned with is to determine the support of m from a knowledge of $u_\infty(\hat{x}; d)$ for $\hat{x}, d \in \Omega$.

The first question to ask is if the support of m is uniquely determined by u_∞. This question is certainly not trivial since in R^2 (in contrast to R^3) it is unknown whether or not the *values* of m are uniquely determined by u_∞. However, in our case we have the following remarkable theorem of Sun and Uhlmann [21] showing that the *support* of m is indeed uniquely determined by u_∞:

THEOREM 3.1. *If u_∞^i is the far field pattern corresponding to n_i and $u_\infty^1(\hat{x}; d) = u_\infty^2(\hat{x}; d)$ for all $\hat{x}, d \in \Omega$ then $n_1 - n_2 \in C^\alpha(R^2)$ for every $\alpha, 0 \le \alpha \le 1$.*

Uniqueness theorems for scattering by an anisotropic medium are essentially unknown, although such results are of obvious importance for problems in microwave medical imagining. The only uniqueness theorem for such problems that we know of is for the case when the medium is a homogeneous dielectric and orthotropic, i.e. $n(x)$ is of the form

$$n(x) = \begin{pmatrix} n_{11} & n_{12} & 0 \\ n_{12} & n_{22} & 0 \\ 0 & 0 & n_{33} \end{pmatrix}$$

where the entries n_{ij} are real constants. In this case it can be shown that the far field pattern determines the support of $m = I - n$ [5].

We now turn our attention to a description of the "simple method" for determining the support of $m = 1 - n$ from a knowledge of the fair field pattern corresponding to (3.5)-(3.7). We first define

$$\Phi(x, y_0) := \frac{i}{4} H_0^{(1)}(k|x - y_0|), \quad y_0 \in D$$

and note that the far field pattern of Φ is given by

$$\Phi_\infty(\hat{x}, y_0) = \frac{e^{i\pi/4}}{\sqrt{8\pi k}} e^{-ik\hat{x}\cdot y_0}.$$

For $g \in L^2(\Omega)$ we define the *Herglotz wave function* with kernel g to be

$$v(x) := \int_\Omega e^{ikx\cdot d} g(d) ds(d).$$

Then it is easily shown [4], [14] that g satisfies the *far field equation*

$$(3.8) \qquad (Fg)(\hat{x}) := \int_\Omega u_\infty(\hat{x}; d) g(d) ds(d) = \frac{e^{i\pi/4}}{\sqrt{8\pi k}} e^{-ik\hat{x}\cdot y_0}$$

if and only if there exists a function w such that, for v the Herglotz wave function with kernel g, w and v satisfy the *interior transmission problem*

$$\Delta_2 w + k^2 n(x)w = 0 \qquad \text{in } D$$
$$(3.9) \qquad \Delta_2 v + k^2 v = 0$$

$$w - v = \Phi(x, y_0) \qquad\qquad x \in \partial D$$

$$(3.10) \qquad \frac{\partial}{\partial \nu_x}(w - v) = \frac{\partial}{\partial \nu_x}\Phi(x, y_0).$$

The main point here is that we can relate the solvability of the integral equation of the first kind (3.8) to the solution of the boundary value problem (3.9), (3.10). The "simple method" is based on showing that as y_0 tends to ∂D, $y_0 \in D$, v becomes unbounded and hence $\|g\|_{L^2(\Omega)}$ becomes unbounded.

The interior transmission problem is a rather unusual boundary value problem in that it relates two different differential equations by their Cauchy data on the boundary. For what is known about such problems we refer the reader to [4], [14] and [19]. In order to justify the basis of the "simple method" for determining the support of m, we must deal with three unpleasant facts: 1) It is not clear that the interior transmission problem has a unique solution; 2) If the interior transmission problem has a unique solution, it is unlikely that v is a Herglotz wave function or, rephrased, the far field equation (3.8) in general does not have a solution (this is a reflection of the fact that the inverse scattering problem is improperly posed); 3) If the interior transmission problem has a unique solution, it is not obvious that v becomes unbounded as y_0 tends to ∂D for $y_0 \in D$.

In order to deal with the above three problems, we introduce a special class of weak solutions to (3.9), (3.10). Let B be a ball centered at the origin with $D \subset B$. Then if v and w satisfy the interior transmission problem, Green's formula and Rellich's lemma imply that

$$(3.11) \qquad w(x) + k^2 \iint_D \Phi(x, \xi)m(\xi)d\xi = v(x), \qquad x \in D$$

$$(3.12) \qquad -k^2 \iint_D \Phi(x, \xi)m(\xi)w(\xi)d\xi = \Phi(x, y_0), \qquad x \in \partial B.$$

Let H be the closure in $L^2(D)$ of the space of entire solutions to the Helmholtz equation. We say that $w \in L^2(D)$, $v \in H$, is a weak solution of the interior transmission problem if v and w satisfy (3.11), (3.12). We then have the following theorem [11]:

THEOREM 3.2. *Let c be a positive constant and assume $\operatorname{Im} n(x) \geq c$ for $x \in D$. Then there exists a unique weak solution w, v of the interior transmission problem. Furthermore,*

$$\lim_{y_0 \to \partial D} \|v(\cdot, y_0)\|_{L^2(D)} = \infty.$$

Proof. (Outline) Let

$$< \varphi, \psi > := \iint_D m(\xi)\varphi(\xi)\overline{\psi(\xi)}d\xi.$$

Then it can be shown that $L^2(D) = H \oplus H^\perp$ where orthogonality is with respect to this sesquilinear form and we can define the bounded projection operator $P : L^2(D) \to H^\perp$. We first consider uniqueness. This is a problem of when two different elliptic partial differential equations can have the same Cauchy data, a problem previously studied by Hans Lewy [17]. In our case, the addition formula for Bessel functions and (3.12) (for $\Phi = 0$) imply that $w \in H^\perp$. Letting T be the integral operator in (3.11), we have $w + PTw = 0$. From $\text{Im } n(x) \geq c$ for $x \in D$ and Green's formula we can now show that $w = 0$ and hence $v = 0$. This proves uniqueness. Existence follows from an examination of $w + PTw = f$ for an appropriately chosen function f and making use of the Fredholm alternative. $\square$

The problem of showing the existence of a unique solution to the interior transmission problem for Maxwell's equations remains an open problem except for the special case when the permittivity is constant [4].

Since every function in H can be approximated in $L^2(D)$ by a Herglotz wave function, the above theorem provides partial answers to the first two of the three problems associated with the interior transmission problem listed above. These two partial answers can now be used to address the third problem and provide the basis of the "simple method" for solving the inverse scattering problem of determining the support of m from u_∞ [3], [11]:

THEOREM 3.3. *For every $\epsilon > 0$ there exists $g(\cdot, y_0) \in L^2(\Omega)$ such that*

1.) $\left\| (Fg)(\hat{x}) - \frac{e^{i\pi/4}}{\sqrt{8\pi k}} e^{-ik\hat{x}\cdot y_0} \right\|_{L^2(\Omega)} < \epsilon$,

2.) $\lim\limits_{y_0 \to \partial D} \| g(\cdot, y_0) \|_{L^2(\Omega)} = \infty$,

3.) $\lim\limits_{y_0 \to \partial D} \| v(\cdot, y_0) \|_{L^2(D)} = \infty$ *where v is the Herglotz wave function with kernel g.*

The support of m is now found by finding a (regularized) solution g of the far field equation for y_0 on some partition P of a rectangle containing D and then determining those points $y_0 \in P$ where $\|g\|_{L^2(\Omega)}$ achieves its maximum (approaching ∂D from inside D). Note that since u_∞ is contaminated by noise and $\|g\|_{L^2(\Omega)}$ tends to infinity as y_0 tends to ∂D, for optimal results the regularization procedure for determining g from the improperly posed far field equation should depend on y_0. In particular, it has recently been shown that if the far field equation is solved by Tikhonov's regularization method where the regularization parameter $\alpha = \alpha(y_0)$ is determined by Morozov's discrepancy principle, excellent reconstructions of the support can be obtained by looking at the level curves of α [11]. An open problem is to prove that such a regularization procedure in fact always yields an approximation to the g whose existence is guaranteed by the above theorem.

4. The simple method for the detection of leukemia in the leg.

As discussed in Section 2, for the problem of detecting leukemia in the upper part of the lower leg the anomaly D is situated in a piecewise constant background medium (i.e. the healthy leg) with line sources as incident fields and near field data. In this case the far field equation (3.8) is replaced by the *modified near field equation*

$$(4.13) \quad \int_{\Gamma} [E^s(x;\xi) - E^s_h(x;\xi)] g(\xi) ds(\xi) = E_\delta(x;y_0) \quad , x \in \Gamma$$

where $\Gamma := \{x : |x| = a\}$ contains D and the discontinuities of the components of the leg in its interior, $E^s_h(x;\xi)$ is the scattered electric field due to a line source through ξ and the healthy leg as scatterer, $E^s(x;\xi)$ is the measured electric scattering data due to a line source through ξ and $E_\delta(x;y_0)$ is the radiating fundamental solution corresponding to the healthy leg with $y_0 \in D$ [6]. Note that in this case E^s and E^s_h is the data for our problem.

In order to apply this modified "simple method" to the problem of detecting leukemia in the leg (subject to the assumptions of Section 2), we have digitized the cross section of the human leg found in [13], p. 114. Synthetic data for $E^s(x;y_m)$, $1 \le m \le M$, $y_m \in \Gamma$, is generated by using the spectral-finite element code of [15] modified for complex wave numbers. The values of n are taken from [18] and are given in the following table:

Medium	$n(x)$
water	1 + .08i
fat/bone	.08 + .02i
muscle	.68 + .32i
healthy marrow	.81 + .46i
proliferated marrow	1.62 + .23i

The circle Γ has radius $a = .07$ meters. Since at 1.6 GHz (the frequency used in our numerical simulations) we have $k = 287$, this means that ka is approximately 20.

To determine the support of $m = 1 - n$ numerically, we discretize (4.13) and solve the resulting discrete problem by regularization using the Morozov discrepancy principle. The details of discretizing (4.13) can be found in [6] but for completeness we provide a brief description here. We approximate the measured scattered field E^s and the background field E^s_h by finite Fourier series as follows:

$$E^s(x;\xi) = \sum_{n=-N}^{N} E_n(\alpha) \exp(in\theta),$$

$$E^s_h(x;\xi) = \sum_{n=-N}^{N} E_{h,n}(\alpha) \exp(in\theta),$$

where $x = a \exp(i\theta)$ and $\xi = a \exp(i\alpha)$ (since we assume that the measurements are carried out on a circle of radius a). Assuming that g can also be approximated by a finite series

$$g(\xi) = \sum_{n=-N}^{N} g_n(y_0)\, e^{in\alpha}$$

and using reciprocity to exchange x and ξ in (4.13) we see that (4.13) may be discretized as

$$2\pi \sum_{n=-N}^{N} (E_{-n}(\alpha) - E_{h,-n}(\alpha))\, g_n(y_0) = E_\xi(\alpha, y_0)$$

where $E_\delta(\alpha, y_0) = E_\delta(\xi, y_0)$.

Using measurements at M data points α_j, $1 \le j \le M$, equally spaced on Γ results in the discrete problem of finding

$$\vec{g}(y_0) = (g_{-N}(y_0),\, g_{-N+1}(y_0), \cdots, g_N(y_0))^T$$

such that

(4.14) $$A\vec{g} = \vec{F}(y_0)$$

where

$$\vec{F} = \vec{f_\delta}/\|\vec{f_\delta}\| \text{ and } \vec{f_\delta} = (E_\delta(\alpha_1, y_0),\, E_\delta(\alpha_2, y_0), \cdots, E_\delta(\alpha_M, y_0))^T,$$

and A is the $M \times (2N+1)$ complex matrix with entries given by

$$A_{l,m} = 2\pi\, (E_{-m}(\alpha_l) - E_{h,-m}(\alpha_l)).$$

Since we always chose $M > 2N+1$ (here in fact $M = 64$ and $N = 30$), equation (4.14) is overdetermined but very ill-conditioned. We find an approximate regularized solution to (4.14) using the Morozov discrepancy principle [22], [11].

In our numerical experiments the measured data is inaccurate due to numerical error in computing the forward problem and due to error deliberately added to the algorithm (we will describe this error later). Thus we actually have at our disposal a matrix A_δ, and we assume knowledge of $\delta > 0$ such that

$$\|A - A_\delta\| \le \delta$$

where $\|.\|$ is the spectral norm. The Morozov technique calculates an approximate solution $\vec{g}_\delta$ of (4.14) such that

$$\|A_\delta\, \vec{g}_\delta - F(y_0)\|^2 = \delta^2 \|\vec{g}_\delta\|^2$$

via Tikhonov regularization. The norm used here is the discrete least squares norm on $\mathbb{C}^M$ and $\mathbb{C}^{2N+1}$ respectively.

The results in this paper are obtained using MATLAB. We perform two types of numerical experiments. In the first experiment we use data that has been corrupted by noise as described in [6]. In the second experiment we compute the data E^s using a domain that differs slightly from that used to compute E_h^s and E_δ. In particular we adjust the outer boundary of the fat layer (water/fat interface) to increase the area of the fat layer. We describe the results of these experiments next.

4.1. Comparison with previous results. Our first set of numerical experiments shows that the discrepancy method of determining g works as well as the old method in [6] with fixed Tikhonov parameter (but without the arbitrary choice of the Tikhonov parameter). In this case we add artificial noise to the data to simulate measurement error by defining

$$(A_\delta)_{l,j} = A_{i,j} \left(1 + \varepsilon\chi_{i,j}\right) \qquad 1 \le l \le M,\ 1 \le j \le N,$$

where ε is a small parameter and $\chi_{i,j}$ is a normally distributed random number in the range [-1, 1]. As in [6] we chose $\varepsilon = 0.013$ which gives approximately 1% relative error in A_δ. We choose

$$\delta = \|A - A_\delta\|$$

so that we ignore discretization error in computing A or A_δ. The grid and distribution of tissue types is shown in Figure 1. In Figures 2 and 3 we show reconstruction from [6] and the corresponding results for the discrepancy method. Clearly this new method works almost as well as our old scheme without having the arbitrary problem of choosing a regularization parameter.

4.2. Stability to domain changes. In this case we move the nodes defining the outer edge of the fat/water interface to increase the area of the fat layer. The increase in thickness is not uniform (since triangles in the mesh must remain non-degenerate which limits node movement in some parts of the grid). The modified grid is then used to compute E^s, whereas the original grid is used to compute E_h^s and E_δ. This simulates a plausible source of error in the application of the method. To simulate an approximate knowledge of the error parameter δ we take

$$\delta = 2\|A - A_\delta\|.$$

In our first experiment of this type (see Figures 4 and 5) we perturb the fat layer to increase the area by 0.5%. Even though this is a small change in the outer edge of the fat layer, it produces an error in A of approximately 1% (similar to the choice $\varepsilon = 0.0133$ in the previous section). As a result, the reconstructions are quite similar in quality to those in Figures 2 and

3. In Figures 6 and 7 we increase the perturbation so that the area is increased by 1.1%. Although the results are degraded slightly, we can still identify the structures present.

In Figures 5 and 7 we have graphed the Tikhonov regularization parameter computed by the Morozov discrepancy method. As was observed in [6], this parameter sometimes gives a better reconstruction than the norm of g itself. However, the difficulty of using this parameter is that it is not sensitive to the absence of change in the scatterer (i.e. only noise on the data). For example, in Figure 5a) the regularization parameter is showing some patterns even though nothing is present. Thus we feel that both the norm of g and the regularization parameter need to be used in this inverse algorithm to have the best possibility of identifying a structure.

We have also found it necessary to normalize the right hand side of (4.14) when using the Morozov technique. This normalization ensures that $\vec{F}$ has unit norm and scales away changes in $\vec{F}$ due to absorption of waves traveling from the interior of the leg. In Figure 8 we show the normalized and unnormalized computation of Target 1 using the perturbed fat layer data as in Figure 6. The need for normalization is clearly demonstrated.

5. Acknowledgement and disclaimer. Effort of authors sponsored by the Air Force Office of Scientific Research, Air Force Materials Command, USAF, under grant number F49620-95-1-0067. The US Government is authorized to reproduce and distribute reprints for governmental purposes notwithstanding any copyright notation thereon. The views and conclusions contained herein are those of the authors and should not be interpreted as necessarily representing the official policies or endorsements, either expressed or implied, of the Air Force Office of Scientific Research or the US Government.

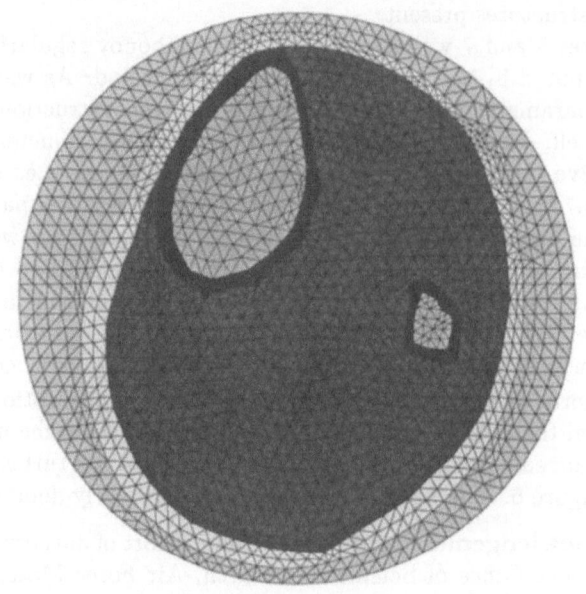

(a) Fine mesh.

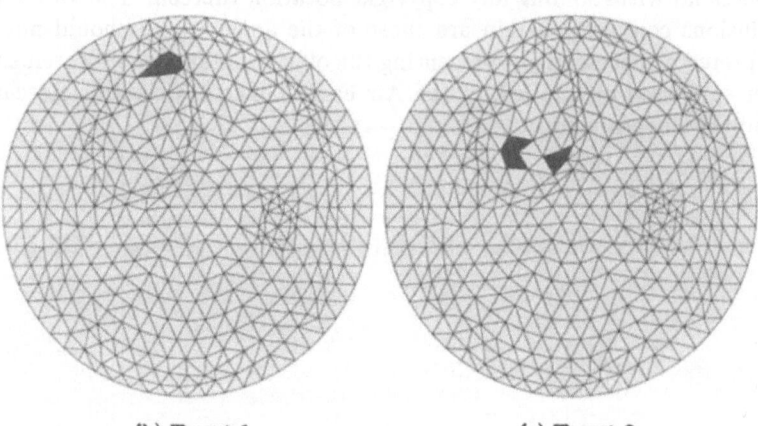

(b) Target 1.   (c) Target 2.

FIG. 1. *The mesh and targets used in this paper. In panel a) the black region is bone and within each bone is marrow. The outermost region is water and there is a fat layer surrounding the muscle of the leg. This mesh is used for all forward calculations and inverse calculations. Panels b) and c) show the region in the marrow of proliferated cells (in black). Target 1 in b) is close to the surface whereas Target 2 in c) is further from the surface and more complex.*

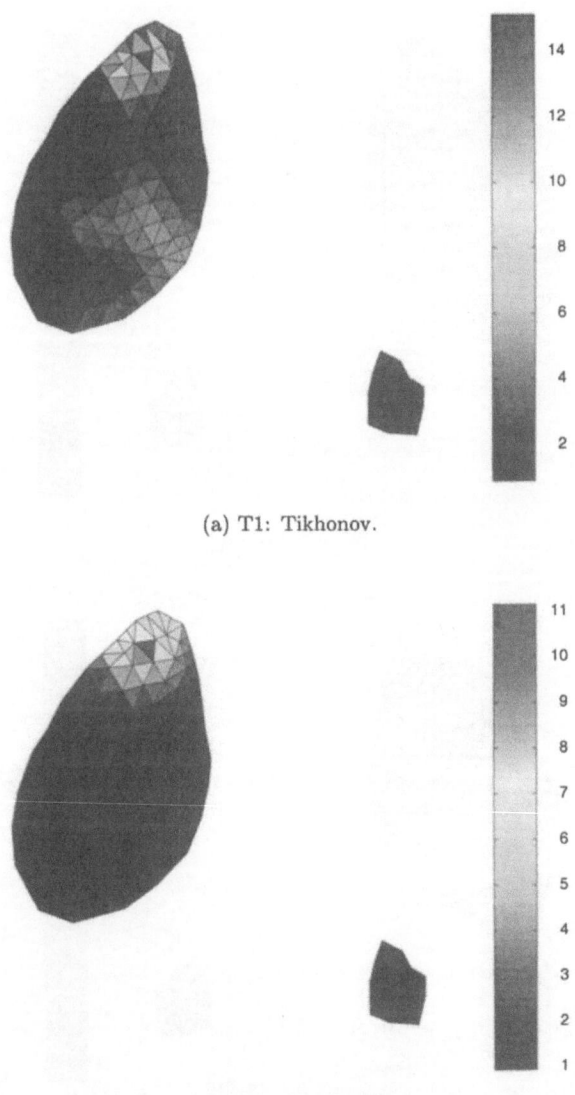

(a) T1: Tikhonov.

(b) T1: Morozov.

FIG. 2. *Here we compare the results of reconstructing Target 1 (denoted T1, see Figure 1 b)) using the straight forward Tikhonov approach from [6] with the Morozov discrepancy approach discussed in this paper. Here $\epsilon = 0.01333$ or approximately 1% relative error in the data. The results are quite similar.*

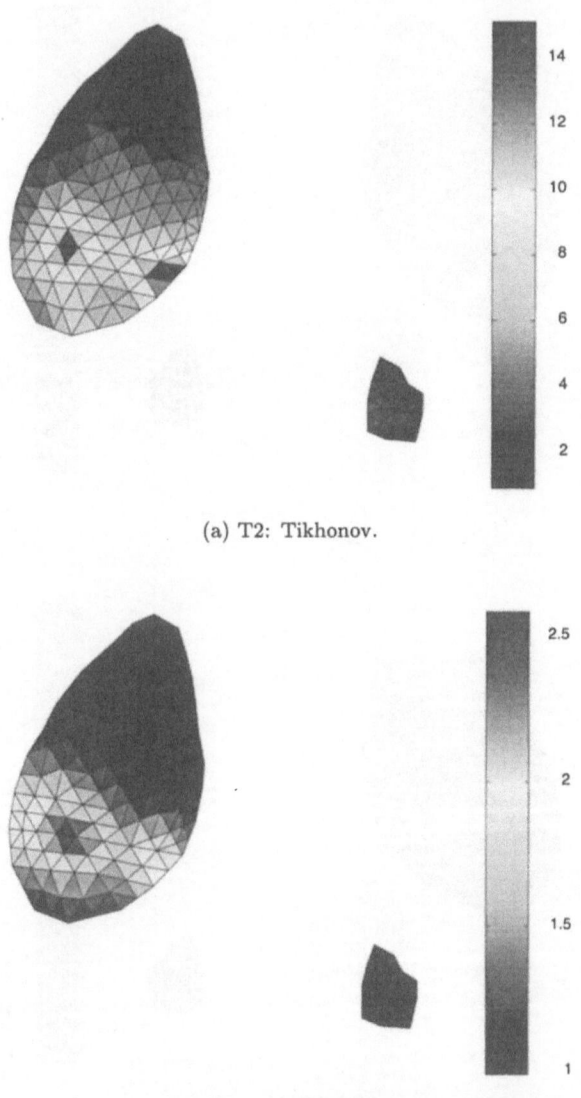

(a) T2: Tikhonov.

(b) T2: Morozov.

FIG. 3. *Here we compare the results of reconstructing Target 2 (denoted T2, see Figure 1 c)) using the straightforward Tikhonov approach from [6] with the Morozov discrepancy approach discussed in this paper. Parameters are the same as for Figure 2.*

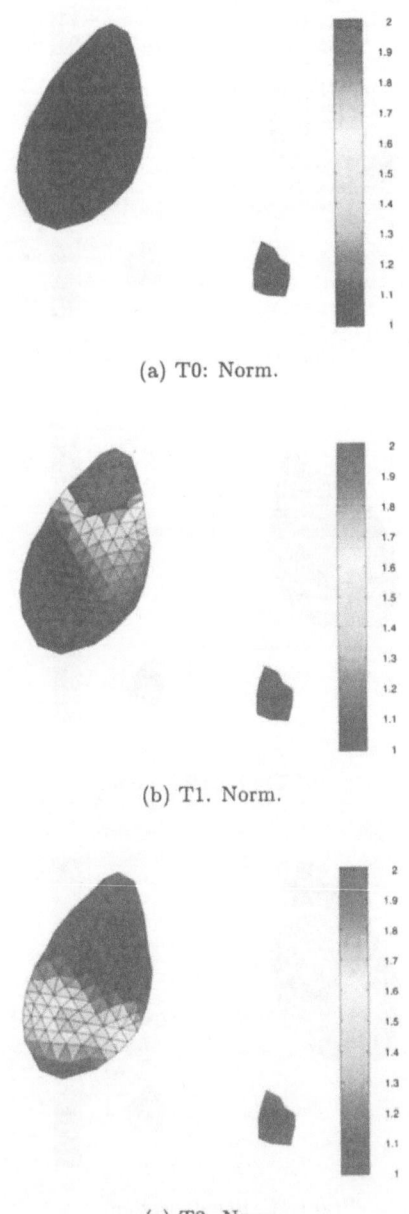

(a) T0: Norm.

(b) T1. Norm.

(c) T2. Norm.

FIG. 4. *Here we compare the results of reconstructing T0 (no target present), T1 (target 1) and T2 (target 2) when the fat area is increased by 0.55%. We show graphs of $1/\|g\|$ (denoted Norm).*

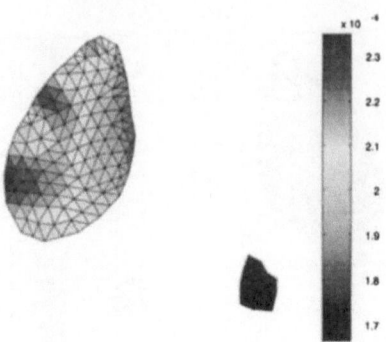

(a) T0. Regularization.

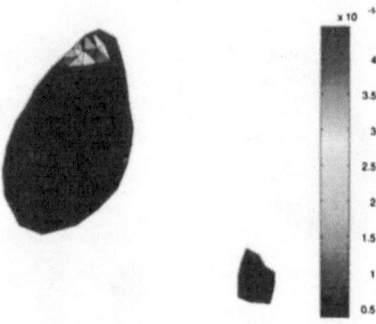

(b) T1. Regularization.

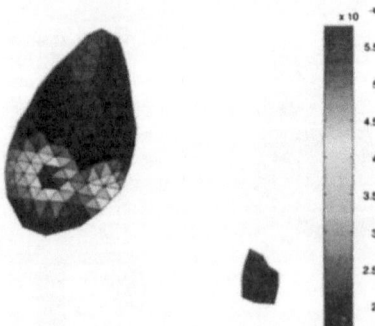

(c) T2. Regularization.

FIG. 5. *Here we compare the results of reconstructing T0 (no target present), T1 (target 1) and T2 (target 2) when the fat area is increased by 0.55%. We show graphs of the Tikhonov parameter determined by the Morozov dicrepancy principle (denoted Regularization).*

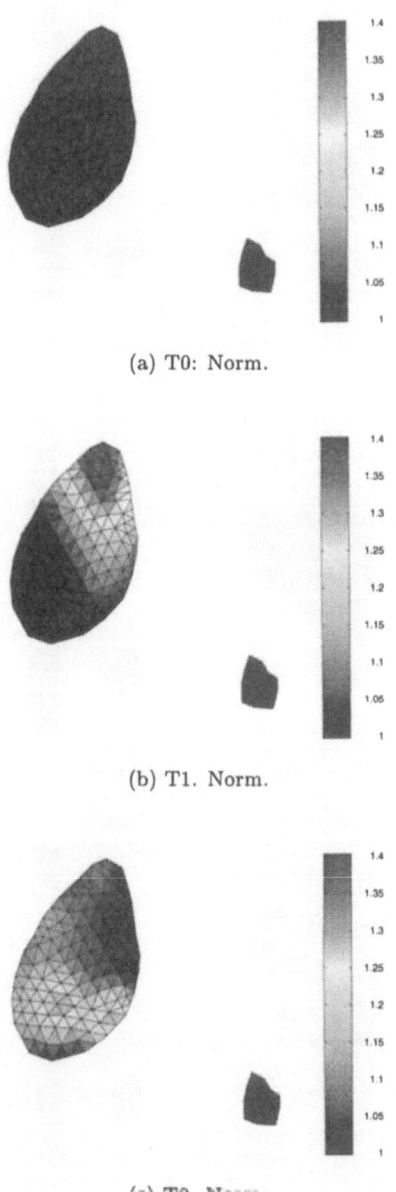

(a) T0: Norm.

(b) T1. Norm.

(c) T2. Norm.

FIG. 6. *Here we compare the results of reconstructing T0 (no target present), T1 (target 1) and T2 (target 2) when the fat area is increased by 1.1%. We show graphs of* $1/\|g\|$ *(denoted Norm).*

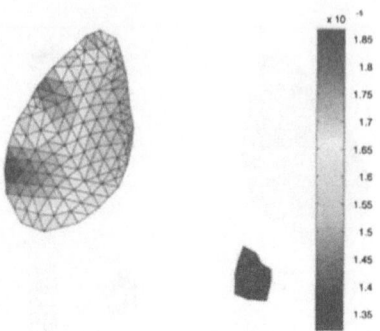

(a) T0. Regularization.

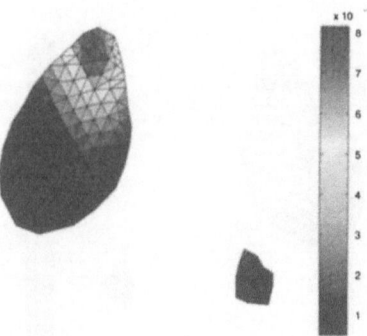

(b) T1. Regularization.

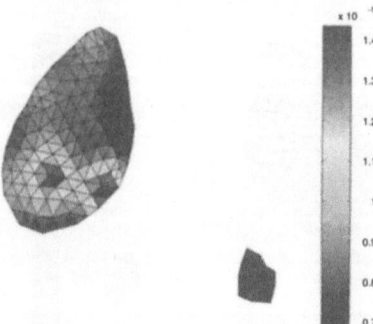

(c) T2. Regularization.

FIG. 7. *Here we compare the results of reconstructing T0 (no target present), T1 (target 1) and T2 (target 2) when the fat area is increased by 1.1%. We show graphs of the Tikhonov parameter determined by the Morozov discrepancy principle (denoted Regularization).*

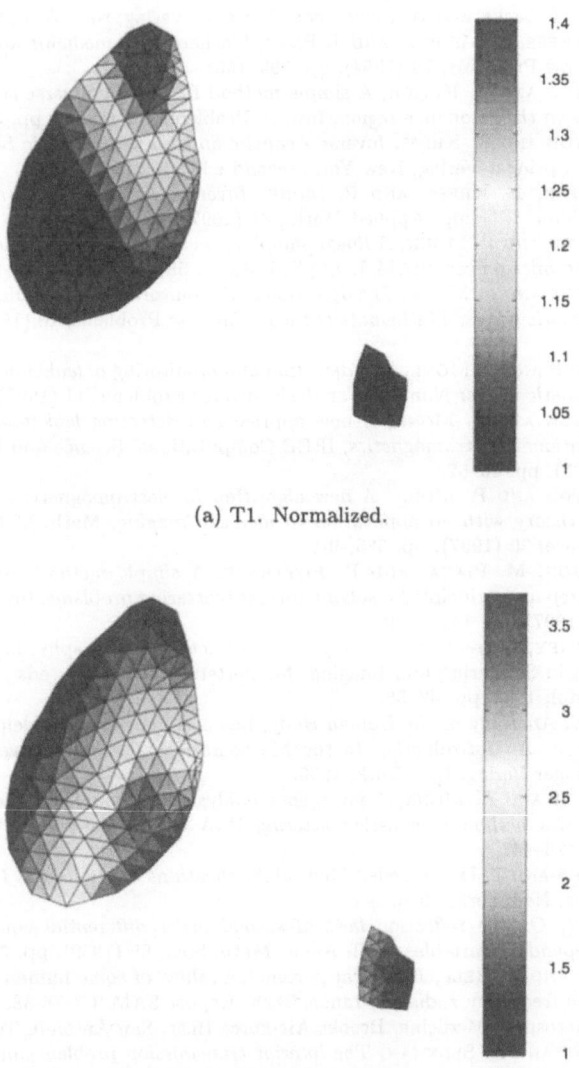

(a) T1. Normalized.

(b) T1. Non-normalized.

FIG. 8. *Here we compare the results of reconstructing Target 1 (T1) when the right hand side of (4.2) is normalized (so $\vec{F} = \vec{f}/\|\vec{f}\|$) above, and when it is not normalized (so $\vec{F} = \vec{f}$) below. Clearly it is necessary to normalize the right hand side.*

REFERENCES

[1] R. ALBANESE, *Wave propagation inverse problems in medicine and environmental health*, in Inverse Problems in Wave Propagation, G. Chavent, G. Papanicolaou, P. Sacks, and W. Symes, eds., Springer-Verlag, New York, 1997, pp. 1–12.

[2] R. ALBANESE, R. MEDINA, AND J. PENN, *Mathematics, medicine and microwaves*, Inverse Problems, 10 (1994), pp. 995–1007.

[3] D. COLTON AND A. KIRSCH, *A simple method for solving inverse scattering problems in the resonance region*, Inverse Problems, 12 (1996), pp. 383–393.

[4] D. COLTON AND R. KRESS, *Inverse Acoustic and Electromagnetic Scattering Theory*, Springer-Verlag, New York, second edition, 1998.

[5] D. COLTON, R. KRESS, AND P. MONK, *Inverse scattering from an orthotropic medium*, J. Comp. Applied Math., 81 (1997), pp. 269–298.

[6] D. COLTON AND P. MONK, *A linear sampling method for the detection of leukemia using microwaves*, SIAM J. Applied Math., 58 (1998), pp. 926–941.

[7] D. COLTON AND P. MONK, *The detection and monitoring of leukemia using electromagnetic waves: Mathematical theory*, Inverse Problems, 10 (1994), pp. 1235–1251.

[8] D. COLTON AND P. MONK, *The detection and monitoring of leukemia using electromagnetic waves: Numerical analysis*, Inverse Problems, 11 (1995), pp. 329–342.

[9] D. COLTON AND P. MONK, *A new approach to detecting leukemia: Using computational electromagnetics*, IEEE Computational Science and Engineering, 2 (1995), pp. 46–52.

[10] D. COLTON AND P. MONK, *A new algorithm in electromagnetic inverse scattering theory with an application to medical imaging*, Math. Methods Applied Science, 20 (1997), pp. 385–401.

[11] D. COLTON, M. PIANA, AND R. POTTHAST, *A simple method using Morozov's discrepancy principle for solving inverse scattering problems*, Inverse Problems, 13 (1997), pp. 1477–1493.

[12] A. DEVANEY, *Current research topics in diffraction tomography*, in Inverse Problems in Scattering and Imaging, M. Bertero and E. Pike, eds., Adam Hilger, Bristol, 1992, pp. 47–58.

[13] H. GRAY, *Anatomy of the Human Body*, Lea and Febiger, Philadelphia, 1959.

[14] A. KIRSCH, *An Introduction to the Mathematical Theory of Inverse Problems*, Springer-Verlag, New York, 1996.

[15] A. KIRSCH AND P. MONK, *Convergence analysis of a coupled finite element and spectral method in acoustic scattering*, IMA J. of Numerical Analysis, 9 (1990), pp. 425–447.

[16] L. LARSEN AND J. JACOBI, eds., *Medical Applications of Microwave Imaging*, IEEE Press, New York, 1986.

[17] H. LEWY, *On the reflection laws of second order differential equations in two independent variables*, Bull. Amer. Math. Soc., 65 (1959), pp. 37–58.

[18] J. PENN AND E. BELL, *Electrical parameter values of some human tissues in the radio frequency radiation range*, Tech. Report SAM-TR-78-38, USAF School of Aerospace Medicine, Brooks Air Force Base, San Antonio, Texas, 1978.

[19] B. RYNNE AND B. SLEEMAN, *The interior transmission problem and inverse scattering from inhomogeneous media*, SIAM J. Math. Anal., 22 (1991), pp. 1755–1762.

[20] J. STAMNES, L. GELIUS, I. JOHANSEN, AND N. SPONHEIM, *Diffraction tomography applications in seismics and medicine*, in Inverse Problems in Scattering and Imaging, M. Bertero and E. Pike, eds., Adam Hilger, Bristol, 1992, pp. 268–292.

[21] Z. SUN AND G. UHLMANN, *Recovery of singularities for formally determined inverse problems*, Comm. Math. Physics, 153 (1993), pp. 431–445.

[22] A. TIKHONOV, A. GONCHARSKY, V. STEPANOV, AND A. YAGOLA, *Numerical Methods for the solution of ill-posed problems*, Kluwer, Dordrecht, 1995.

IMAGE RECONSTRUCTION FROM EXPERIMENTAL DATA IN DIFFUSION TOMOGRAPHY

MICHAEL V. KLIBANOV*, THOMAS R. LUCAS* , AND
ROBERT M. FRANK*

Abstract. The authors have recently introduced a novel imaging algorithm for optical/diffusion tomography, the "Elliptic Systems Method" (ESM). In this article the performance of the ESM is analyzed for experimental data. Images are obtained for the case of a single source and seven (7) detector locations, an unusually limited number of source/detector pairs. These images are verified by numerical simulation. A new approach to data fitting (at the detectors) is introduced.

1. Introduction. The problem of Diffusion Tomography (DT) consists of imaging small abnormalities hidden in a diffuse background medium using measurements of an output radiation on the boundary of the medium. Propagation of radiation in such a medium is governed by the diffusion Partial Differential Equation (PDE), which is a parabolic PDE in the case of time resolved data. Applications of DT include thermal imaging, imaging using electromagnetic radiation (since Maxwell's system can be often reduced to a parabolic equation), etc.

In the past several years a great amount of attention has been paid to optical imaging in a turbid media, because of its potential applications to medical imaging, c.f. [1-3],[5], [12], and [13]. Two examples of turbid media are biological tissues and murky water. The source of radiation in this case is an ultrafast laser pulse of about 100 femtoseconds duration (1 femtosecond = 10^{-15} second, and 1 picosecond (ps) = 10^{-12} second).

The dominating factor of light propagation in a turbid media is photon scattering. Absorption also takes place, but in biological tissues, for example, the absorption coefficient is much less than the scattering coefficient [2]. Photons passing through a turbid medium experience many random scattering events, thus propagating in a diffuse-like manner. The problem of optical imaging which we are considering is to reconstruct one small perturbation of at least one optical parameter of the medium. This leads to the solution of our Inverse Scattering Problem (ISP). This small perturbation characterizes both the presence (if any) of an abnormality (such as a tumor) and the correction to the initial guess about the background medium.

Optical imaging of the female breast gives a good example. Cancerous tumors absorb light more than normal tissues. This absorbing contrast is from 2:1 to 4:1 [16]. Nevertheless, since the sizes of these tumors are small as compared with the size of the breast, they can be considered as low contrast objects which form small perturbations of the background

*Department of Mathematics, University of North Carolina at Charlotte, Charlotte, NC 28223.

medium. Rigorously speaking, the L_2 norm of the unknown perturbations (due to the presence of tumors) of the corresponding coefficients of the diffusion equation (see below) is much smaller than the L_2-norm of the "known part" of this coefficient. This "known part" describes the reference medium. Thus, the perturbation approach should be applicable in the image reconstruction algorithm.

A number of numerical methods have been proposed for DT c.f. [1-3], [5], and [13]. The vast majority of these algorithms are based on a perturbation approach. Some authors also propose to use conventional computed tomography [5]. The most crucial step of the perturbation approach, is the solution of a linearized ISP. This solution is used then for an update of the unknown coefficient(s). The previously developed numerical methods reduce the linearized ISP to an ill-posed integral equation of the first kind with respect to the perturbation of the sought coefficient(s) of the diffusion PDE.

Recently we have proposed a novel approach to the image reconstruction algorithms, the "Elliptic Systems Method" [8]. The ESM is also a version of the perturbation approach, requiring only a single source location, in the case of time resolved data. However, unlike previously developed methods, the ESM reduces the linearized problem to a well-posed boundary value problem for a coupled system of elliptic PDEs. Therefore, solution of this system by the Finite Element Method (FEM) amounts to the factorization of a sparse, well-conditioned matrix, which can be done very rapidly. The number N of PDEs in this system serves as the regularization parameter for this (originally ill-posed) problem. Usually, $N \leq 4$, since images degrade for larger N, see [8] for a discussion of this topic. This paper is partly expository and partly original. In the expository part (sections 4, 5, and the Appendix) we follow [8] describing the ESM, its numerical implementation and formulating a "limiting" theorem. A new element is introduced in section 5 which offers a different method of data fitting at the detectors. By this method we fit the data using integrals of Legendre polynomials, rather than these polynomials themselves, as in [8]. This approach has some advantages over [8]. In the original part of the paper (sections 2, 3, and 6) we present, for the first time, results of imaging experimental data using the ESM. We also verify these results using numerical simulation, thus presenting new images from simulated data. A few remarks must be made about the connections between the theoretical derivation of the ESM and the specific numerical results obtained in this paper. The theoretical derivation of the ESM was performed for the case where the domain of interest Ω had a sufficiently smooth boundary; the coefficients of the parabolic PDE were also assumed to be sufficiently smooth. In addition, the abnormality(ies) were assumed to be much smaller (in L_2-norm, for example) than the known part of the sought coefficient, in order to justify the perturbation approach. We must point out, however, that we have worked with the experimental data which we had. In this case some requirements of the theoretical derivation were violated (see section 2 for details). Namely:

1. Although our current code works with the 2-dimensional problem only, the data was collected in a 3-D region, which was a rectangular prism.

2. The 2-D cross-section of this prism was a square. Thus, its boundary was not smooth, which naturally poses a question about singularities of solutions of the corresponding PDEs near the boundary. We avoid this issue by the use of simple cut-off near the boundary. More sophisticated methods can be applied, but they are beyond the scope of this article.

3. In the experiment the inclusion was a black absorber. So formally, within this inclusion, the value of the unknown absorption coefficient was infinite, and the diffusion coefficient was zero. Therefore, the perturbation cannot be considered small, and the parabolic equation is not valid within such an abnormality. As a point of clarification, we observe that in the practical scenario of medical imaging, for example, light passes through the tumors, and they are certainly not black absorbers.

4. We worked with the case of a single source and only 7 detector locations, whereas the ESM ideally would use many detector locations (though still a single source) to obtain good interpolation of the boundary data.

Nevertheless, we decided to apply the ESM to these data. We observed that the perturbation approach converged and good estimates for the locations of the inclusions were obtained. This is consistent with the common observation that frequently numerical methods work successfully for ranges of parameters which are broader than those guaranteed by the theory.

In section 2 we describe the experiment. In section 3 we describe the forward and inverse problems. Sections 4 and 5 are devoted respectively to the derivation of the ESM and its numerical implementation. In section 6 we present the imaging results from the experimental data and verify them through simulations. We discuss obtained results in section 7.

2. Experiment. In order to model light propagation in tissue, people often use Intralipid Solution, which is a conventional model of an isotropically scattering turbid medium. Below we consider only isotropically scattering mediums. Three optical parameters characterizing a turbid medium are the speed of light c in the medium, the transport mean free path l_t, and the mean absorption length l_a. The transport scattering and absorption coefficients are respectively $\mu'_s = 1/l_t$, and $\mu_a = 1/l_a$. l_t is the mean distance between two consecutive collisions of photons. l_a is the mean distance in which the intensity of light decays by the factor e. In low absorbing media, such as biological tissues, $l_a >> l_t$. Let $d(\Omega, x_0)$ be the maximal distance between the source position x_0 and points of the domain Ω filled with a turbid medium. Then the optical thickness of Ω is the number of mean free paths across Ω, $d(\Omega, x_0)/l_t$.

The experimental data was delivered to us by the well established experimental group led by R.R. Alfano (City College of CUNY, New York). The turbid medium used was uniform Intralipid Solution with $l_t \approx 2.7$ mm, $l_a \approx 500$ mm and $c = 0.225 \frac{mm}{ps}$. The Intralipid Solution was in a container with transparent walls. The container was a rectangular prism whose horizontal cross-section was a 60 mm × 60 mm square and the height was 90 mm. The light source and detectors were located on the boundary of the vertical mid-plane.

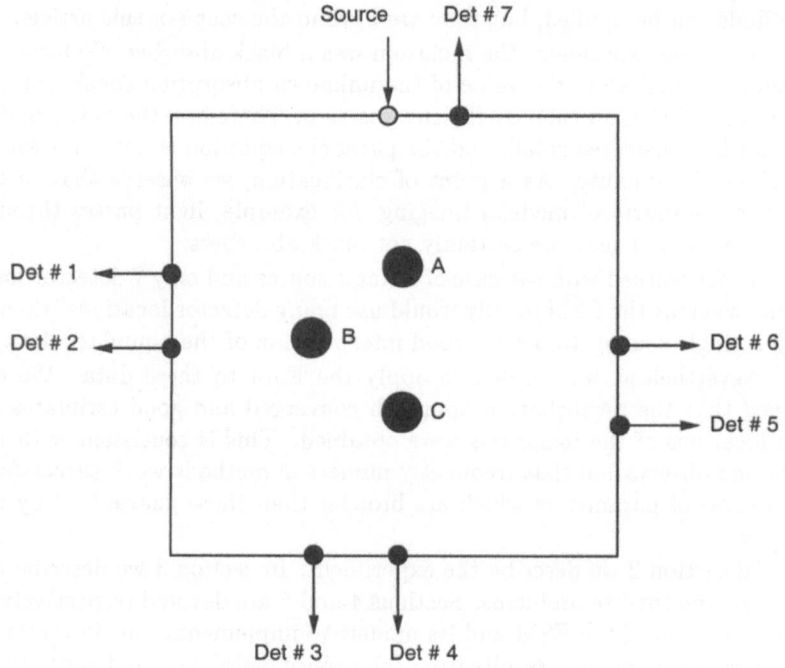

FIG. 1. *Source/detectors/ inclusions configuration in the experiment.*

The inclusion was a black painted rod of 4.8 mm diameter and 90 mm height. Thus, this inclusion was a black absorber with $l_t = l_a = 0$ inside. This abnormality was consecutively placed in three (3) different positions $A, B,$ and C. The center of each inclusion was 10 mm off the center of the square, see Fig. 1. We will refer to these inclusions as A, B, and C.

Let $\Omega \subset R^2$ be the square cross-section. We number its sides counterclockwise, with the left side being #1. On each of sides 1, 2 and 3 two optical fibers were placed, one at the midpoint and the other separated by a distance of 12.5 mm. On the backscattering side #4, however, only one fiber was placed. Below we will call these optical fibers "detectors" and number these detectors as shown on Fig. 1. The coordinates of the source and detectors are given in Table 1.

The light source was an ultrafast laser with pulses of 100 femtoseconds

TABLE 1
Coordinates of the light source and detectors, in mm.

Source/detectors	(x, y) coordinates
Source	$(0, 30)$
#1	$(-30, 12.5)$
#2	$(-30, 0)$
#3	$(-12.5, -30)$
#4	$(0, -30)$
#5	$(30, -12.5)$
#6	$(30, 0)$
#7	$(12.5, 30)$

duration at a wavelength of 625 nanometers. The time resolved output light intensity was collected by optical fibers using streak camera on the time interval $t \in (0, 2000)$ ps. The time between two consecutive readings was 1 ps. The output light was measured for the following scenarios:

1. For the case when the inclusion was absent, i.e. the response of the homogeneous reference medium, and

2. Separately, for the cases of the inclusions A, B and C.

As an example of the measured output intensities, Fig. 2 displays 4 time resolved curves on detector #4, which was opposite the source. The solid curve represents the response of the reference medium; the dotted, dash-dotted, and dashed curves represent the intensities of the cases A, B, and C respectively.

Images from the same experimental data were obtained earlier by the CUNY group using a different image reconstruction algorithm [3]. However, unlike ourselves, this group had treated the data for the A, B and C cases as coming from the *same* inclusion using three (3) different source positions and seven (7) detectors per source, which were obtained by three consecutive rotations by $\pi/2$ of the entire medium.

3. Forward and inverse problems. Since we have implemented only the 2-dimensional version of the ESM, we will, for the sake of simplicity, focus on 2-dimensional space. 3-D theory is very similar [8]. In [8] we have assumed that the whole space R^2 was filled with a turbid medium, thus the parabolic Cauchy problem was considered. However, in the case when the detectors are placed on the boundary of the turbid medium (as above), different boundary conditions should be used. These are the so-called "extrapolated boundary conditions" [4], which are described below.

Let $\Omega \subset R^2$ be a bounded domain with the boundary $\partial\Omega$. In sections 3 and 4 we will assume that Ω is a convex domain with a sufficiently smooth boundary $\partial\Omega$. Let $G \subset R^2$ be a bigger domain, $\Omega \subset G$, such that for any point $x \in \partial G$ there exists a unique point $x' \in \partial\Omega$ such that the outward

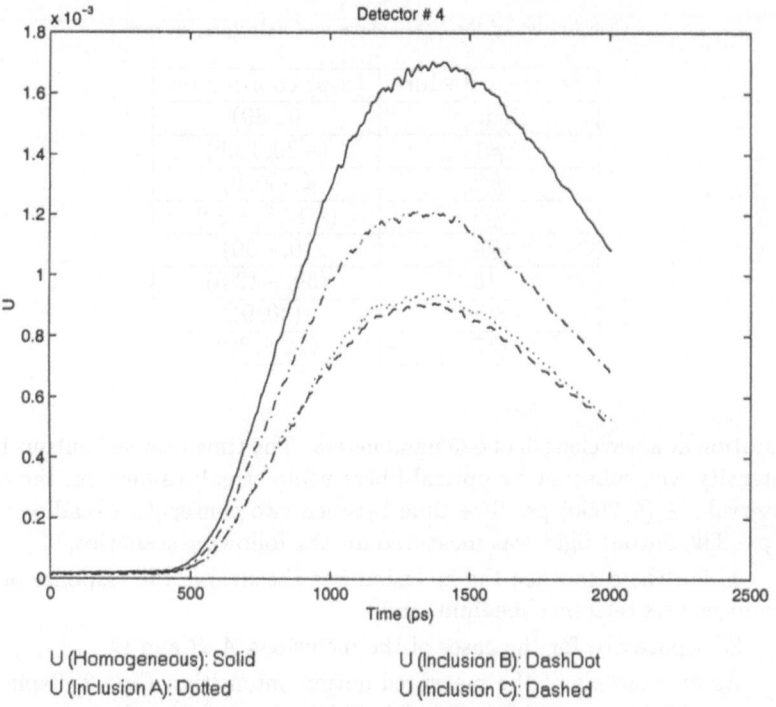

U (Homogeneous): Solid   U (Inclusion B): DashDot
U (Inclusion A): Dotted   U (Inclusion C): Dashed

FIG. 2. *Experimental data for detector #4 (opposite the source).*

distance in the normal direction is given by

$$(3.1) \qquad |x - x'| = \alpha \, l_t$$

where α is a positive constant. Then ∂G is an extrapolated boundary of the domain Ω. The importance of this extrapolated domain G is due to the fact that zero Dirichlet boundary conditions are assigned on ∂G. The constant α depends on the mismatch of the speed of light in the turbid medium and air. The speed of light in a vacuum is about $0.3 \ \frac{mm}{ps}$. The speed of light in water is about $0.225 \ \frac{mm}{ps}$. Assuming that the speed of light in air and in the Intralipid solution are the same as respectively in a vacuum and in water, we find that the mismatch of the light speeds in our case is $1.33 = 0.3/0.225$. Hence, one should take $\alpha \approx 1.33$. We used $\alpha = 1.1$, which provides good quality images.

Let $u(x, t)$ be the light intensity with the light source located at $\{x_0\}$. Then the function u satisfies the diffusion/parabolic equation for $x \in G$, $t \in (0, T)$, where $T = $ const. > 0,

$$(3.2) \qquad u_t = div \, (D(x) \, \nabla u) - a(x) \, u, \quad (x, t) \in G \times (0, T)$$

(3.3) $$u(x, 0) = \delta(x - x_0).$$

The extrapolated boundary condition is

(3.4) $$u|_{\partial G} = 0.$$

In (3.2) $D(x)$ is the diffusion coefficient and $a(x)$ is the absorption term [13]

(3.5) $$D(x) = \frac{c\,l_t(x)}{3}, \; a(x) = \frac{c}{l_a(x)}.$$

Note that a more general expression for the diffusion coefficients is $D(x) = [3(\mu_s'(x) + \mu_a(x))]^{-1}$ [13]. However, in low absorbing media (such as we are dealing with) $\mu_s'(x) \gg \mu_a(x)$, leading to formula (3.5). Further, since within a black absorber $l_t = l_a = 0$, then within the rod used in the experiment

(3.6) $$D(x) = 0, \; a(x) = \infty.$$

We assume for purposes of our theoretical derivation that $D \in C^{k+1+\gamma}(\overline{\Omega})$, $a \in C^{k+\gamma}(\overline{\Omega})$, and these coefficients can be extended into $G\backslash\Omega$ in such a way that $D \in C^{k+1+\gamma}(\overline{G})$ and $a \in C^{k+\gamma}(\overline{G})$, $\gamma \in (0,1)$, the integer $k > 0$ will be chosen later (see Theorem A3 in the Appendix), and $C^{k+\gamma}$ are the Hölder spaces. If the boundary ∂G is sufficiently smooth and both functions $D(x)$ and $a(x)$ are known, than the parabolic forward problem (3.2)-(3.4) has the unique solution $u \in C^{2k+\gamma, k+\frac{7}{2}}(\overline{G} \times [\varepsilon, T])$ for any $\varepsilon \in (0, T)$ [9].

To pose the ISP we assume that the light intensity $u(x, t)$ is measured at a set of points $\{x_i\}_{i=1}^m \in \partial\Omega$ over a time interval (T', T). We note that at each detector's location x_i the data for $t < t(x_i) = |x_i - x_0|/c$ is just noise. This is due to the fact no light arrives at the point x_i before $t = t(x_i)$. On the other hand, the diffusion equation (3.1) is formally valid for all $t \in (0, T)$. This is a well-known inconsistency between the diffusion equation and the actual time requiring for light to reach the detectors. Fortunately, however, both our results as well as previous results from imaging experimental data [3] show that images can be successfully obtained. Thus, we assume that

$$T' > T(\Omega, x_0) = \frac{\max_{x \in \overline{\Omega}} |x - x_0|}{c}.$$

In the specific case of the experimental data we are dealing with $T(\Omega, x_0) = \frac{\sqrt{60^2 + 30^2}}{0.225}$, or 298 ps.

We interpolate the detectors readings over the entire boundary $\partial\Omega$, thus obtaining the function $\varphi(x, t)$,

(3.7) $$u|_{\partial\Omega} = \varphi(x, t), \; x \in \partial\Omega, \; t \in (T', T).$$

Note that the accuracy of the function φ depends, in part, on the distances between neighboring detectors. If these distances are sufficiently small, i.e. if there are sufficiently many detectors placed on $\partial\Omega$, then the function φ is given with a good accuracy (subject to the noise in the data). If, however, these distances are not small, as in the case of the experimental data above with seven (7) detectors, then the accuracy of the function φ could be poor away from the detectors.

In our currently implemented code only the function $a(x)$ in (3.2) is treated as an unknown function. However, the ESM can be extended to work for the case where either the diffusion coefficient $D(x)$, or both functions $D(x)$ and $a(x)$ are unknown. This extension of our method is not yet implemented and forms a subject of our future effort. Thus, we will focus on the determination of the function $a(x)$. In the experimental data the difference in the absorption coefficient between the medium and the object is infinite, as the object is a black absorber.

Inverse scattering problem. *Given the function $D(x)$ for $x \in G$, the function $a(x)$ for $x \in G\backslash\Omega$ and the function $\varphi(x,t)$ in (3.7), determine the function $a(x)$ for $x \in \Omega$.*

In the current numerical implementation of the ESM we solve a coupled elliptic system of the 4th order, see below and [8]. In the future we plan to solve a coupled second order elliptic system and corresponding results will be discussed in forthcoming publications. In order to solve this 4th order system, the Neumann boundary conditions on $\partial\Omega$ are also required. One way to do this would be to solve the boundary value problem for the equation (3.1) in the domain $G\backslash\Omega$ using the boundary conditions (3.4) and (3.7). However, this would lead to additional computations. For this reason we have chosen a simpler way which is described below. In order to check the validity of this approach, we compared images from simulated data for two cases: (i) using the approach described below, and (ii) computing the normal derivative $\frac{\partial u}{\partial n}|_{\partial\Omega}$ using the solution of the forward problem. For the given ranges of parameters the images were very similar (not shown in this paper).

For $x \in \partial\Omega$ let $n = n(x)$ be the outward unit normal vector on $\partial\Omega$. Then

$$(3.8) \qquad \frac{\partial u}{\partial n}(x,t) = \lim_{s \to 0} \frac{u(x+sn,t) - u(x,t)}{s}$$

Because of (3.1) $(x + \alpha l_t n) \in \partial G$. Hence (3.4) implies $u(x + \alpha l_t n, t) = 0$. If the number αl_t is sufficiently small, then (3.6) implies

$$\frac{\partial u}{\partial n}(x,t) \approx -\frac{1}{\alpha l_t}u(x,t), \text{ for } x \in \partial\Omega, \ t \in (T',T).$$

Thus, because of (3.5) we will assume that

$$(3.9) \qquad \frac{\partial u}{\partial n} = -\frac{1}{\alpha l_t}u = -\frac{1}{\alpha l_t}\varphi(x,t), \ x \in \partial\Omega, \ t \in (T',T)$$

Thus, (3.9) is equivalent to the boundary condition of the third kind $\frac{\partial u}{\partial n} + \sigma u = 0$, for a positive constant σ.

4. Elliptic systems method (ESM).

4.1. Perturbation approach versus the given experimental data.

We describe the ESM only for the above case when the function $a(x)$ in (3.2), (3.5) is unknown. Since the ESM is a version of the perturbation approach, we assume that $a(x) = a_0(x) + h(x)$ where the function $a_0(x)$ is given. This function describes a good guess about the background medium. The perturbation term $h(x)$ is unknown; it describes both inclusions and the correction to the background medium. In medical applications, however, the imaging of small "tumor-like" inclusions is often of more interest than the imaging of the background media.

Following the general idea of the perturbation approach, we will assume that the function $h(x)$ is much smaller than the background function $a_0(x)$, leading to a linearization with respect to $h(x)$ on each consecutive iteration of the ESM. For example, one might assume that

$$(4.1) \qquad \|h\|_{L_2(\Omega)} << \|a_0\|_{L_2(\Omega)}.$$

As an example we will consider the case of a single inclusion of circular shape with radius r. For simplicity, assume that

$$h(x) = \begin{cases} h_0(x) = \text{const.} > 0, & \text{within inclusion} \\ 0, & \text{otherwise} \end{cases}$$

Also, assume that $a_0(x) = \text{const.} = a_0 > 0$ (as in the case of the above experiment). Then in the 2-dimensional case (4.1) means that

$$(4.2) \qquad \frac{\|h\|_{L_2(\Omega)}}{\|a_0\|_{L_2(\Omega)}} = \frac{\pi r^2 h_0}{A(\Omega)|a_0|} << 1,$$

where $A(\Omega)$ is the area of the domain Ω. However, in the case of a black absorber $h_0 = \infty$, see (3.5), (3.6). Hence, the condition (4.2) is not valid for our specific set of the experimental data. Nevertheless, convergence of consecutive Newton-like iterations of the ESM was observed, and correct locations of inclusions were obtained. This falls outside of our current theory.

4.2. Three steps of the ESM.

In this subsection we assume for theoretical purposes that $x_0 \notin \overline{\Omega}$. In practice we sometimes drop this condition.

Step 1: Elimination of $h(x)$. Let the function $u_0(x,t)$ be the solution of the boundary value problem (3.2)-(3.4) with $a = a_0$. That is,

$$(4.3) \qquad u_{0t} = \text{div}\,(D(x)\nabla u_0) - a_0(x)u_0, \quad (x,t) \in G \times (0,T)),$$
$$u_0(x,0) = \delta(x - x_0),$$
$$u_0|_{\partial G} = 0.$$

Let $v(x,t) = u(x,t) - u_0(x,t)$. Then for $(x,t) \in G \times (0,T)$, using (3.2)-(3.4),

$$(4.4) \qquad v_t = div\,(D(x)\nabla v) - a_0(x)v - h(x)(u_0 + v)$$
$$v(x,0) = 0, \; v|_{\partial G} = 0.$$

Note that by the maximum principle $u_0 \geq const. > 0$ for $(x,t) \in \overline{\Omega} \times [0,T]$. Hence, for $(x,t) \in \Omega \times (0,T)$ we can consider the normalized solution

$$(4.5) \qquad H(x,t) = \frac{v}{u_0} = \frac{u}{u_0} - 1.$$

Substitute $v = Hu_0$ into (4.4). We obtain

$$\left[H_t - div\,(D\nabla H) - 2D\frac{\nabla u_0}{u_0}\nabla H + h(1+H) \right] u_0$$
$$= -H\left[u_{0t} - div\,(D\nabla u_0) + a_0 u_0 \right].$$

By (4.3) the right-hand-side of this equality equals to zero. Hence the equation for the function H is

$$H_t = div\,(D\nabla H) + 2D\frac{\nabla u_0}{u_0}\nabla H - h(x)(1+H), \quad (x,t) \in \Omega \times (0,T).$$

Linearization of this equation with respect to h entails dropping the product hH. Hence the equation for H, linearized with respect to the perturbation $h(x)$, is

$$(4.6) \qquad H_t = div\,(D\nabla H) + 2D\frac{\nabla u_0}{u_0}\nabla H - h(x), \quad (x,t) \in \Omega \times (0,T).$$

To establish the initial condition for $H(x,0)$, we assume that coefficients a, a_0, and D satisfy the conditions of Conjecture A in Appendix. Thus, by this conjecture

$$(4.7a) \qquad H(x,0) = 0, \; x \in \overline{\Omega}$$

$$(4.7b) \qquad \text{and } H \in C^{4,2}\left(\overline{\Omega} \times [0,T] \right).$$

Now, in order to eliminate the perturbation term $h(x)$ from the equation (4.6), we differentiate this equation with respect to t using the fact that $\frac{\partial}{\partial t}h(x) \equiv 0$. Let $p(x,t) = \frac{\partial H}{\partial t}$. Then (4.7) leads to

$$(4.8) \qquad H(x,t) = \int_0^t p(x,\tau)\,d\tau$$

Thus, (4.7) and (4.8) lead to an integro-differential equation for the function p:

$$(4.9a) \quad Lp := p_t - div\,(D\,(x)\,\nabla p) - 2D\,(x)\,\frac{\partial}{\partial t}\left[\frac{\nabla u_0}{u_0}\int_0^t \nabla p\,(x,\tau)\,d\tau\right] = 0$$

$$(4.9b) \qquad\qquad p \in C^{2,1}\,(\overline{\Omega} \times [0,T])$$

There are two boundary conditions for the function $p\,(x,t)$, but no initial conditions. To derive these boundary conditions, we use (3.7) and (3.9). By (3.7)

$$(4.10) \qquad p|_{\partial\Omega} = \frac{\partial}{\partial t}\left(\frac{\varphi\,(x,t)}{u_0\,(x,t)}\right) = \varphi_2\,(x,t) \ , \ t \in (T',T)$$

This gives the first boundary condition. Let $s = -\,(\alpha l_t)^{-1}$. Then (3.9) implies

$$\frac{\partial u}{\partial n} = su \ \text{and} \ \frac{\partial u_0}{\partial n} = su_0 \ \text{for} \ x \in \partial\Omega.$$

Hence

$$\frac{\partial}{\partial n}\left(\frac{u}{u_0}\right)\bigg|_{\partial\Omega} = \left[\frac{1}{u_0}\frac{\partial u}{\partial n} - \frac{u}{u_0^2}\frac{\partial u_0}{\partial n}\right]\bigg|_{\partial\Omega}$$

$$= \left[s\frac{u}{u_0} - s\frac{u}{u_0}\right]\bigg|_{\partial\Omega} \equiv 0.$$

Thus, the second boundary condition for the function $p\,(x,t)$ is

$$(4.11) \qquad\qquad \frac{\partial p}{\partial n}|_{\partial\Omega} = 0 \ , \ t \in (T',T)\,.$$

We will consider the function $p\,(x,t)$ on a subinterval $(T_0,T_F) \subseteq (T',T)$. Once the function p is computed, then the perturbation $h\,(x)$ can be reconstructed from (4.6) by

$$(4.12) \quad h\,(x) = \frac{1}{T_F - T_0}\int_{T_0}^{T_F}\left[div\,(D\,(x)\,\nabla H) + 2D\frac{\nabla u_0}{u_0}\nabla H - H_t\right]dt,$$

where the function H is determined by (4.8). We take the average value in (4.12) because in practical computations the integrand has a t dependence. Other variations for the reconstruction of $h\,(x)$ from H are also possible, and will be reported on in later papers.

Therefore we will now focus on the solution of the boundary value problem (4.9)-(4.11). In fact, this is a boundary value problem for a parabolic integro-differential equation with data on the lateral surface $\partial\Omega \times (T',T)$.

An elegant numerical method for solving similar problems, but without the presence of integrals, was developed and tested by R. Lattes and J. L. Lions in [11]; also see relevant publications in [6,7]. It should be noted that the method of [11] has not been applied in the context of finding x-dependent coefficients of PDEs. By [11] one should, in a certain sense, solve a boundary value problem for the 4th order equation $L^*Lp = 0$ with the boundary data (3.9). Here L^* is the operator formally adjoint to L.

However, the technique of [11] is not applicable in our case, since the operator L contains Volterra-like integrals from 0 to t, whereas the data (3.9) are not given for $t < T'$. In addition, the term $\nabla u_0 / u_0$ has a singularity as $t \to 0$. For example, if $D \equiv \text{const.}$, then $\nabla u_0 / u_0 \approx (x - x_0) / (2Dt)$ as $t \to 0$. Hence, a different approach should be taken to find the function $p(x, t)$.

Step 2: Coupled system of elliptic PDEs. Let $(T_0, T_F) \subset (T', T)$ be a subinterval of the time interval (T', T). We will now consider the expansion of the function $p(x, t)$ into a generalized Fourier series with respect to an orthonormal basis in $L_2(T_0, T_F)$ and obtain a coupled elliptic system with respect to the x- dependent coefficients of this expansion.

Let $\{\tilde{a}_k(t)\}_{k=1}^{\infty}$ be an orthonormal basis in $L_2(0, T)$, such that all functions $\tilde{a}_k(t)$ are real valued and analytic as functions of the real variable t for $t > 0$. Because the function $p(x, t)$ is analytic with respect to t for $t > 0$ [Lemma A1 in the Appendix] and because of (4.9b), it can be represented as a generalized Fourier series with respect to t on the interval $(0, T)$,

$$p(x, t) = \sum_{k=1}^{\infty} \tilde{a}_k(t) \, \tilde{Q}_k(x), \text{ for } (x, t) \in \Omega \times (0, T).$$

The functions $\tilde{Q}_k(x) \in C^2(\overline{\Omega})$ are generalized Fourier harmonics, leading to

$$\tilde{Q}_k(x) = \int_0^T p(x, t) \, \tilde{a}_k(t) \, dt.$$

Following the Galerkin method, we assume that

$$p(x, t) \approx \sum_{k=1}^{N} \tilde{a}_k(t) \, \tilde{Q}_k(x), \text{ for } (x, t) \in \Omega \times (0, T),$$

where $N \geq 1$ is an integer. Reorthogonalize the functions $\tilde{a}_k(t)$ on (T_0, T_F). Then we obtain an orthonormal basis $\{a_k(t)\}_{k=1}^{\infty}$ in $L_2(T_0, T_F)$. Hence

(4.13a) $$p(x, t) \approx \sum_{k=1}^{N} a_k(t) \, Q_k(x), \text{ for } (x, t) \in \Omega \times (0, T)$$

(4.13b) $$\text{where } Q_k(x) = \int_{T_0}^{T_F} p(x, t) \, a_k(t) \, dt.$$

Now we want to obtain a coupled elliptic system with respect to the functions $Q_k(x)$. First, consider the boundary conditions for these functions on $\partial\Omega$. Denote

$\beta(x) = (\beta_1(x), \beta_2(x), ..., \beta_N(x))$, for $x \in \partial\Omega$, where

$$(4.13c) \qquad \beta_k(x) = \int_{T_0}^{T_F} \varphi_2(x, t) a_k(t) \, dt, \quad 1 \leq k \leq N.$$

Also, introduce the N-dimensional vector valued function $Q(x) = (Q_1(x), ..., Q_N(x))$. Multiply both sides of (4.9a) by $a_k(t)$ for $k = 1, ..., N$, and integrate with respect to t over the interval (T_0, T_F). Since the functions $\{a_k(t)\}$ are orthonormal on (T_0, T_F), by using (4.9) and (4.13), we obtain the following elliptic boundary value problem

$$(4.14a) \qquad A(Q) := div(D(x)\nabla Q) - \sum_{j=1}^{2} B_j(x) \frac{\partial Q}{\partial x_j} - CQ = 0$$

$$(4.14b) \qquad Q|_{\partial\Omega} = \beta(x), \quad \frac{\partial Q}{\partial n}|_{\partial\Omega} = 0$$

where $Q \in C^2(\overline{\Omega})$. The $N \times N$ matrices $B_j \in C^{2+\gamma}(\overline{\Omega})$ depend on the function u_0, and are given by

$$(B_1, B_2)_{ks} = -2D(x) \int_{T_0}^{T_F} a_k(t) \frac{\partial}{\partial t} \left[\frac{\nabla u_0}{u_0} \int_0^t a_s(\tau) \, d\tau \right] dt$$

$$= -2D(x) a_k(t) \frac{\nabla u_0}{u_0} \int_0^t a_s(\tau) \, d\tau \, |_{T_0}^{T_F} + 2D(x)$$

$$\int_{T_0}^{T_F} a_k'(t) \frac{\nabla u_0}{u_0} \int_0^t a_s(\tau) \, d\tau dt$$

by integration by parts. The elements of the $N \times N$ matrix C are

$$(4.15) \qquad c_{ks} = \int_{T_0}^{T_F} a_s'(t) \, a_k(t) \, dt; \quad k, s = 1, ..., N.$$

Remarks: (i) In our setting N is a regularization parameter for our originally ill-posed inverse problem. Hence, unlike the forward boundary value problems, we cannot allow $N \to \infty$. In fact, N is often no greater than 4 or 5, see section 6.

(ii) In the ideal theoretical situation the data φ is given without noise and $N = \infty$. In the realistic scenario, however, the function φ is given with noise, and one takes $N < \infty$. In addition, the values of the function φ for $t < T'$ are not used. Hence, in solving the system (4.14) with noisy data and small N the goal is to get a good approximation for the first

N generalized Fourier harmonics $Q_k(x)$ of the function $p(x, t)$, and from this approximation to obtain good resulting images. So far, our extensive testing of the ESM for realistic ranges of parameters has fulfilled this goal; see section 6 and [8].

(iii) When conventional methods are used to solve the original inverse scattering problem, it gives the appearance of being data rich and computationally intensive, c.f. [1-3], [13]. This is because of the use of many source terms, with each source term having a number of associated detectors, each of which has hundreds of time readings. In the ESM, as developed here, this is no longer true, since only one source term is used, and the data at each detector is reduced to a small number of generalized Fourier coefficients related to the normalized solution $H(x, t)$. These rapid one-dimensional calculations at each detector also serve to remove much of the noise, where as mentioned above the number of terms N serves as a regularization parameter.

Step 3: Numerical scheme for the problem (4.14). Boundary value problems such as (4.14) are central to the ESM. Indeed, a major novelty of (4.14) is that it represents a *differential* rather than the conventional integral form of the resulting system. Therefore if this problem is well-posed, one can apply methods for its solution from the well established theory of numerical solutions to partial differential equations. This leads to the factorization of large, sparse matrices, with rapid solution times. The problem (4.14), can be solved by at least two approaches. One approach is to simply solve the second order system (4.14a) using only the first boundary condition (4.14b). However, we have chosen the fact that this problem can be formulated as an overdetermined system with the two boundary conditions (4.14b), rather than just one. This leads to solving an associated 4th order elliptic system. We have chosen to do this because we can rigorously prove that this problem is well-posed. On the other hand, we have not yet proven the well-posedness of the second order system (4.14a) when using only the Dirichlet boundary condition (4.14b). This topic will be studied in our forthcoming publications.

Let A^* be the operator formally adjoint to the operator A, $A^*(Q) := div\,(D(x)\nabla Q) + \sum_{j=1}^{2} \frac{\partial}{\partial x_j}(B_j^T(x)Q) - C^T Q$ where the superscript "T" denotes the transpose of a matrix. We consider the boundary value problem for the following elliptic system of the 4th order

$$(4.16) \qquad (A^*A)(P) = 0, \; P|_{\partial\Omega} = \beta(x), \; \frac{\partial P}{\partial n}|_{\partial\Omega} = 0.$$

If, for example $D \equiv 1$, then $A^*A = (\nabla^2)^2 +$ terms with derivatives up to the third order. Here $(\nabla^2)^2$ is the biharmonic operator, $(\nabla^2)^2 = \sum_{j=1}^{2} \frac{\partial^2}{\partial x_j^2}\left(\sum_{k=1}^{n} \frac{\partial^2}{\partial x_k^2}\right)$. It can easily be established that the problem (4.16) is equivalent to the following minimization problem:

$$\|A\,(P)\|^2_{L_2(\Omega)} \to \min; \quad \text{for } P|_{\partial\Omega} = \alpha\,(x)\,, \ \frac{\partial P}{\partial n}|_{\partial\Omega} = 0, \ P \in H^2\,(\Omega)\,.$$

Hence in solving (4.16), we find a *quasi-solution* of the overdetermined problem (4.14). Well-posedness of the problem (4.16) follows from

THEOREM 4.1. [8]. *Let Ω be a convex bounded domain with its boundary $\partial\Omega \in C^\infty$. Let the components $\beta_i\,(x)$ of the vector valued function $\beta\,(x)$ satisfy the condition $\beta_i\,(x) \in C^{4+\gamma}\,(\partial\Omega)$ for $1 \le i \le N$. Then there exists an unique solution $P\,(x) = (P_1\,(x)\,,...,\,P_N\,(x))$ with $P_i \in C^{4+\gamma}\,(\overline{\Omega})$ of the problem (4.16), and $\|P\|_{4+\gamma} \le M\,\|\beta\|_{4+\gamma}$, where the positive constant M depends only on the domain Ω and the operator A, and $\|P\|_{4+\gamma} =$*

$$\left[\sum_{i=1}^{N} \|P_i\|^2_{C^{4+\gamma}(\overline{\Omega})}\right]^{\frac{1}{2}}.$$

5. Numerical implementation. We have solved the system (4.16) using the FEM, for a more detailed description see [8]. Tests were conducted on a Silicon Graphics Indigo (SGI) with one processor. The implementation presented here has the following basic features:

1. It is designed to solve the above inverse problem for the case when the function $a\,(x)$ is unknown. Let $a\,(x) = a_0\,(x) + h\,(x)$, where the given function a_0 is our good guess about the background medium, and the unknown function h is a small perturbation of a_0. The core of the inverse solver is the solution of the 4th order coupled elliptic system (4.16). Given the solution of (4.16), the code recovers $h\,(x)$ from (4.12) and readily updates a_0 as $a_0\,(x) := a_0\,(x) + h\,(x)$. Next, it solves the new parabolic problem (4.3) and updates the normalized boundary conditions at the detectors. Then it solves the problem (4.16) again. This describes one iteration of the ESM.

2. Having the above applications in mind, we look for regions, where the absorption is larger than that of the background medium. In those regions of Ω where the computed function $h\,(x) < 0$, the code has an option to take $h\,(x) := 0$.

3. We observed in our tests (below) that the quality of the images was poor near the boundary of Ω, and especially near the corners. We attributed this to the non-smoothness of the boundary of a square and to the possibility of singularities in the solutions at the corners. To deal with this issue, we simply made cut-off of images in a neighborhood of the boundary of Ω. Other methods of dealing with this are possible, but they are not discussed here.

4. In the case of simulated data, we have followed a suggestion of [14] modeling a small target as a binary function (*const.* > 0 within the target and 0 elsewhere), which, however varies smoothly between this *const.* and 0 within a region whose size is much smaller than the size of the target

(such a structure of targets is a user choice, and *not* a restriction of the applicability of the ESM).

5. To obtain the values of the boundary conditions $\beta(x_i) = Q(x_i)$ at the detectors, we apply the following procedure. Let $(T_0, T_F) \subset (T', T)$ be the above mentioned time interval over which the function $p(x, t)$ is analyzed, see step 2 in section 4 and (4.13b). Let $\{a_n(t)\}_{n=1}^{\infty}$ be the set of Legendre polynomials which are orthonormal on (T_0, T_F). Let

$$\varphi_1(x_i, t) = \frac{\varphi(x_i, t)}{u_0(x_i, t)} - 1.$$

and K be an integer. Then we approximate the function $\varphi_1(x_i, t)$ as

$$(5.1a) \qquad \varphi_1(x_i, t) \approx \sum_{n=1}^{K} C_n^i \int_0^t a_n(\tau) \, d\tau, \quad t \in (T', T)$$

where the coefficients C_n^i of this expansion are found from the least squares fit of $\varphi_1(x_i, t)$ over the interval (T', T) by these integrals of Legendre polynomials. Thus

$$(5.1b) \qquad \varphi_2(x_i, t) = \frac{\partial}{\partial t}\left(\frac{\varphi(x_i, t)}{u_0(x_i, t)}\right) \approx \sum_{n=1}^{K} C_n^i a_n(t)$$

The approximation (5.1a) of φ_1 by integrals of Legendre polynomials, rather than with fitting φ_2, the partial derivative of φ_1 with respect to time, with Legendre polynomials themselves (using integration by parts [8]) as in (5.1b) is a new element in our code, which was not presented in [8]. The disadvantage of using (5.1b) alone is that an arbitrary displacement is added to the curves recovered by $\int_0^t \varphi_2(x_i, \tau) \, d\tau$ between T' and T. The disadvantage of the new scheme (5.1a) is that while $\varphi_1(x_i, t)$ may be fitted accurately between T' and T, the derivative function $\varphi_2(x_i, t)$ may not be fitted well. We have found the last effect to be minor compared with the first. We have found that (5.1a) provides images of better quality than previously reported in [8]. Thus, (4.13) and (5.1) imply that

$$C_n^i = Q_n(x_i), \quad \text{and } \beta(x_i) = (Q_1(x_i), ..., Q_N(x_i)).$$

Note, that if $K < N$, then the boundary conditions for the functions $Q_{K+1}(x), ..., Q_N(x)$ is zero, since $Q_{K+1}(x_i) = ... = Q_N(x_i) = 0$ for all detectors x_i.

6. Let $u^e(x_i, t)$ and $u_0^e(x_i, t)$ denote the functions obtained in the experiment, whereas $u^s(x_i, t)$ and $u_0^s(x_i, t)$ denote the comparable functions obtained in numerical simulations. The solid line in Fig. 3 displays the time dependence of the function

$$\varphi_1^e(x_4, t) = \frac{u^e(x_4, t)}{u_0^e(x_4, t)} - 1$$

for the detector #4 for the C inclusion where time is given in picoseconds. One can clearly see that this function is rather noisy for $t < 500$ ps. The same observation was made for all other detectors/inclusions.

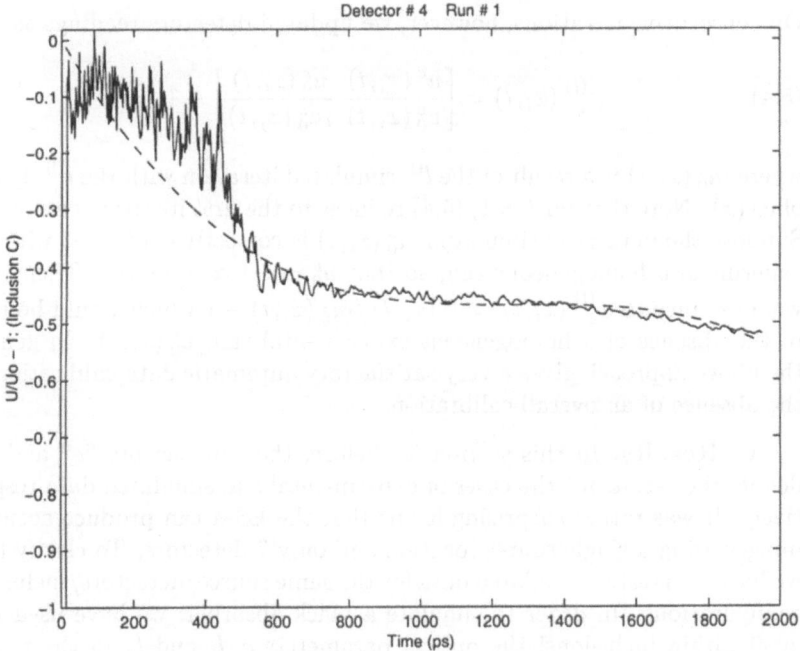

FIG. 3. *Function $u^e/u_0^e - 1$ (solid line) and its fit (dashed line) for detector #4, case C.*

For this reason we have taken $(T', T) = (550, 2000)$ ps and $(T_0, T_F) = (600, 1950)$ ps, on the first iteration of the ESM. We have also taken $K = 3$ and $N = 4$ for this iteration. The dashed line in Fig. 3 displays our fit (5.1) of the function $\varphi_1^e(x_4, t)$ with $K = 3$ over (T', T). On all consecutive iterations of the ESM we took $(T', T) = (650, 750), (T_0, T_F) = (600, 700), K = 3$, and $N = 1$. This was the choice for all numerical results presented in this paper.

7. Having the values $\beta(x_i)$ at the detectors, we interpolate them over the entire boundary $\partial\Omega$ to obtain the boundary condition $\beta(x)$. In the case of 7 detectors, as in the experiment, we use linear interpolation over the total boundary of the square Ω. In the case of 52 detectors, however as in [8], we use interpolation by C^2 cubic splines, over the four sides considered separately.

8. We used the value $\alpha = 1.1$ in (3.1). Thus we took $G = \{x_1, x_2 : |x_1|, |x_2| < 33\}$.

9. On the first iteration of the ESM we have consistently taken

$$(5.3) \qquad \varphi_1^{(1)}(x_i, t) = \varphi_1^e(x_i, t) = \frac{u^e(x_i, t)}{u_0^e(x_i, t)} - 1$$

On consecutive iterations, however, we updated detectors readings as

$$(5.4) \qquad \varphi_1^{(l)}(x_i, t) = \left[\frac{u^e(x_i, t)}{u_0^e(x_i, t)} \frac{u_0^s(x_i, t)}{{}_l u_0^s(x_i, t)}\right] - 1,$$

where ${}_l u_0^s(x_i, t)$ is a result of the lth simulated iteration with the lth update of $a_0(x)$. Note that for $l = 1$, (5.4) reduces to the first iteration case, (5.3). Suppose the numerical simulation $u_0^s(x_i, t)$ is correctly calibrated with the experimental homogeneous run, so that $u_0^e(x_i, t) \approx u_0^s(x_i, t)$. Then (5.4) would reduce to $\varphi_1^{(l)}(x_i, t) = u^e(x_i, t) / {}_l u_0^s(x_i, t) - 1$ which would be used in the absence of a homogeneous experimental run, $u_0^e(x_i, t)$. In general the above approach gives a very satisfactory automatic data calibration, in the absence of an overall calibration.

6. Results. In this section, as before, the superscripts "e" and "s" denote the results for the cases of experimental and simulated data respectively. It was rather surprising for us that the ESM can produce accurate images using a single source location and only 7 detectors. To clarify this, we have also tested simulated data for the same source/detectors/inclusions configurations. In order to simulate a black absorber, we have used $h = 2000$ within inclusions; the optical parameters c, l_t and l_a in the rest of the medium were the same as those of the experiment. We have also introduced multiplicative Gaussian noise in the data $u^s(x_i, t)$ with standard deviation $\sigma = 0.01$.

We have observed that the first iteration of the ESM provides rather good locations for inclusions. On the subsequent iterations these locations improved somewhat, and artifacts are decreased. We have also observed that the images became stable after 3 iterations of the ESM, which effectively means convergence. In all the tests presented below we stopped the iterative process on the 4th iteration. Observe that convergence was achieved despite the fact that the sufficient convergence conditions were not satisfied, see (4.2). The total CPU time required for these 4 iterations was just 11.4 minutes: (i) 5 minutes for the solution of the forward problem on the first iteration of the ESM for $t \in (0, 2000)$ ps, (ii) 1.5 minutes for the solution of the forward problem for $t \in (0, 750)$ ps on each of the 3 remaining iterations, (iii) 1.17 minutes for the solution of the elliptic system (4.16) with 1024 quadratic finite elements and the number of equations $N = 4$ on the first iteration of the ESM, and (iv) 0.25 minutes for the solution of the system (4.16) for 1024 quadratic finite elements, but with $N = 1$ for the 3 remaining iterations. This timing result compares favorably with competing methods [8] by orders of magnitude.

While the locations of the inclusions were always imaged with an accuracy, which we consider as good given only 7 detectors, the values of the function $h(x)$ within the inclusions were always imaged poorly, similarly to [8]. To help the reader to read our images better, we note that each small square is 3.75 mm × 3.75 mm on these images, that is the overall square of size 60 mm by 60 mm has 16 small squares on each side, as is seen in the figures.

Test A^e. Fig. A^c displays the correct image. We have not shown the values of $h(x)$ as $h(x) = \infty$ within a black absorber. The maximal value of the function $h(x)$ was reported at $(0.625, 13.125)$ mm as compared with the actual center at $(0, 10)$ mm. Fig. A_2^e displays the obtained image. In this Figure consider an ellipse within which the values of the imaged function $h(x)$ were 90% or more than those of the maximal value of the computed h. We will call such ellipses "most absorbing ellipses" (MAE). The axis of MAE on Fig. A_2^e are about 4.5 mm and 3.7 mm, which is close to the diameter of the rod (4.8 mm). Fig. A_3^e shows the 3-D view of the image.

Test A^s. Fig. A_2^s displays the image and shows an obvious similarity with Fig. A_2^e. Even the artifacts are similar. The maximal value of $h^s(x)$ was reported at $(-0.625, 9.375)$ mm.

FIG. A^c. *Correct image of inclusion A.*

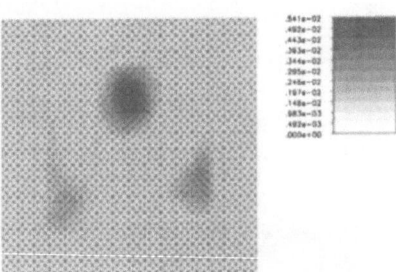

FIG. A_2^e. *Image of the experimental data after 4 iterations.*

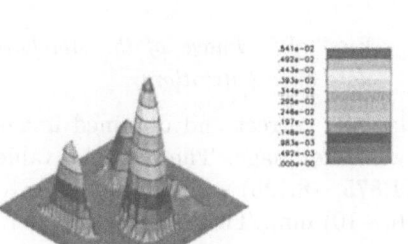

FIG. A_3^e. *3-dimensional view of the experimental data after 4 iterations.*

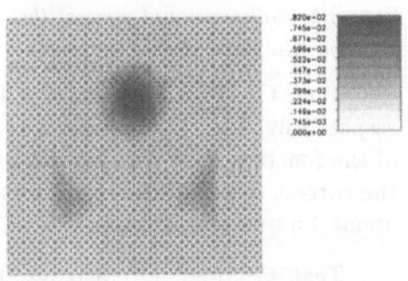

FIG. A_2^s. *Image of the simulated data after 4 iterations.*

Test B^e. Figs. B^e and B_2^e show the correct and obtained image respectively. The maximal value of the function $h(x)$ was reported at $(-8.125, -1.875)$ mm, which is very close to the correct center of $(-10, 0)$ mm. The axis of the MAE were about 7.5 mm and 4.9 mm. Fig. B_3^e displays the 3-D view of the image.

Test B^s. Fig. B_2^s shows the obtained image. The maximal value of the function $h(x)$ was reported at $(-6.875, 1.875)$ mm. The similarity of Figs. B_2^e and B_2^s is apparent. Again, the artifacts are similar.

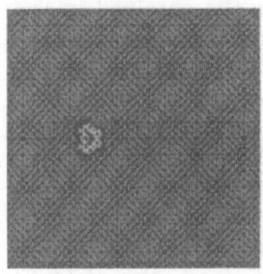

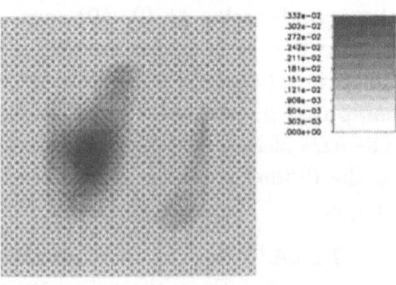

FIG. B^c. *Correct image of inclusion B.*

FIG. B_2^e. *Image of the experimental data after 4 iterations.*

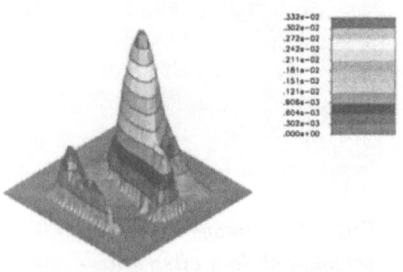

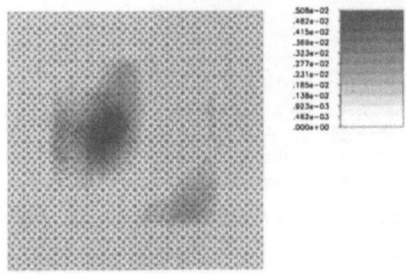

FIG. B_3^e. *3-dimensional view of the experimental data after 4 iterations.*

FIG. B_2^s. *Image of the simulated data after 4 iterations.*

Test C^e. Figs. C^e and C_2^e display the correct and obtained image respectively. Fig. C_3^e shows a 3-D view of the image. The maximal value of the function $h(x)$ was reported at $(1.875, -8.125)$ mm which is close to the correct center of the inclusion at $(0, -10)$ mm. The axis of MAE were about 4.5 mm and 3.8 mm.

Test C^s. Fig. C_2^s displays the obtained simulated image, which is similar to Fig. C_2^e. The maximal value of the function $h(x)$ was reported at $(1.875, -8.125)$ mm, which is exactly the same as in the case C^e.

TABLE 2

Correct and imaged A, B, and C inclusions for both experimental and simulated data, in mm.

Inclusion	Correct center	Point of $\max[h^e(x)]$	Point of $\max[h^s(x)]$
A	$(0, 10)$	$(0.625, 13.125)$	$(-0.625, 9.375)$
B	$(-10, 0)$	$(-8.125, -1.875)$	$(-6.875, 1.875)$
C	$(0, -10)$	$(1.875, -8.125)$	$(1.875, -8.125)$

FIG. C^c. *Correct image of inclusion C.*

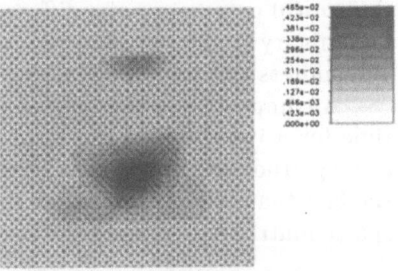

FIG. C_2^e. *Image of the experimental data after 4 iterations.*

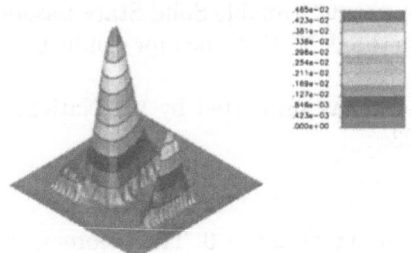

FIG. C_3^e. *3-dimensional view of the experimental data after 4 iterations.*

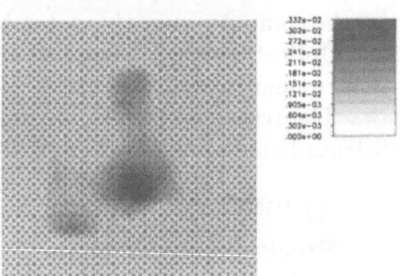

FIG. C_2^s. *Image of the simulated data after 4 iterations.*

The results of these tests are summarized in Table 2.

7. Discussion. We have demonstrated that the ESM works well for a specific experimental data. The locations of the inclusions were imaged with good accuracy given such a limited number of source/detectors pairs and violation of the convergence condition (4.2). Furthermore, images from simulated data were very similar to those obtained from the experimental data, and even the artifacts were very similar. However, values of the absorption coefficient within inclusions were imaged poorly, similarly with [8]. It is possible that this is due to a current code limitation on the number N of terms Q_n; currently $N \leq 4$. Note that the number N of equations for this system is a regularization parameter for the inverse problem under consideration. On the other hand, the number n of a generalized Fourier harmonics Q_n is similar to the frequency. It is well known from physics that it is impossible to extract high frequency harmonics from a noisy signal. We also note that the majority of current imaging algorithms in diffusion tomography image values of the function $h(x)$ within inclusions poorly. Likewise, it is clear from Fig. 3 that accurate data at the boundary can be extracted for only a few functions, Q_n.

Another question can be posed about the sensitivity of the ESM to a proper knowledge of the reference medium. The point is that in a practical setting it is obviously impossible to have accurate advance knowledge of the optical parameters of a diffuse medium. Currently we have made some preliminary studies of this topic, which show a rather positive answer. We will address this issue with more details in our forthcoming publication.

An important advantage of the ESM is its speed. The total CPU time for 4 iterations of the ESM was only 11.4 minutes, which is orders of magnitude less than many existing algorithms. This is mainly due to the fact that the ESM amounts to the factorization of a well-conditioned, sparse matrix representing the resulting elliptic system.

8. Acknowledgments. The authors are grateful to the experimental group led by Professor R. R. Alfano (City College of CUNY, New York) for providing the experimental data [3] and supporting in part the effort for the computer runs through a NASA grant "Tunable Solid State Lasers and Optical Imaging." The authors also thank Y. V. Ivanov for conducting some numerical experiments.

The research of Klibanov and Lucas was supported by the National Science Foundation grant DMS-9704923.

APPENDIX

Precise behavior of the function $u(x, t)$ **as** $t \to 0$. The theorems of this section were proven in [8]. We use these results to establish (4.7a). First consider the parabolic Cauchy problem, rather than the boundary value problem (3.2), (3.3). Let $u(x, t)$ be the solution to the Cauchy problem

$$u_t = div\left(D(x)\nabla u\right) - a(x)u, \quad (x, t) \in R^n \times (0, T), \quad n = 2, 3,$$

$$u(x, 0) = \delta(x - x_0).$$

It is well known that the function $u(x, t)$ decays exponentially as $t \to 0$. In this section we show that $u(x, t)$ can be "regularized" in a certain sense. In particular, for the case of $D = const.$, one obtains

$$u(x, t) = \frac{\exp\left[\frac{-|x - x_0|^2}{4Dt}\right]}{(2D\sqrt{\pi t})^n} w(x, t),$$

where $w(x, 0) = 1$.

Consider the hyperbolic Cauchy problem

(A.1) $$v_{tt} = \text{div } (D(x)\nabla v) + a(x)v, \ x \in R^n, \ t > 0$$

$$v|_{t=0} = 0, \ v_t|_{t=0} = \delta(x - x_0)$$

The structure of the solution of this problem was studied carefully in [14] in the 3-D case and in [17] in the 2-D situation. In accordance with these results, we impose somewhat *excessive* smoothness conditions on the coefficients D and a. Functions u and v are connected through the following analog of the Laplace transform (see formula (7.131) in [10])

(A.2) $$u(x,t) = \frac{1}{2\sqrt{\pi}t^{3/2}} \int_0^\infty \exp\left[-\frac{y^2}{4t}\right] yv(x,y)dy$$

The following lemma follows immediately from (A.2).

Lemma A1. *The function* $u(x,t)$ *is analytic with respect to* t *as a function of the real variable for* $t > 0$.

In the hyperbolic case, $\sqrt{D(x)}$ is the speed of sound in the media, and $v(x,t) = 0$ for $t < \tau(x,x_0)$, where $\tau(x,x_0)$ is the travel time from the point $\{x_0\}$ to the point $\{x\}$. The function $\tau(x,x_0)$ satisfies the eikonal equation $|\nabla_x \tau|^2 = \frac{1}{D(x)}$, see [14]. This equation determines rays along which the first arrival signal propagates from $\{x_0\}$ to $\{x\}$. Also, these rays are geodesic lines in the Riemann metrics $d\tau = \frac{1}{\sqrt{D(x)}}\sqrt{(dx_1)^2 + (dx_2)^2 + (dx_3)^2}$. In the case $D = const.$, these rays are simply straight lines.

Definition. The family of geodesic lines generated by the function $D(x)$ is regular in $\overline{\Omega}$ if for any pair x, $x_0 \in \overline{\Omega}$ there exists a unique geodesic line $L(x,x_0)$ connecting $\{x\}$ with $\{x_0\}$.

In what follows, we will always assume the regularity of the family of geodesic lines $L(x,x_0)$ *in* $\overline{\Omega}$.

Let $H(z)$ be the Heaviside function. That is, $H(z) = 1$, for $z \geq 0$ and $H(z) = 0$ for $z < 0$. Since the source location $\{x_0\}$ is fixed in our inverse problem, we denote $\tau_0(x) = \tau(x,x_0)$. The following theorem follows from Theorem 4.1 and Lemma 3.4 of [14].

Theorem A1. *In* (A.1) *let* $n = 3$, $D \in C^{l+4}(R^n)$, $a \in C^l(R^n)$ *with* $l \geq 2s+7$ *and* $s \geq -1$. *Then* $\tau_0(x) \in C^{l+3}(\overline{\Omega})$, *and, for* $x \in \overline{\Omega}$, *the function* $v(x,t)$ *can be represented in the form*

$$v(x,t) = \frac{H(t)}{2\pi} \sum_{k=-1}^s \sigma_k(x)H_k(t^2 - \tau_0^2(x)) + v_s(x,t),$$

where

$$H_{-1}(t^2 - \tau_0^2(x)) = \frac{\delta(t - \tau_0(x))}{\tau_0(x)},$$

$$H_k(t^2 - \tau_0^2(x)) = \frac{[t^2 - \tau_0^2(x)]^k}{k!}H(t - \tau_0(x)), k \geq 0,$$

Further, the functions $\sigma_k(x) \in C^{l-2k}(\overline{\Omega})$, *and for* $s \geq 1$ *the derivatives* $D_x^\alpha D_t^\beta v_s \in C\{t \geq \tau(x), x \in \overline{\Omega}\}$, *where* $\alpha = (\alpha_1, \alpha_2, \alpha_3)$, $|\alpha| = \alpha_1 + \alpha_2 + \alpha_3, |\alpha| + \beta \leq s - 1$ *and* $|\alpha| \leq l - 2s - 3$. *Furthermore, the function* $v_s(x,t)$ *can be represented in the form* $v_s(x,t) = H_{s+1}(t^2 - \tau_0^2(x))v_s^0(x,t)$, *where the function* v_s^0 *is bounded for* $(x,t) \in \{\tau_0(x) \leq t \leq Z, x \in \overline{\Omega}\}$ *for any* $Z > \tau_0(x)$.

In addition, it follows from the standard energy estimates for the hyperbolic Cauchy problems that for $x \in \overline{\Omega}$ and $t \to \infty$ $|D_x^\beta v_s(x,t)| \leq e^{kt}$, for a $k = const. > 0$. For the ESM, we need the function $u \in C^{4+\gamma, 2+\frac{7}{2}}(R^n \times [\varepsilon, T])$ (for every $\varepsilon \in (0,T)$) and the function $w \in C^{4,2}(\overline{\Omega}_T)$. This, in combination with Theorem A1, imposes smoothness conditions on the coefficients $D(x)$ and $a(x)$ in the following theorem:

Theorem A2. *Let* $n = 3, x_0 \notin \overline{\Omega}$, *and in Theorem A.1* $s = 5, l = 17$. *Further, let* $D(x) \geq const. > 0$ *in* R^3, $D = const.$ *outside of* Ω, $D \in C^{21}(R^3)$, $a \in C^{17}(R^3)$, *and suppose all of these derivatives of the functions* $D(x)$ *and* $a(x)$ *are bounded in* R^3. *In addition, let the family of the geodesic lines generated by the function* $D(x)$ *be regular in* $\overline{\Omega}$. *Then the solution of the parabolic Cauchy problem (1.1)* $u \in C^{4+\gamma, 2+\frac{7}{2}}(R^3 \times [\varepsilon, T])$ *(for every* $\varepsilon \in (0,T)$) *has the form*

$$(A.3) \qquad u(x,t) = \frac{\exp\left[-\frac{\tau_0^2(x)}{4t}\right]}{t^{\frac{3}{2}}} w(x,t),$$

where the function $w \in C^{4,2}(\overline{\Omega}_T)$, $w(x,0) > 0$ *in* $\overline{\Omega}$ *and* $w(x,0)$ *does not depend on* $a(x)$.

Now consider the 2-D case. We use a 2-D analog of Theorem A1, see [17, p. 228]. Namely,

$$v(x,t) = \frac{\sigma_{-1}(x)}{\sqrt{t^2 - \tau_0^2(x)}} + \sqrt{t^2 - \tau_0^2(x)}\left[\sum_{k=0}^s \sigma_k(x)H_k\left(t^2 - \tau_0^2(x)\right) + v_s(x,t)\right]$$

where the function $v_s(x,t)$ has the same properties as above. Hence, the following result can be proven similarly to Theorem A2.

Theorem A3. *Let* $n = 2$, *and the rest of conditions of Theorem A2 be satisfied. Then the function* $u \in C^{4+\gamma, 2+\frac{7}{2}}(R^2 \times [\varepsilon, T])$ *(for every* $\varepsilon \in (0,T)$) *has the form*

$$u(x,t) = \frac{\exp\left[-\frac{\tau_0^2(x)}{4t}\right]}{t} w(x,t),$$

where the function w *satisfies the same conditions as those listed in Theorem A2.*

Now consider the boundary value problem (3.2)-(3.4). Obviously, an analog of the formula (A.2) is valid. Hence, Lemma A1 is also valid for this problem. However, we have yet proven analogs of Theorems A2 and A3 for the boundary value problem (3.2)-(3.4), unlike the case of the Cauchy problem. We anticipate that such results can be proven using the construction of the Green's function for the parabolic operator [9]. Thus, we make the following

Conjecture A. *Theorems A2 and A3 are valid for the boundary value problem (3.2)-(3.4) for the case of a "reasonable" (in certain sense) domain G.*

REFERENCES

[1] S. ARRIDGE AND M. SCHWEIGER, 1995, *Sensitivity to prior knowledge in optical tomographic reconstruction*, Proc. of SPIE, **2389**, 378-388 (SPIE - The Society of Photo-Optical Instrumentation Engineers).

[2] R. BARBOUR, H. GRABER, J. CHANG, S. BARBOUR, S. KOO, AND R. ARONSON, 1995, *MRI-guided optical tomography*, IEEE Comp. Sci. Eng., **2**, #4, 63-77.

[3] W. CAI, B. DAS, F. LIU, M. ZEVALLOS, M. LAX, AND R. ALFANO, 1997, *Time-resolved optical diffusion tomographic image reconstruction in highly scattering turbid media*, Proc. Natl. Acad. Sci. USA, **93**, 13561-13564.

[4] K. CASE AND P. ZWEIFEL, 1967, *Linear Transport Theory*, Addison-Wiley Publishing Company, London.

[5] S. COLAK, D. PAPAIOANNOU, G. HOOFT, M. VAN DEN MARK, H. SCHOMBERG, J. PAASSCHENS, J. MELISSEN, AND N. VAN ASTEN, 1997, *Tomographic image reconstruction from optical projections in light-diffusing media*, Applied Optics, **36**, 180-213.

[6] M. KLIBANOV AND F. SANTOSA, 1991, *Computational quasi-reversibility method for Cauchy problems for Laplace's equation*, SIAM J. Appl. Math., **51**, 1655-1675.

[7] M. KLIBANOV AND RAKESH, 1992, *Numerical solution of a time-like Cauchy problem for the wave equation*, Math. Methods in Appl. Sci., **15**, 559-580.

[8] M. KLIBANOV, T. LUCAS, AND R. FRANK, 1997, *A fast and accurate imaging algorithm in optical/diffusion tomography*, Inverse Problems, **13**, 1341-1361.

[9] O. LADYZENSKAYA, V. SOLONNIKOV, AND N. URALCEVA, 1968, *Linear and Quasilinear Equations of Parabolic Type*, AMS, Providence, RI.

[10] M. LAVRENTIEV, V. ROMANOV, AND S. SHISHATSKII, 1986, *Ill-Posed Problems Of Mathematical Physics And Analysis*, AMS, Providence, RI.

[11] R. LATTES AND J.-L. LIONS, 1969, *Method of Quasi-Reversibility: Applications To Partial Differential Equations*, Elsevier, New York.

[12] *Mathematics and Physics of Emerging Biomedical Imaging*, 1996, National Research Council, Institute of Medicine, National Academic Press, Washington, D.C.

[13] M. O'LEARY, D. BOAS, B. CHANCE, AND A. YODH, 1994, *Images of inhomogeneous turbid media using diffuse photon density waves*, in Optical Society of America Proc. "Advances In Optical Imaging And Photon Migration," **21**, 106-115.

[14] V. ROMANOV, 1987, *Inverse Problems of Mathematical Physics*, VNU Press, Uthrecht.

[15] L. SOURIAU, B. DUCHÊNE, D. LESSELIER, AND R. KLEINMAN, 1006, *Modified gradient approach to inverse scattering for binary objects in stratified media*, Inverse Problems, **12**, 463-481.

[16] A. WOLBRAST, 1993, *Physics of Radiology*, Appleton & Lange, Norwalk, CT.

[17] V. YAKHNO, 1990, *Inverse Problems for Differential Equations of Elasticity*, Novosibirsk, Nauka (in Russian).

Major theorems [...]

REFERENCES

THE APPLICATION OF THE X-RAY TRANSFORM TO 3D CONFORMAL RADIOTHERAPY*

ROBERT Y. LEVINE†, EUGENE A. GREGERSON† AND MARCIA M. URIE‡

Abstract. The inversion problem for γ-ray ($\geq 1MeV$) conformal radiotherapy is analyzed using the mathematics of tomographic reconstruction. It is shown that the delivered dose can be approximated by the dual attenuated x-ray transform of the filtered beam profile function. The number of intensity-modulated beams required for dose conformation to a tumor is derived. The sampling requirement is at most $(2\pi r_{max}W_{max} + 5/2)$ beams for a 2D tomotherapy geometry, where r_{max} and W_{max} are the maximum spatial extent and frequency, respectively, of the radiation dose. We generalize this 'Bow Tie' solution to 3D, suggesting a sufficient beam number given by $(\Delta\omega/2\pi W_{max})(2\pi r_{max}W_{max}+5/2)^2$, where $\Delta\omega$ is the frequency resolution of the beam front modulation. The matrix inversion implicit in this bound suggests a criterion for beam orientation selection. Beam angles should be chosen such that the SVD inversion to beam profiles is non-singular for the entire configuration of beams. The natural Hilbert space metric among beam profiles provides another criterion for choosing beam angles. The total squared intensity at each beam angle (in the ρ - metric) is used as a ranking of beam orientations to maximize the overlap between the sampled and continuous beam profile functions. The measure is displayed relative to 3D tissue space contours to define an optimum subset of beams. The formalism is applied to real brain and prostate tumor data consisting of radiologist-generated tumor and organ-at-risk contours, prescribed dose and dose limits, and CT images.

1. Introduction. Radiotherapy planning involves the choice of a finite number of intensity-modulated radiation beams at various orientations relative to the patient. The goal is the delivery of a lethal dose to the tumor volume (TV) with minimum leakage to organs-at-risk (OARs) and normal tissue. The inversion problem in treatment planning is the determination of a map from a known conforming (prescribed) dose function in tissue space to beam functions in profile space. The profile space in three dimensions is a tangent plane attached to each point on the sphere, which represents the beam front at the corresponding orientation relative to the center of rotation of the beam gantry. In the early 1980s, it was realized that a powerful mathematical abstraction, namely the theory of projections [1], approximates the relationship between intensity-modulated beams and delivered dose [2]–[7]. The theory has been applied to the inversion of prescribed dose functions to beam profiles which 'reconstruct' the dose function tomographically [8]–[10]. Alternative approaches to the inversion problem involve simulated annealing [11]–[16], interval linear feasibility search by iterative projection [17, 18], linear programming [19, 20], constrained random search [21, 22], and iterative projection and deconvolution [23]–[27]. Reviews are found in Refs.[28] and [29].

*This work was sponsored by the Department of the Air Force under contract F19628-90-C-0002. Please direct correspondence to R.Y. Levine (bob@spectral.com).

†MIT Lincoln Laboratory, Lexington, MA 02173.

‡Department of Radiation Oncology, University of Massachusetts Medical Center, Worcester, MA 01655.

In recent years techniques have been developed to deliver intensity-modulated radiation dynamically without the use of patient-specific compensators. The instruments typically modulate radiant energy by variation of the beam dwell-time in a particular trajectory through the tissue [30]-[34]. Intensity-modulated treatments have been verified with multileaf collimators [31, 32], [35]-[40] for OAR shielding and TV dose escalation of head/neck [41] and prostate tumors [42]. It has been suggested that reduced post-treatment Prostate-specific Antigen (PSA) levels were achieved by safely escalating dose on prostate tumors [43, 44]. Alternative methods for beam modulation include variable-width blocks [45, 46] and special-purpose wedge placement [47].

While it is clearly established that dose-based inversion is computable with modern machines, the required beam number and orientations are not completely understood. A large variety of inversion algorithms, beam numbers, and orientations have been reported in the conformal inversion literature. Direct comparisons are difficult due to the differences in beam parameterizations and dose computation algorithms; as well as the diverse set of phantoms and real data cases to which the algorithms have been applied. In an example with 3D intensity-modulated multileaf collimators (MLC), Webb used simulated annealing with 21 beams in various 3D configurations on a cylindrically symmetric phantom [14]. The maximum tumor control probability (TCP) was found for beams in a coplanar configuration parallel to the phantom axis. In a similar 2D study on a variety of real data cases, Webb found 32 uniformly spaced beams sufficed for conformal treatment [11]. Webb's results, with a 3D beam/dose model similar to the parameterization in Section 2, motivated our investigation of the x-ray transform. Holmes and Mackie reported $2\% - 4\%$ dose errors in the inversion with six modulated beams on a 2D circular phantom [9]; whereas Bortfeld, et al., employed nine beams on 3D iteratively-improved reconstructions of dose with non-opposing beams for an irregularly shaped nasopharynx tumor [7]. Censor, et al., employed 12 intensity-modulated fan beams for 2D brain tumor treatment derived by a feasible solution search algorithm [17]. Recently, Boyer reported good 3D conformation to a prostate tumor using nine coplanar modulated beams [29]. A six beam plan was recently employed to safely escalate a conforming prostate dose to $81Gy$ with a dynamic multileaf collimator [42]. Recently, the Memorial Sloan-Kettering group suggested heuristic guidelines to allow less than five beams [48], numbers advocated by Brahme from biological response models [49]. In our simulations we have obtained good results with 12-14 beams on brain and prostate tumors using the iterative backprojection of the 3D x-ray transform [50]. Alternative parameterizations of beams and dosages; for example, beam weights and wedge placement, have also resulted in a variety of beam numbers and configurations. In one case Niemierko, et al., employed 5-6 beams determined from a random search algorithm on the beam weights of 188 initial beams [21, 22]. By starting with a large number of beams the Neimierko algorithm effectively performed beam orientation selection as a result of beam weight computation. Censor, et al., tested a similar approach to beam selection in the reduction of a 24 beam plan to six and two beams [17]. Another approach to the beam placement problem is simply to allow the therapist to choose beam orientations followed by an optimization on the beam weights [48]. Using a beam wedge parameterization on real 2D data, Morrill, et al., employed 36 equally-spaced beams with linear programming [19, 20] and simulated annealing inversion [12]. In Ref. [16] the weights of 54 non-coplanar beams from 27 directions are adjusted for optimization of a prostate tumor dose. The beam number was reduced to an equivalent thirteen beam plan by removal of low-weighted beams. It should be noted that the beam numbers employed in these examples arise as much from constraints on clinical implementation and algorithm run time as from dose conformation goals.

It seems clear that the properties of the objective function space depend critically on whether beam weights or orientations are the independent variables. In the former case, the space may be unimodal, while it may contain many local minima in the space of beam angles [51]. Also, the response of different tissues to radiation is a critical component to optimum beam numbers. For example, Mohan, et al., have suggested that, due to tissue response models, a large lung tumor should be treated with a few beams directly intersecting the tumor, whereas conformal multiple beam treatment is effective for prostate tumors [52]. A more relevant measure of treatment effectiveness may be biological response, rather than conformation to a prescribed dose [53]–[58]. For this reason inversion algorithms have been developed in which TV and OAR response models are used to define optimization goals [12, 16, 21, 22, 59].

Planning for realistic treatment constrains the inversion problem in a variety of ways. Current treatment protocols involve less than 20 beams, which is possibly an extreme undersampling of the optimum continuous beam profile function. Beam placement is limited by the physical geometry of the patient and treatment machines. Complications in the dose model, which are not expressed in a simple backprojection of the filtered beam profiles, include photon absorption, electron scatter effects, dose build-up at interfaces, and non-planar patient contours. In this paper we show that, even with these dose model complications, the delivered dose in 3D γ-ray therapy can be approximated by the dual attenuated x-ray transform of the filtered beam profiles. In addition, we consider the number of beams necessary to implement 3D inversion from a prescribed dose function using the theory of projections. Beam number and placement criteria, generalizing 2D results based on the Radon transform, are derived from the 3D attenuated x-ray transform. The 2D case is appropriate for treatment about a cylindrical axis, whereas 3D beam configurations assume a center of spherical rotation. The latter geometry includes treatment at oblique angles to the gantry axis of rotation. We derive a quantitative relationship among beam number, profile function sampling, dose function radius, and spatial frequency content. Also, an algorithm is suggested for beam selection in the case of extreme undersampling; that is, when the number of beams is much smaller than the sampling threshold. The condition number for the (matrix) inversion to a specific set of beams is proposed as a possible measure of the configuration effectiveness. This criterion for an undersampled profile also applies to treatments constrained by a limited range of angles. An alternative angle-dependent measure, the ρ-metric, is derived from the overlap between sampled and continuous beam profile functions. The beam number and selection criteria were obtained following derivations in Refs. [60] (electron microscopy) and [61] (limited-angle CT) for 2D tomographic reconstruction. The beam sampling and selection algorithms are easily generalized for photon absorption, electron scatter, and a non-planar tissue-air interface. These results require that transverse electron transport prior to dose delivery is less than the width of the beam build-up region. The beam build-up effect, due to photon-electron conversion at the tissue-air interface, is not invertible by projections. However, the effect can be approximated by a cutoff of delivered dosage at an appropriate skin depth in the tissue.

Section 2 contains the beam/dose model including effects of photon attenuation, electron scatter, beam build-up, and non-planar tissue-air contours. It is assumed that homogeneous tissue is described by a single γ-ray ($\geq 1.0 MeV$) absorption constant μ ($\simeq \mu_{H_2O} = .03 cm^{-1}$). The beam/dose model, taken from Ref. [9], is expressed in terms of the 3D dual attenuated x-ray transform relating delivered dose and the set of beam profiles. In Section 3 the mathematics of the 3D x-ray transform is applied to obtain a beam number sufficiency condition which generalizes the 2D 'Bow Tie' solution in CT [60, 62]. Treatment about a cylindrical axis, for which the 2D sampling bound is appropriate, is compared to the 3D case of treatment about a sphere centered on the tumor. The relevant function spaces for radiotherapy inversion, physical (tissue) and profile (beam), each admit a natural metric which measures the overlap of two functions. It is shown in Section 4 that the appropriate metric for comparison of delivered dosages from sampled and complete beam profiles is the ρ-metric from CT [61]. For an input prescribed dose, the metric is employed in a display of ranked beam orientations around the patient. The filtering of projected dosages, required for the inversion of the dual attenuated x-ray transform, results in unphysical negative beam intensities [4, 5]. To compensate for the non-negativity constraint, we developed a 3D generalization of an iterative inversion algorithm suggested in Ref. [9]. The iterative inversion algorithm is described in Section 5. A complete dose model, including beam build-up effects, can be incorporated in the 3D backprojection step of the inversion algorithm based on the x-ray transform.

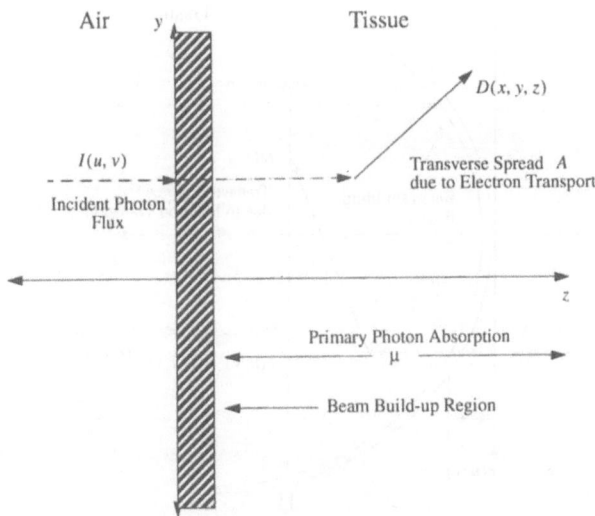

FIG. 1. *Definition of parameters for convolutional dose model in Ref. [9].*

In Section 6 real patient data sets of brain and prostate tumors are analyzed. The inversion algorithm inputs are radiologist-generated TV/OAR contours and prescribed dosages. Radiotherapy plans resulting from iterative inversion of the x-ray transform are judged for dose conformation as a function of beam number and orientation. The results, as summarized in the conclusion, indicate that the proposed algorithms may aid the radiophysicist in the initial choice of beam configurations. The appendix contains a discussion of accelerator calibration for multiple intensity-modulated beams.

2. Dose/Beam model. An approximate radiotherapy dose model is based on a 3D convolutional relationship between photon fluence and delivered electron dose [9]. As described in Figure 1, a beam of intensity $I(x, y)$ $[J/cm^2]$ is incident along the z-direction on tissue with a uniform absorption constant μ $[cm^{-1}]$. The primary photon energy per unit mass $T(x, y, z)[J/cm^3 kgm]$, called the TERMA (Total Energy Released per Mass), at a depth z is given by

$$(1) \qquad\qquad T(x, y, z) = \frac{\mu}{\rho} e^{-\mu z} I(x, y),$$

where ρ is the tissue density $[kgm/cm^3]$. Following the model in Refs. [9] and [10], the dose at (x, y, z), $D(x, y, z)$, is a convolution of the TERMA with a 3D kernel $\tilde{A}$ given by

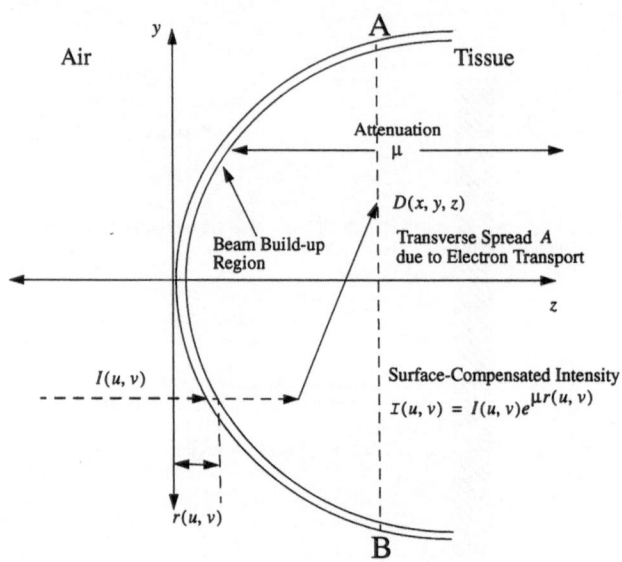

FIG. 2. *Non-planar tissue-air interface described by distance $r(u, v)$ to the patient contour along the (u, v) ray. Definition of compensated beam intensity $\mathcal{I}(u, v)$.*

$$D(x, y, z) = \frac{\mu}{\rho} \int_x \int_y \int_0^t e^{-\mu w} I(u, v) \bar{A}(x - u, y - v, z - w) \, du \, dv \, dw$$

$$(2) \qquad = \frac{\mu}{\rho} e^{-\mu z} A * I \quad (0 \leq z \leq t),$$

where t is the tissue thickness, $z = 0$ is the location of a planar air-tissue interface, $*$ represents 2D convolution, and A is a (z, t)-dependent kernel defined by

$$(3) \qquad A(x, y; z, t) = \int_0^t e^{\mu(z - w)} \bar{A}(x, y, z - w) \, dw.$$

It is interesting to consider the (z, t)-dependence of the scattering kernel in Eq. (3) as determined by phantom measurements [9]. For a ^{60}Co source ($1MeV$), the kernels for different depths diverge at 1% of the peak value, which corresponds to off-axis ($x = y = 0$) distances of about $2cm$. Higher energy beams ($\sim 24MeV$) have greater depth dependence with variable-depth kernels diverging at about the 10% to peak value level for off-axis distances of $2.5cm$. The phantom thickness (t) dependence of the scattering kernels is negligible for both the $24MeV$ and ^{60}Co sources. Accordingly, following the approximation in Ref. [9], we drop the (z, t)-dependence in the kernel $A(x, y; z, t)$. Part of this depth dependence is recovered by including a multiplicative beam build-up function in Eq. (2).

Under reasonable assumptions on A in Eq. (2), it is possible to correct $I(u,v)$ for a non-planar tissue-air interface [63]. Figure 2 contains the model of a *convex* patient contour including the beam build-up region, incident photon flux, and intersection of the $z = w$ plane at points **A** and **B**. Assuming that the patient contour is described by $r(u,v)$, which is the distance to the tissue along a ray emanating from (u,v), the depth to the transverse plane through (u,v,w) is given by $d = w - r(u,v)$. Therefore, the TERMA appearing in Eqs. (1) and (2) is given by

$$(4) \qquad T(u,v,w) = \frac{\mu}{\rho} e^{-\mu w} e^{\mu r(u,v)} I(u,v).$$

In addition, if the electron spread A is smaller than the beam build-up thickness, thereby not adding to the uncertainty due to beam build-up effects, we can ignore the effect of the transverse convolution at the endpoints **A** and **B**. The dose is then approximated by

$$(5) \qquad D(x,y,z) = \frac{\mu}{\rho} e^{-\mu z} \int du dv A(x - u, y - v) \mathcal{I}(u,v),$$

with, $\mathcal{I}(u,v) = I(u,v) e^{\mu r(u,v)}$, as a beam intensity which is compensated for a non-planar tissue-air interface. The beam build-up region at the tissue-air interface corresponds to a non-equilibrium condition between the gamma-ray beam and secondary electrons which deliver the dose [28, 64]. The effect can be modeled in Eq.(5) by the replacement of $e^{-\mu z}$ by a function $h(z)$ with a peak at the build-up depth and an exponential tail. Build-up effects, which can be significant for high energy beams (depths $\sim 5cm$ for $24MeV$ beams), are easily incorporated in the iterative inversion in Section 5.

Figure 3 contains the geometry for the mapping between tissue and profile spaces which approximates 3D radiation therapy. The mapping to 2D planar modulated beams is a model of treatment with a gantry-mounted multileaf collimator at oblique angles to the patient axis. An orthonormal triad $(\vec{\theta}_1, \vec{\theta}_2, \vec{\theta})$ defines the vector $\vec{\theta}$ normal to the 2D beam front with basis vectors $(\vec{\theta}_1, \vec{\theta}_2)$ in the plane $\theta^{\perp}$. The plane $\theta^{\perp}$ transverse to the beam is described by coordinates (u,v) such that a ray emanating from the point (u,v) enters the tissue at a distance $r(\vec{\theta}, u, v)$. Assuming the beam rotates on a sphere of radius R_0, a point $\vec{x} \in \mathcal{R}^3$ in the tissue is parameterized by

$$(6) \qquad \vec{x} = u\vec{\theta}_1 + v\vec{\theta}_2 + (R_0 - t)\vec{\theta},$$

where t is the distance from $\vec{x}$ to the orthogonal projection onto $(u,v) \in \theta^{\perp}$. The depth in the tissue to $\vec{x}$ is given by

$$(7) \qquad d = (t - r(\vec{\theta}, u, v)) = R_0 - \vec{x} \cdot \vec{\theta} - r(\vec{\theta}, \vec{x} \cdot \vec{\theta}_1, \vec{x} \cdot \vec{\theta}_2).$$

We are interested in the dose delivered from a set of modulated beams with an intensity profile $I(\vec{\theta}, u, v)$ denoting the fluence emitted from the

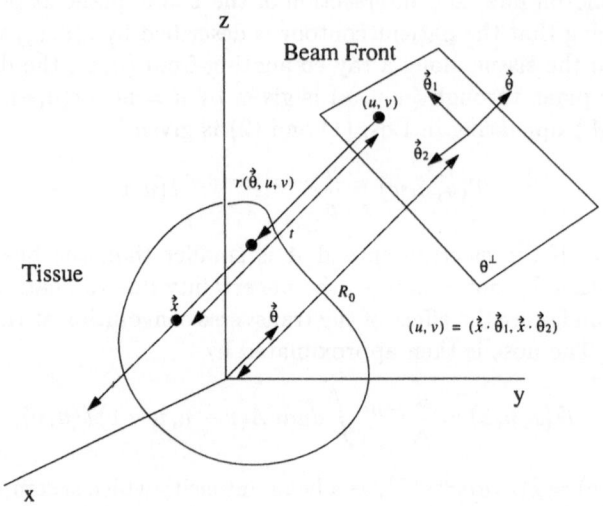

FIG. 3. *Coordinate systems relating tissue $\vec{x}$ and profile $(\vec{\theta}, \vec{y}), \vec{y} \in \theta^\perp$, Hilbert spaces for 3D dose/beam model.*

point (u, v) on the beam front oriented at an angle $\vec{\theta} \in \mathcal{S}^2$. Assuming $r(\vec{\theta}, u, v)$ is the distance to the patient contour from the (u, v) point on the $\vec{\theta}$-oriented beam, define the compensated intensity profile function

$$(8) \qquad \mathcal{I}(\vec{\theta}, u, v) = I(\vec{\theta}, u, v)e^{\mu r(\vec{\theta}, u, v)}.$$

Integration of Eq. (5) over all beam angles then results in the expression $(\vec{x} \in \mathcal{R}^3)$

$$(9) \quad D(\vec{x}) = \frac{\mu e^{-\mu R_0}}{\rho} \int d\vec{\theta} e^{\mu \vec{x} \cdot \vec{\theta}} \int du dv A(\vec{x} \cdot \vec{\theta}_1 - u, \vec{x} \cdot \vec{\theta}_2 - v) \mathcal{I}(\vec{\theta}, u, v)$$

$$(10) \qquad = \frac{\mu e^{-\mu R_0}}{\rho} \int_{\mathcal{S}^2} d\vec{\theta} e^{\mu \vec{x} \cdot \vec{\theta}} (A * \mathcal{I})(\vec{\theta}, \vec{x} \cdot \vec{\theta}_1, \vec{x} \cdot \vec{\theta}_2),$$

where R_0 is the distance of each beam center from the center of rotation and $*$ denotes the 2D convolution across the beam front. From the mathematical theory of 3D projections, Eq. (10) can be written

$$(11) \qquad D = \frac{\mu e^{-\mu R_0}}{\rho} P^\sharp_{-\mu}(A * \mathcal{I}),$$

where $P^\sharp_{-\mu}$ is the dual attenuated x-ray transform [1, 65, 66].

The goal of radiotherapy planning is the inversion of Eq. (11); that is, the computation of the beam profile $\mathcal{I}$ from a desired prescribed dosage D. In the limit of high beam energy, the tissue absorption constant μ approaches the water value of $.03cm^{-1}$ corresponding to a tissue half-value length of about $33cm$. Treatment of tumors through tissue depths generally much less than $30cm$ suggests the zeroth order approximation to Eq. (11) in the limit $\mu = 0$. In Section 5 we implement an iterative inversion algorithm in which corrections for absorption and beam build-up effects have been incorporated in the backprojection step. Formally, in the limit $\mu = 0$, the dual x-ray transform has an inverse given by [1]

$$(12) \qquad (P^{\sharp})^{-1} = JP$$

where P is the 3D x-ray transform, which for a function f on $\mathcal{R}^3$ yields a profile function Pf given by

$$(13) \qquad Pf(\vec{\theta}, \vec{y}) = \int_{-\infty}^{\infty} f(\vec{y} + t\vec{\theta})dt, \ \vec{y} \in \theta^{\perp}.$$

The operator J in Eq. (12) is a 2D filter, linear in spatial frequency, which is applied to each beam front. Denoting by $\hat{h}(\vec{\eta})$ the Fourier transform of h, the beam front filter at angle $\vec{\theta}$ is given by

$$(14) \qquad \widehat{Jf}(\vec{\eta}) = \frac{|\vec{\eta}|}{2\pi^2} \hat{f}(\vec{\eta}), \ \vec{\eta} \in \theta^{\perp}.$$

Equations (12) - (14) generalize a standard 2D Radon transform inversion, employed in CT, known as the backprojection of filtered projections [1, 67]. Substitution of Eq. (12) into Eq. (11), followed by deconvolution of electron scatter effects, results in the formal linear inversion,

$$(15) \qquad \mathcal{I} = \frac{\rho e^{\mu R_0}}{\mu} A^{-1} * JPD.$$

Equation (15) is easily generalized for photon absorption in the case of 2D cylindrical symmetry, which is appropriate for treatment with a standard linac gantry. For what is effectively a 2D inversion, P and J are replaced by the *attenuated* Radon transform [65],

$$(16) \qquad P_{\mu}f(\vec{\theta}, \vec{y}) = \int_{-\infty}^{\infty} f(\vec{y} + t\vec{\theta})e^{-\mu t}dt, \ \vec{y} \in \theta^{\perp},$$

and the filter [1]

$$(17) \qquad \widehat{J_{\mu}f}(\vec{\eta}) = \begin{cases} \frac{|\vec{\eta}|}{2\pi^2} \hat{f}(\vec{\eta}) & |\vec{\eta}| \geq \mu \\ 0 & |\vec{\eta}| < \mu \end{cases},$$

respectively. In our experiments we have found that the water absorption length of $33cm$ for MeV photons has little effect in the 2D filter because μ^{-1}

is less than the frequency space sampling resolution. Therefore, absorption effects are incorporated in only the x-ray transforms P_μ and $P^\sharp_{-\mu}$. In the next section, Eq. (15), which is the 3D-equivalent of the formulation in Refs. [9] and [10], is applied to the derivation of a sufficient beam number for 3D radiotherapy. While beam sampling and selection criteria can be defined within the zeroth order approximation of $\mu = 0$, the modulated beam fronts are obtained from the iterative backprojection algorithm in Section 5 using higher order effects such as scattering and beam build-up. However, a clinically acceptable delivered dose calculation requires Monte Carlo or Differential Pencil Beam (DPB) simulation [64] with the derived beam profiles as input. A mathematical treatment of the dose operator, which generalizes the dual x-ray transform to a fan beam description, is found in Ref. [23].

It is interesting to consider the benefits of radiotherapy with the three-dimensional geometry in Figure 3. There are, of course, many situations in which beam trajectories at oblique angles to the patient axis avoid OARs and maximize the projected tumor area to the beam front. Inversion to an effective configuration should include these beam directions among the candidate solutions. A less tumor-specific benefit is suggested intuitively by the 3D backprojection of a projected prescribed dose function $p(\vec{x})$ approximated by

$$(18) \qquad P^\sharp(P(p(\vec{x}))) \sim \frac{1}{r^2} * p.$$

The corresponding relationship in two dimensions, involving the Radon and dual Radon transforms, is given by $R^\sharp(R(p(\vec{x}))) \sim \frac{1}{r} * p$. The $1/r^2$ - convolution in Eq. (18) suggests a higher degree of dose conformation to the tumor and resulting lower exposure to surrounding normal tissues. Geometrically, this corresponds to the spread of radiation into a larger volume with the corresponding reduction in the (per kilogram) dosage.

3. Beam number criteria. In this section the theory of projections is applied to obtain a sufficient number of beams necessary to implement Eq. (15); that is, to obtain a 'reconstruction' of the delivered dose from the set of beam profiles. The beams will be oriented on a sphere relative to a center of rotation in the patient (most likely the TV center). The derived beam number is a generalization of the well-known 2D 'Bow Tie' condition, appropriate for CT reconstruction, to the case of a gantry which is not constrained to rotate about a cylindrical axis of symmetry.

The Projection-Slice Theorem for the 3D x-ray transform, relating the Fourier transforms of the 2D projection and original 3D function, is given by [1]

$$(19) \qquad \widehat{Pf}(\vec{\theta}, \vec{\eta}) = \sqrt{2\pi}\hat{f}(\vec{\eta}), \ \vec{\eta} \in \theta^\perp.$$

Application of Eq. (19) to the Fourier transform of Eq. (15), with the substitution of Eq. (14) for J, results in the expression,

$$(20) \qquad \widehat{A * \mathcal{I}}(\vec{\theta}, \vec{\eta}) = \frac{\rho e^{\mu R_0}}{\mu (2\pi)^{3/2}} |\vec{\eta}| \hat{D}(\vec{\eta}), \ \vec{\eta} \in \theta^{\perp}.$$

Following the derivation of the 2D 'Bow Tie' condition in Ref. [60] for electron microscopy, expand the delivered dose and its Fourier transform in spherical harmonics,

$$(21) \qquad D(r, \theta, \phi) = \sum_{l=0}^{\infty} \sum_{m=-l}^{+l} g_{lm}(r) Y_{lm}(\theta, \phi),$$

and

$$(22) \qquad \hat{D}(W, \Theta, \Phi) = \sum_{l=0}^{\infty} \sum_{m=-l}^{+l} G_{lm}(W) Y_{lm}(\Theta, \Phi),$$

where W is the magnitude of the 3D frequency space vector $W\vec{\Theta}$. From the Funk-Henke Theorem for spherical harmonics $\{Y_{lm}\}$ [1],

$$(23) \qquad \int_{S^2} e^{i\sigma \vec{\theta} \cdot \vec{\omega}} Y_{lm}(\vec{\omega}) d\vec{\omega} = \frac{(2\pi)^{3/2} i^l}{\sqrt{\sigma}} J_{(l+1/2)}(\sigma) Y_{lm}(\vec{\theta}),$$

where $\vec{\theta}$ denotes the unit vector $\vec{\theta} = (\theta, \phi) \in S^2$; and the expression for the Fourier transform in spherical coordinates,

$$(24) \qquad \hat{D}(W\vec{\Theta}) = \int r^2 dr d\vec{\theta} e^{i2\pi r W \vec{\theta} \cdot \vec{\Theta}} D(r\vec{\theta});$$

we have by the substitution of Eqs. (21) and (22),

$$(25) \qquad G_{lm}(W) = \frac{(2\pi) i^l}{\sqrt{W}} \int r^{3/2} dr J_{(l+1/2)}(2\pi r W) g_{lm}(r).$$

In Eqs. (23) and (25) J_n is the nth order Bessel function.

We are interested in the reconstruction of the filtered delivered dose $|\vec{\eta}| \hat{D}(\vec{\eta})$, $\vec{\eta} \in \Theta^{\perp}$, from the finite sampling of beams $\widehat{A * \mathcal{I}_\Theta}$. From Eq. (20), the beam at angle $\vec{\Theta}_j$, $j = 1, \ldots, M$ determines frequency space samples of $\hat{D}(\vec{\eta})$ evaluated at $\vec{\eta} \in \Theta_j^{\perp}$. More specifically, substitution of $\vec{\eta} = W\vec{\Theta}_{jk}$, $\vec{\Theta}_{jk} \in \Theta_j^{\perp}$, $k = 1, \ldots, N_j$ into Eq. (20) yields

$$(26) \qquad \widehat{A * \mathcal{I}_{\Theta_j}}(W\vec{\Theta}_{jk}) = \frac{\rho e^{\mu R_0}}{\mu (2\pi)^{3/2}} W \hat{D}(W\vec{\Theta}_{jk}).$$

The geometry suggested by Eq. (26) is shown in Figure 4 in which samples on an annulus of radius W, from the beam oriented at angle $\vec{\Theta}_j$, are proportional to the corresponding frequency space samples of $\hat{D}(\vec{\eta})$.

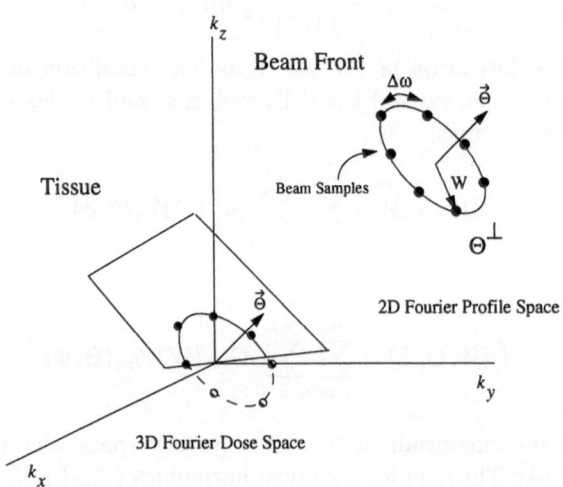

FIG. 4. *Geometry of 3D delivered dose reconstruction from filtered beam profiles.*

The set $\{G_{lm}(W)\}$ completely determines the delivered dosage $\hat{D}(\vec{\eta})$ at a radial spatial frequency of W. Hence, the condition for beam reconstruction of $\hat{D}$ is the inversion of the matrix equation ($j = 1, \ldots, M; k = 1, \ldots, N_j$) for $\{G_{lm}(W)\}$ given by

$$(27) \qquad W\hat{D}(W\vec{\Theta}_{jk}) = W\sum_{lm} G_{lm}(W)Y_{lm}(\vec{\Theta}_{jk}).$$

Define the matrix $Y_{(jk)(lm)} \equiv Y_{lm}(\vec{\Theta}_{jk})$ to obtain

$$(28) \quad (Y \bullet (WG))_{(jk)} = W\sum_{lm} Y_{(jk)(lm)}G_{lm} = (A * \mathcal{I})_{\Theta_j}(W\vec{\Theta}_{jk}).$$

Equation (28) represents N_j equations (the number of samples in the plane $\Theta_j^{\perp}$) among the unknowns $\{G_{lm}(W)\}$. Assuming a maximum value l of L in the expansion of Eq. (22), the corresponding number of unknowns is $\sum_{l=0}^{L}(2l + 1) = (L + 1)^2$, which results in a consistency bound (number equations $\geq$ number unknowns) given by

$$(29) \qquad \sum_{j=1}^{M} N_j \geq (L + 1)^2.$$

The number of samples obtained from each beam for the reconstruction of $\hat{D}$ at frequency W is determined by the 2D Fourier transform of the beam front evaluated on the annulus of radius W. A resolution length of $\Delta\omega$ in frequency on the beam front suggests a number of samples $N_j(W) = 2\pi W/\Delta\omega$. Substitution into Eq. (29) results in a beam number threshold given by

$$(30) \qquad M \geq \frac{\Delta\omega}{2\pi W}(L+1)^2.$$

A sufficient (maximum) value of L can be estimated by assuming a maximum extent of the delivered dosage r_{max} with the property $g_{lm}(r) = 0$ for $r \geq r_{max}$ in Eq. (21). Therefore, at a spatial frequency of W, a bound $2\pi r_{max} W$ exists for the Bessel function argument in Eq. (25). From the Debye approximation [1, 60, 62], $J_n(x) \simeq 0$, for $n \geq (x+2)$, substituted into Eq. (25), we have $G_{lm}(W) \simeq 0$ for $l \geq L \equiv (2\pi r_{max} W + 3/2)$. Therefore, from Eq. (30) a sufficient number of beams for dose reconstruction at frequency W is given by

$$(31) \qquad M \geq \frac{\Delta\omega}{2\pi W}(2\pi r_{max} W + 5/2)^2.$$

In the limit $2\pi r_{max} W \gg 1$, Eq. (31) is written

$$(32) \qquad M \geq (r_{max}\Delta\omega)2\pi r_{max} W_{max},$$

where W_{max} is the maximum spatial frequency in the delivered dose.

It is interesting to contrast the beam number criterion for the 2D and 3D cases. Two-dimensional reconstruction results in a bound

$$(33) \qquad M \geq (2\pi r_{max} W_{max} + 5/2),$$

which is known as the 'Bow Tie' condition in CT [62] and electron microscopy [60]. This beam number estimate has been applied to radiotherapy inversion with beams rotating about a cylindrical axis [7, 10]. The 2D inversion bound differs from the 3D bound because only one dose sample is obtained in frequency space from each beam. Therefore, the factor $r_{max}\Delta\omega$, dependent on the *beam* frequency space resolution, does not appear in the 2D bound. Note that the matrix inversion, although never actually computed, is more complex in the cylindrical case due to a required evaluation at each (W, Z) coordinate, rather than each radial spatial frequency W alone.

Evaluation of Eq. (32) with typical radiotherapy parameters provides an estimate of a sufficient number of 3D treatment beams. For example, a 5.0cm tumor ($r_{max} = 2.5cm$) with a dose resolution of $5.0mm (W_{max} = 2.0cm^{-1})$ results in a bound of $32r_{max}\Delta\omega$ beams. Note that if the expected condition $r_{max}\Delta\omega \simeq 1$ is satisfied, the bound is equal to the 2D case. It is interesting to speculate that 3D beam numbers can be reduced by increasing the multileaf collimator spatial frequency sampling, thereby reducing

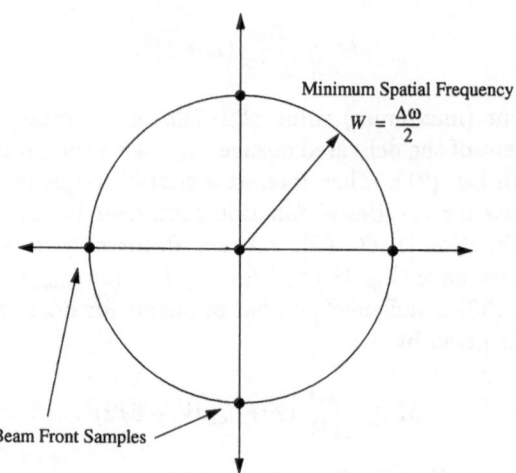

FIG. 5. *Two-dimensional Fourier space of modulated beam front at the minimum spatial frequency with $W = (\Delta\omega)/2$.*

$\Delta\omega$. This property does not exist for 2D configurations. The lower limit of spatial frequency sampling occurs at $W = (\Delta\omega)/2$, shown in Figure 5, for which there are four samples per beam. In this case with $r_{max}\Delta\omega \simeq 1$ a beam number $M \geq 13$ is obtained for the minimum spatial frequency reconstruction. Note that the bounds in Eqs. (31) and (33) are sufficient bounds based on the conservative Debye approximation. An alternative estimate results from Eq. (30) directly, where L is obtained from a spherical harmonic expansion of the dose function. The beam number bounds suggest that large irregularly-shaped tumors require more beams for the dose reconstruction.

4. Beam selection criteria. The sufficient number of beams for dose conformation generally exceeds numbers employed in present-day treatments. In this section we consider beam *selection* criteria; that is, algorithms which determine a subset of beam orientations for treatment. The view of treatment as a reconstruction of the delivered dose from filtered beam profiles, embodied in the matrix inversion of Eq. (28), suggests that the generalized inverse of Y must be well-defined. This corresponds to the condition that the matrix

$$(34) \qquad (Y^\dagger Y)_{(lm)'(lm)} = \sum_{j=1}^{M} \sum_{k=1}^{N_j} Y_{(lm)'}^*(\vec{\Theta}_{jk}) Y_{(lm)}(\vec{\Theta}_{jk}),$$

is non-singular. Therefore, the eigenvalues of $(Y^\dagger Y)$ in Eq. (34) define a selection criterion at each frequency W for the set of beam angles $\{\vec{\Theta}_j, j = 1, \ldots, M\}$. Note that the bounds in Section 3, suggesting an approximately equal beam number for 2D and 3D therapy, depend on the non-singularity of $Y^\dagger Y$. For example, if the samples on the frequency space annulus are not independent, the inversion could require a single sample for each beam and a consequent 3D beam number of $(2\pi r_{max} W + 5/2)^2$. We found for the real data cases in Section 6, however, that the 2D and 3D beam numbers are comparable.

An additional selection criterion, which ranks individual beams, is based on the metrics in the Hilbert spaces of dose and beam functions. The *direct* inner products in physical and profile spaces between two dose functions f, g and beam functions $\underline{f}, \underline{g}$ are defined by

$$(35) \qquad [f, g] = \int_{\mathcal{R}^3} d\vec{x} f(\vec{x}) g(\vec{x}),$$

and

$$(36) \qquad \langle \underline{f}, \underline{g} \rangle = \int_{S^2} d\vec{\theta} \int_{\theta^\perp} d\vec{y} \underline{f}(\vec{\theta}, \vec{y}) \underline{g}(\vec{\theta}, \vec{y}),$$

respectively [61]. The dual x-ray transform is the *metric space dual* of the x-ray transform relative to the direct metrics. The duality relationship is expressed in the equation

$$(37) \qquad [f, P^\sharp \underline{g}] = \langle Pf, \underline{g} \rangle,$$

where f and $\underline{g}$ are functions in physical and profile spaces, respectively. Assuming a continuous profile function $\underline{f}(\vec{\theta}, \vec{y}), \vec{y} \in \theta^\perp$, define the *sampled profile function* $\underline{f}_s$ by

$$(38) \qquad \underline{f}_s(\vec{\theta}, \vec{y}) = \sum_{j=1}^{M} \delta(\vec{\theta} - \vec{\theta}_j) \underline{f}(\vec{\theta}_j, \vec{y}).$$

We are interested in maximizing the direct inner product between the delivered dose functions corresponding to continuous and sampled beam profiles. Substitution of Eq. (11) (with $\mu = 0$) into the direct inner product, and application of Eq. (37), yields

$$(39) \qquad [D, D_s] = (\frac{\mu e^{-\mu R_0}}{\rho})^2 \langle PP^\sharp(A * \mathcal{I}), (A * \mathcal{I})_s \rangle.$$

Assuming that the beam profiles are derived from a prescribed dosage p on $\mathcal{R}^3$ in Eq. (15), we have

$$(40) \qquad (A * \mathcal{I}) = \frac{\rho e^{\mu R_0}}{\mu} JPp,$$

which upon application of Eq. (12) in Eq. (39), results in the expression

$$(41) \qquad [D, D_s] = \langle Pp, (JPp)_s \rangle = \sum_{j=1}^{M} F_j,$$

where F_j, $j = 1, \ldots, M$ is a measure assigned to beam angle $\vec{\theta}_j$ given by

$$(42) \qquad F_j = \int_{\theta_j^\perp} d\vec{y} Pp(\vec{\theta}_j, \vec{y}) J Pp(\vec{\theta}_j, \vec{y}).$$

Equation (42) suggests that an appropriate beam angle-dependent measure of dose conformation is given by

$$(43) \qquad F(\vec{\theta}) = \int_{\theta^\perp} dy Pp(\vec{\theta}, \vec{y}) J Pp(\vec{\theta}, \vec{y}),$$

which is a quantity we display in a spherical plot in the tissue coordinate system. Equation (43) is a measure of an individual beam contribution to the dose reconstruction embodied in Eq. (11). The condition number of the matrix in Eq. (34) is a measure of the dose reconstruction assigned to the entire set of beams.

5. Iterative inversion algorithm. The formal inversion of the prescribed dose, given in Eq. (15), results in unphysical negative beam profiles. A successful method of applying prior constraints, which has been employed by a number of groups [7, 9, 17, 24, 59], is to project iteratively onto acceptable beam profiles. In this paper we use an algorithm, suggested by Holmes and Mackie [9] for 2D tomotherapy, which adds a projected residual dose iteratively onto each beam profile before application of the constraint.

Define the positivity constraint operator C for a profile function $\underline{f}$ by

$$(44) \qquad C\underline{f}(\vec{\theta}, \vec{y}) = \begin{cases} \underline{f}(\vec{\theta}, \vec{y}) & \underline{f} \geq 0 \\ 0 & otherwise \end{cases}.$$

The appropriate iterative inversion algorithm is written in terms of operators $P_{-\mu}^\sharp$, P_μ, and J in Eqs. (11),(16), and (14), respectively. In the 0th order iteration define

$$(45) \qquad \begin{aligned} D^{(0)} &= p \\ i^{(0)} &= C\left[A^{-1} * J P_\mu D^{(0)}\right], \end{aligned}$$

where p is the prescribed dosage on $\mathcal{R}^3$. The jth iteration then requires the projection of a residual dose $R^{(j)}$ as summarized in the sequence of operations,

$$(46) \qquad D^{(j)} = P_{-\mu}^\sharp(A * i^{(j-1)}),$$

$$(47) \qquad R^{(j)} = p - D^{(j)},$$

and

$$(48) \qquad i^{(j)} = C\left[i^{(j-1)} + A^{-1} * J P_\mu R^{(j)}\right].$$

The stopping criterion is a sufficiently small change in the residual $\sqrt{[R^{(j)}, R^{(j)}]}$ versus iteration. The beam profile at the jth iteration, corrected for a non-planar tissue-air interface, is then given by

$$(49) \qquad \mathcal{I}^{(j)} = \frac{\rho e^{\mu R_0}}{\mu} i^{(j)}.$$

The implementation of Eqs. (45) and (48) requires the application of a high frequency cutoff on the beam front filter J in Eq. (14) to remove projected noise and sampling artifacts [7, 10]. As suggested in Ref. [10], we apply a 2D Butterworth lowpass filter given by

$$(50) \qquad R(|\vec{\eta}|) = \frac{1}{\sqrt{1 + (|\vec{\eta}|/|\vec{\eta}_0|)^{2M}}},$$

with cutoff $|\vec{\eta}_0|$ and taper parameter M. The choice of $|\vec{\eta}_0|$ reflects a trade-off in the delivered dose; small $|\vec{\eta}_0|$ with a less conforming uniform TV dose, and large $|\vec{\eta}_0|$ with a more variable dose in the TV. For the examples in Section 6 we found that some experimentation with $|\vec{\eta}_0|$-values was desirable to obtain an optimum delivered dose. As determined by the TV/OAR Dose-Volume Histograms [68, 69], we found optimum results with $|\vec{\eta}_0|$-values given by $W_{bmax}/6$, where W_{bmax} is the highest spatial frequency in the beam, and with M of ten in Eq. (50). These filter parameters are comparable to the 1D lowpass filters used in Ref. [10].

The computer implementation of Eqs. (47)-(48) requires rapid sums along projected $(P_{-\mu})$ and backprojected $(P^{\sharp}_{-\mu})$ rays, beam front filtering (J), and positivity constraint decisioning $(\mathcal{C})$. Higher-order dose models are easily incorporated into the iterative algorithm. For example, the depth-dependent scattering kernel in Eq. (3) can be used in Eq. (47) with a generalized backprojection operator $P^{\sharp}_{-\mu}$ which includes effects of tissue heterogeneities $(\mu = \mu(\vec{x}))$, air gaps, and beam build-up. This is implemented computationally by attenuating the incident fluence with the factor

$$(51) \qquad e^{-\sum_i \mu_i \Delta x_i} h(\vec{x}_i),$$

where μ_i, Δx_i, and $h(\vec{x}_i)$ are the extinction coefficients, step sizes, and beam build-up function at the ith location $\vec{x}_i$ along the ray. Air gap effects are incorporated by defining μ_i and the delivered dosage D to be zero for voxels in air. These effects are important for tumors near the lung, esophagus, and sinuses. In Eq. (48) the projection operator $P_\mu f$ is generalized by accumulating terms $\sum_i e^{\mu_i \Delta x_i} f(\vec{x}_i)$ along the ray projected onto the beam front. The beam front deconvolution filter is the inverse of A in Eq. (10) in which the depth dependence is averaged. Figure 6 contains the flow-chart of the iterative algorithm for conformal inversion to a 3D beam configuration with a higher-order dose model.

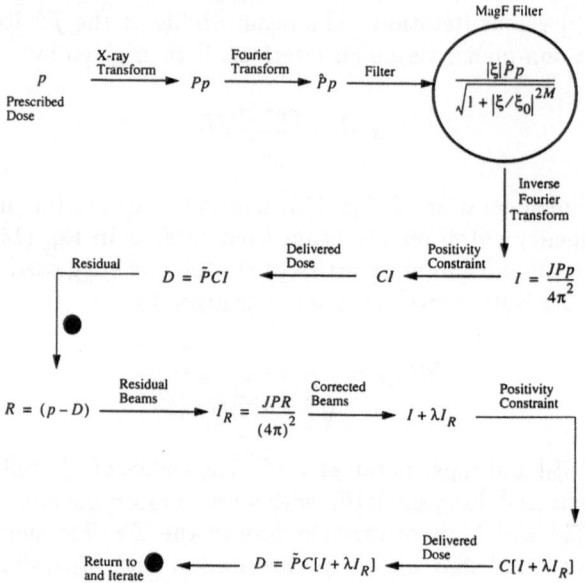

FIG. 6. *Three-dimensional dual X-ray transform inversion by iterative backprojection algorithm.*

Dose prescriptions on OARs are often expressed as a percent OAR tissue volume allowed above a tolerance dose. We have incorporated this type of dose prescription into the iterative inversion algorithm by reducing the delivered dosage on OAR voxels to the tolerance dose level (in Eq. (47)). This procedure, which has had negligible effect on algorithm run-time, is applied after Eq. (48) in the iterative program. An alternative approach is simply to prescribe an extremely small dosage to the OARs, effectively blocking rays which intersect the OAR volume. Plans of this type are similar to the all-none method [5].

6. Results. In this section a computer implementation of the iterative algorithm in Figure 6 is applied to 3D conformal treatments of brain and prostate tumors. The fore-brain tumor data set was obtained from the University of Massachusetts Medical School, and the rear-brain and prostate data sets were obtained from Massachusetts General Hospital. Each data set consisted of radiologist-generated contours of TVs and OARs, CT data, and prescribed TV/OAR dosage and dose limits. In all of the computations a cone beam geometry was assumed with an SAD (source-to-axis distance) of $100cm$ and SCD (source-to-calibration point-distance) of $64cm$. Using the calibration algorithm derived in Appendix A and UMASS tissue maximum ratio (TMR) tables, the absolute beam intensities were derived for a Varian $6MeV$ machine. The filter, derived from a parallel beam geometry,

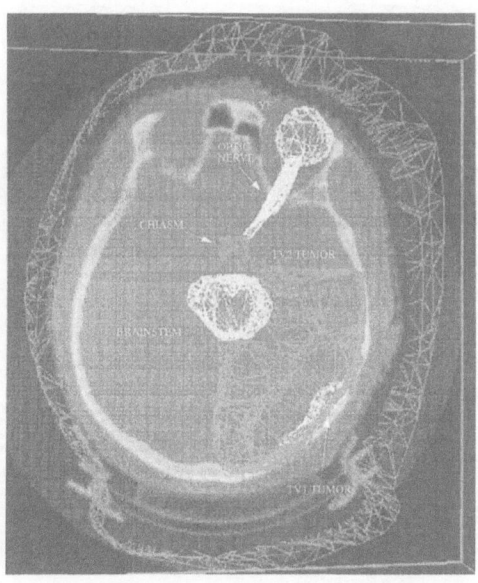

FIG. 7. *Tiled TV1, TV2, and OAR contours for MGH rear-brain tumor. Structures generated by contour tiling algorithm.*

was applied to the cone beam case with a cutoff of about 1/6 the highest spatial frequency in the beam. The tissue voxels, which were located in space by derived skin contours, were assumed to uniformly attenuate the radiation fluence with the water attenuation constant μ. In all cases the algorithms converged in less than 10 iterations; as judged by the direct metric of the dose residual R^j in Eq. (48). Five iterations of the inversion algorithm to 162 beams at a $5.0mm$ dose resolution required six and seventeen minutes on an SGI Indigo2 Extreme ($100MHz$) for the brain and prostate tumors, respectively.

6.1. Rear-brain tumor. Figure 7 contains tiled radiologist-generated contours of a rear-brain tumor and surrounding OARs superimposed on original CT data. The data consisted of 57 2D 320x320 CT images with slice separations of about $2.5mm$. The radiologist contoured the brain data into two TV regions; TV1, the area of obvious tumor and TV2, the area of likely cancerous penetration. Prescribed dosages of $90Gy$ and $65Gy$ were given for TV1 and TV2, respectively. The relevant OARs in Figure 7; brainstem, retina, and optic nerve/chiasm have maximum allowed dosages of $20Gy$, $10Gy$, and $10Gy$, respectively, with 0% dose tolerance. Note that the close proximity of the brainstem to the TV regions, and the prominence of the eye and optic nerve/chiasm are severe limitations to beam configurations. Figure 8 contains an axial slice through the 3D prescribed dose

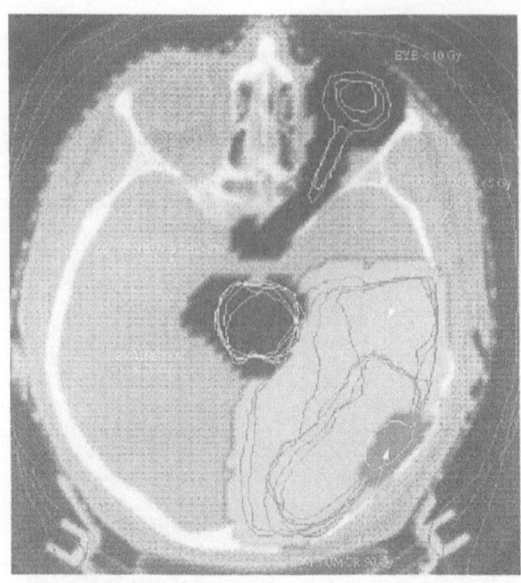

FIG. 8. *Prescribed TV1, TV2, and OAR dosages for MGH rear-brain tumor.*
Three-dimensional contours and CT data in axial slice. Structures generated by contour
fill algorithm.

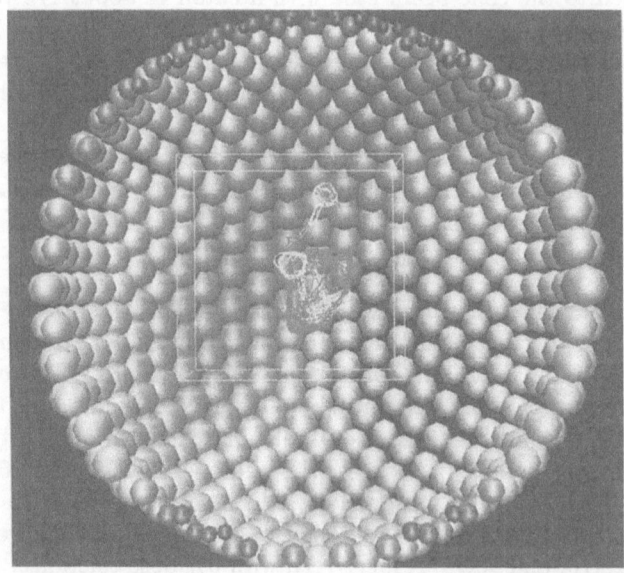

FIG. 9. *Rho-metric ranking of beam orientation for MGH rear-brain tumor (highest*
ranked beams in red).

functions for TV regions and OARs obtained with a contour fill algorithm. A display of the ρ - metric ranking in Eq. (43) for 481 beams relative to the brain center is shown in Figure 9. Note that the higher ranked beams, represented by red spheres, lie roughly along a tilted annulus ranging from the lower-left to upper-right of the 3D image. A display of high-ranked beam directions overlaid on the corresponding beams-eye-view of the contours, shown in Figure 10, indicates that the optimum beam orientations allow maximum TV coverage and avoidance of proximate OARs. Figure 11 contains a plot of the dose residual versus iteration for 481 beam inversion at a dose resolution of $3.0mm$.

Figure 12 contains isodose contours at 10% increments overlaid on CT data and the prescribed dose in an axial slice through the eye, optic nerve, chiasm, and brainstem with the full 481 beam treatment shown in Figure 9. Note the excellent dose conformation around proximate OARs, and delivery of a uniform two-level prescribed dose to the TVs. The $65Gy$ and $90Gy$ tiled isodose surfaces with enclosed rendered tumors are shown in Figure 13. The delivered dose resulting from 33 candidate beams, displayed in the same axial slice, is shown in Figure 14. Although dose contours are clearly degraded relative to the plan in Figure 12, there is acceptable OAR shielding and TV dose conformation.

We also considered treatment plans for TV1 ($65Gy$) alone. Figure 15 contains the side-by-side comparison of 8 and 33 beam treatments using the delivered dosage in the same axial slice as in Figure 12. The results indicate that as few as eight beams are sufficient to deliver a conforming dose. Due to present-day limitations in beam modulation technology and time constraints, a physicist is often limited to 4-5 beams in the design of a radiotherapy plan. Figure 16 contains the results of a four-field plan for the brain tumor with multileaf collimator blocking (no intensity modulation), in which an experienced planner (MU) chose the four-beam configuration for input to the iterative algorithm. The delivered dosage is compared with the 12 highest ranked beams (from a 481 beam plan) with intensity modulation. The isodose curves are shown on CT data in an axial slice through the tumor. The superior dose conformation and TV coverage, indicated by the $65Gy$ isodose surface and curves in Figure 16, suggest significant normal tissue shielding with beam intensity modulation.

6.2. Prostate tumor. Figure 17 contains a tiling of TV and OAR contours for a prostate tumor surrounded by the rectum and bladder. The data consisted of 50 2D 320x320 CT images with a slice separation of $8.0mm$. The prescribed tumor dose was $65Gy$, and the rectum and bladder had dose limits of $20Gy$ with 5% dose tolerance. The ρ-metric ranking of 84 physically accessible beams ($\leq 30°$ off vertical) is shown in Figure 18. Note that the standard lateral beams, allowing a clear trajectory to the tumor between the bladder and rectum, are highly ranked. The lateral beams are among the fourteen highest ranked beams in Figure 19. Figures 20 and 21 contain the 10% isodose curves, overlaid on CT data and the prescribed dose, in axial and sagittal slices with an 84 beam treatment. The inversion algorithm was run with a $5.0mm$ dose resolution. Figures 22 and 23 contain the corresponding isodose curves in the same axial and sagittal slices with a 24 beam treatment. The dose degradation relative to the 84 beam plan is seen in the normal tissue exposure at less than 20% of the peak dosage for comparable TV and OAR conformation.

6.3. Fore-brain tumor. Figures 24 and 25 contain contours of a large, irregular fore-brain tumor embedded among the highly sensitive optic nerves. As seen in the sagittal view in Figure 24, the tumor extends back to the brainstem and down into the lower brain. The dose prescription was for $65Gy$ on the TV, $\leq 10Gy$ on the optic nerves and chiasm, and $\leq 20Gy$ on the eyes and brainstem. The inversion was computed with a $3.0mm$ delivered dose resolution and 0% dose tolerance above the dose limit for all OARs. Figure 26 contains the ρ-metric ranking of 481 accessible beams, indicating optimum beam placement on the tumor side above the eye. A comparison of delivered dosage in the axial slice for 481 and 33 candidate beams is shown in Figures 27 and 28. Note that dose conformation around OARs is not significantly degraded; but rather, the smaller number of beams creates low level normal tissue exposure. Figure 29 contains the $65Gy$ isodose surface with 123 candidate beams indicating complete 3D coverage of the tumor.

Fore-brain tumor treatment planning requires air gap corrections due to the proximity of the sinuses to the tumor. We did not include these corrections in the dose calculations above, as can be seen in the air gap contours in Figures 27 and 28. Recently, the correction has been included in our codes; and preliminary results indicate a minor effect on beam numbers and orientations in the fore-brain tumor case.

7. Conclusion. The traditional gantry-based beam delivery in radiotherapy, as well as the obvious cylindrical dimensions of the human body, suggest 2D conformal inversion. However, particularly in the brain, the locations of TV and OARs may favor treatment at oblique angles to the patient axis. In this paper we generalize to 3D earlier work on 2D conformal radiotherapy involving inversion of the dual Radon transform. In a model with two-dimensional modulated beam fronts oriented on a sphere centered on the tumor; and tissue described uniformly by a single absorption constant, it is shown that the appropriate projection operator relating beams and dose is the attenuated x-ray transform. The mathematics of x-ray transform inversion, involving a generalized Projection-Slice Theorem, relates the required beam number to the number of spherical harmonics in the delivered dose. A rough estimate of the number of spherical harmonics, in terms of dose dimensions and spatial frequency content, is obtained from the Debye approximation for Bessel functions. It is shown that 3D inversion does not require more beams than 2D treatment; and in fact, beam numbers may be reduced by the adjustment of beam modulation sampling. About 30 beams are sufficient for typical tumor dimensions ($5.0cm$) and dose resolution ($5.0mm$), although this number is significantly reduced for targets with small spherical harmonic content. The interpretation of dose functions and beam profiles as existing in Hilbert spaces with a direct metric suggests a natural angle-dependent measure of beam orientation. The measure, which is the 3D generalization of the ρ-metric in CT, maximizes the overlap between the delivered dosages of continuous and sampled beam profiles. An additional beam selection criterion, depending only on the set of angles in the beam configuration, is the condition number of the matrix inversion implicit in the spherical harmonic expansion of the dose.

The inversion of the dual attenuated x-ray transform is implemented by iterative projection of filtered beam profiles to solutions satisfying the non-negativity constraint. Radiotherapy plans are derived from radiologist-generated TV/OAR contours and prescribed dosages for brain and prostate tumors. The plans are evaluated by examining delivered dose contours and isosurfaces. We found that OARs, such as the optic nerves or rectum, could be included in the dose prescription before the iterative inversion to profiles. This has the effect of protecting the OARs by avoiding conflicted beams; that is, beams in which OARs and TVs project into the same point. Overall, we found that beam number thresholds of less than 20 were sufficient to satisfy dose conformation goals in the brain and prostate cases. The degradation in plans with fewer numbers of beams consists primarily of low level normal tissue exposure.

Areas of future work include an investigation of the conditioning of the 3D inversion as determined by the eigenvalues of the matrix in Eq. (34). This will provide apriori information of the effectiveness of beams under limited-angle or preferred direction constraints. We are also implementing higher order dose models into the backprojection step of the iterative algorithm, including the effects of air gaps, beam build-up, and scattering. Finally, the 2D beam front filter has been modified to account for the cone geometry of most accelerator beams. It is expected that available beam numbers will increase dramatically as dynamic, intensity-modulated multileaf collimators become widespread in therapy clinics. The algorithms presented in this paper may assist the planner in the initial design of these more complicated beams and beam configurations.

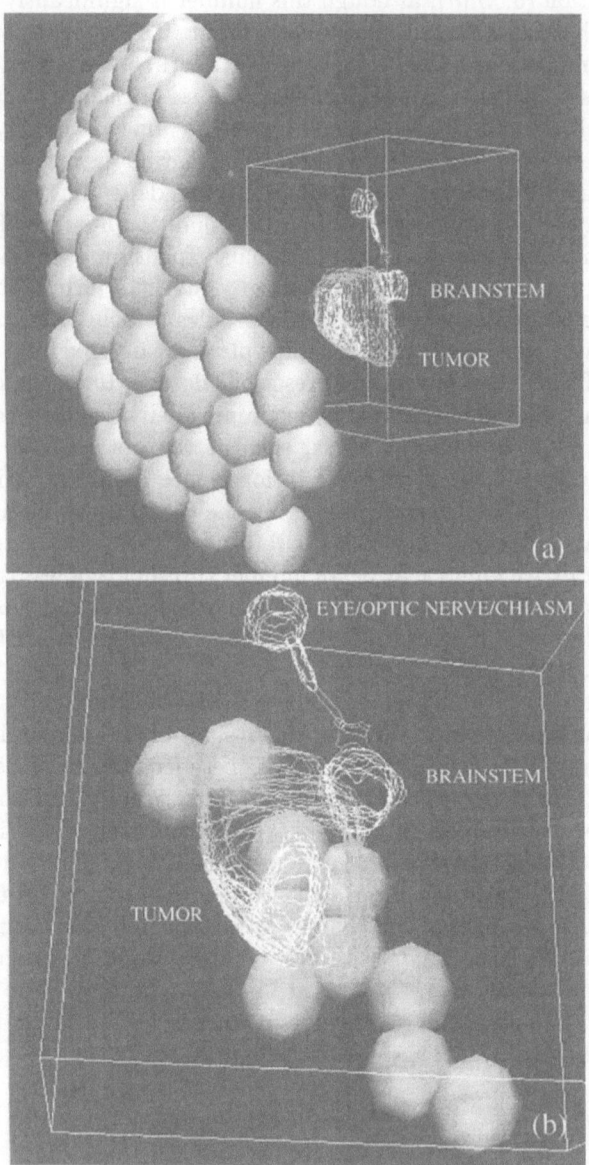

FIG. 10. *Rho-metric ranking for MGH rear-brain tumor beams relative to TV/OAR contours. (a). Forty-four highest ranked beams in side view. (b). Ten highest ranked beams in beams-eye-view.*

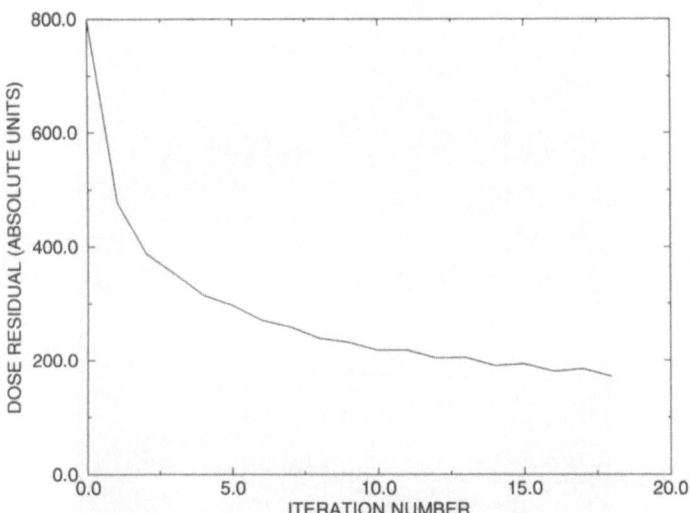

FIG. 11. *Dose residual versus iteration for 481 beam inversion of MGH rear-brain tumor.*

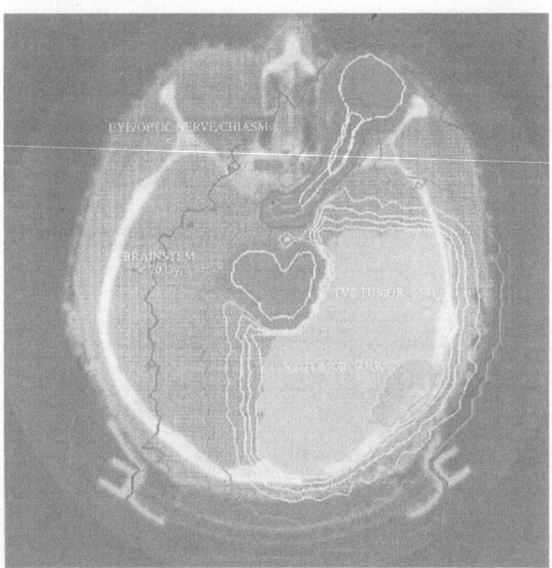

FIG. 12. *Ten percent isodose contours for 481 beam treatment of MGH rear-brain tumor. Two-level TV1 and TV2 dose prescription shown in solid color.*

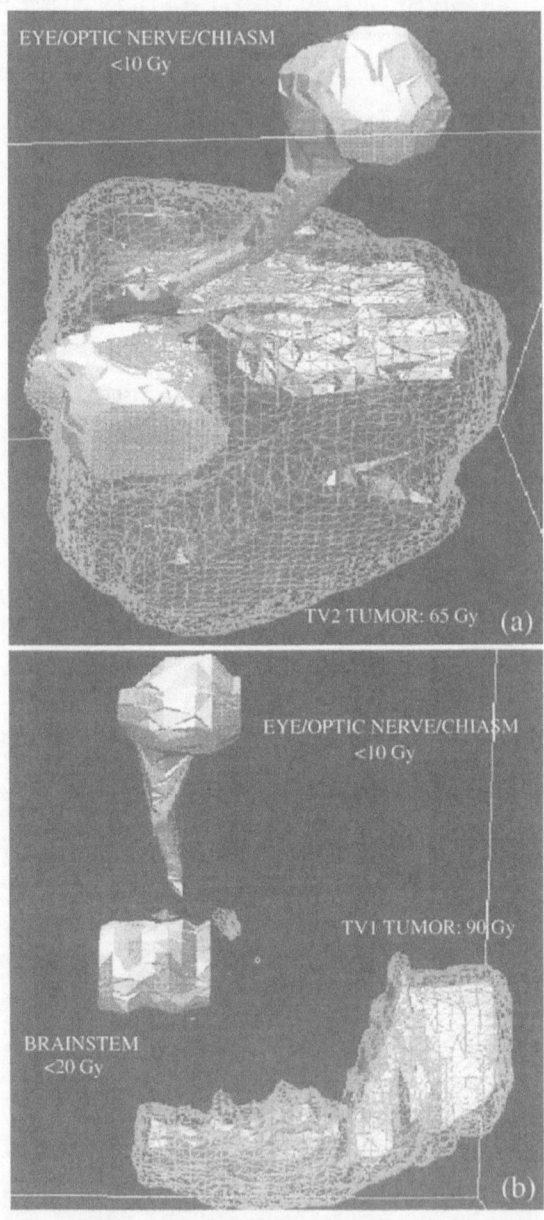

FIG. 13. *Isodose surface for MGH rear-brain tumor with 481 beam treatment. (a). 65Gy isodose surface. (b). 90Gy isodose surface.*

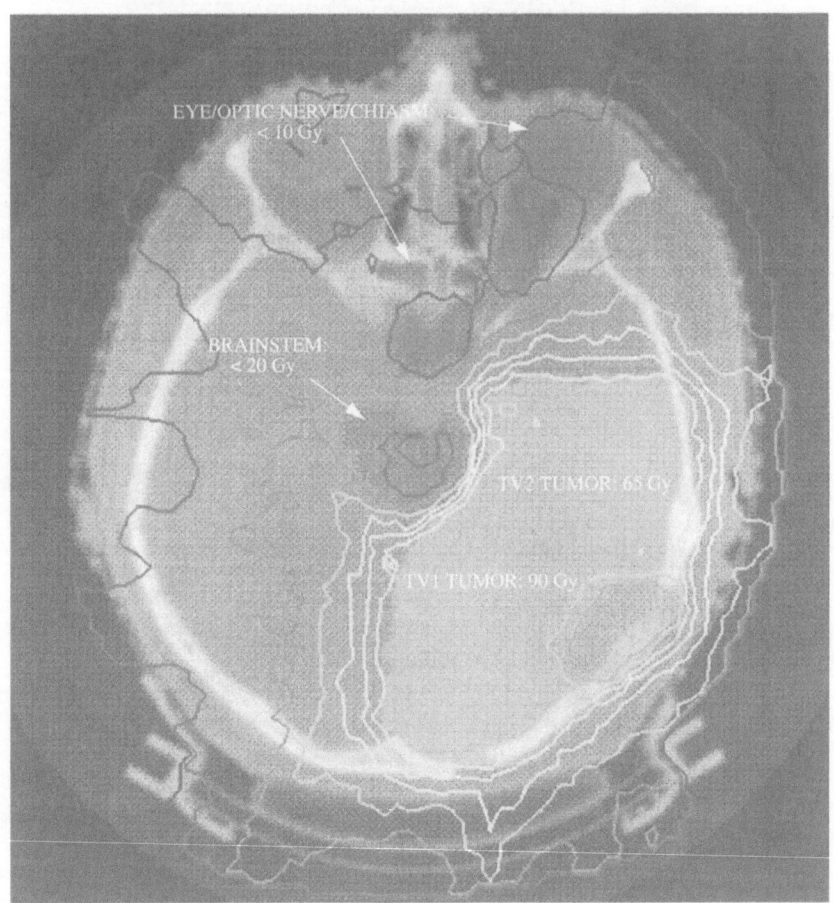

FIG. 14. *Ten percent isodose contours for 33 beam treatment of MGH rear-brain tumor. Two-level TV1 and TV2 dose prescription shown in solid color.*

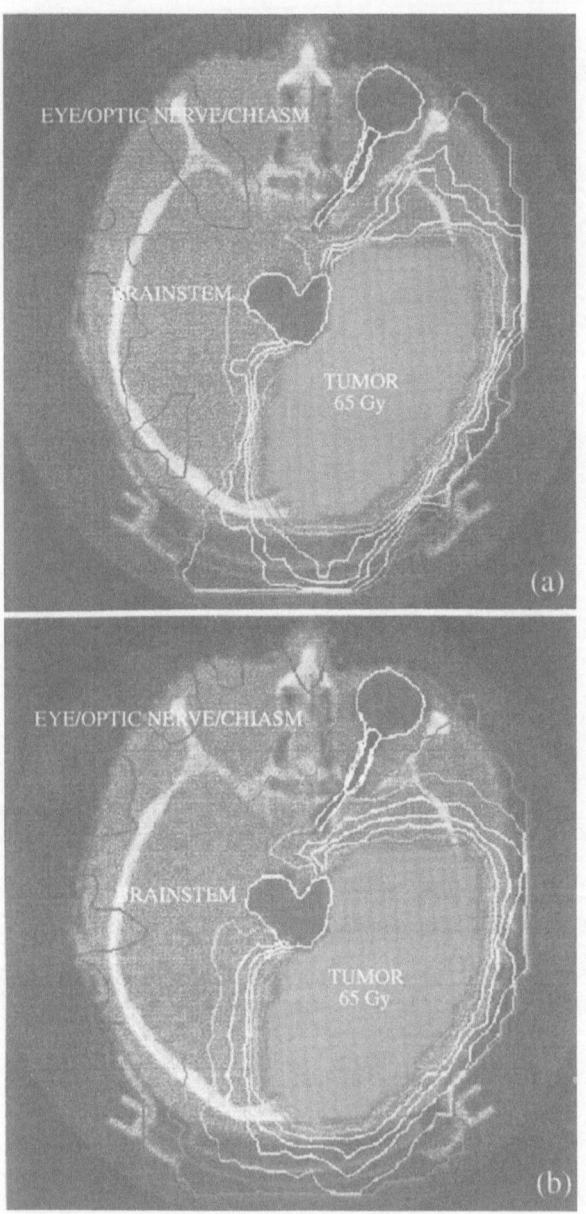

FIG. 15. *Comparison of TV2 region dose conformation with (a) thirty-three and (b) eight modulated beams. Prescribed dose in solid color.*

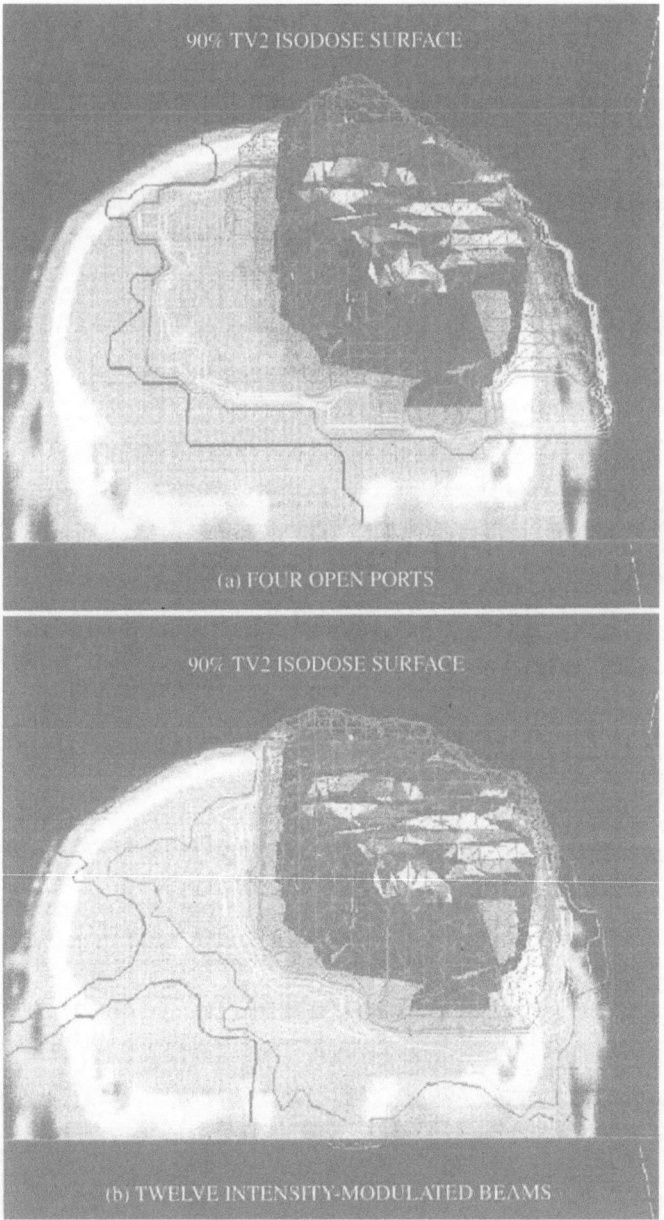

FIG. 16. *Delivered dosage for MGH rear-brain tumor with (a) four open ports and (b) twelve intensity modulated beams. Ninety percent TV2 peak dose surface and ten percent isodose contours in axial slice with rendered tumor.*

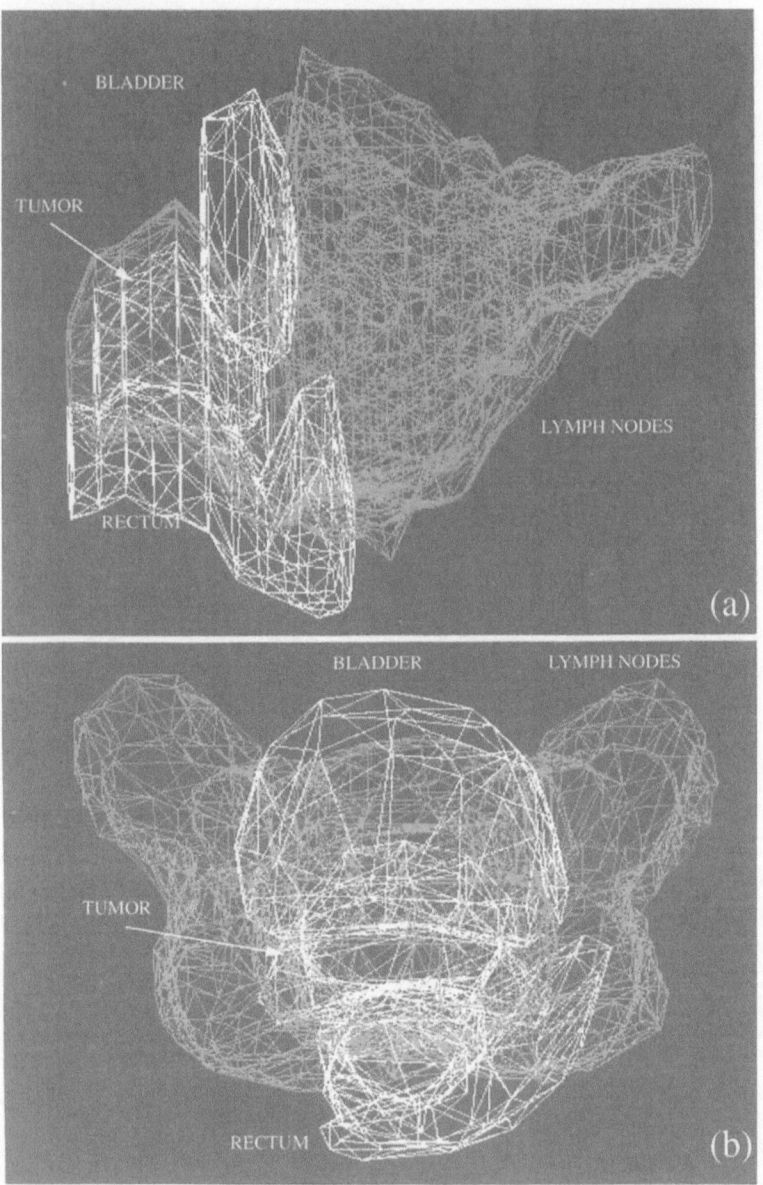

FIG. 17. *(a). Sagittal and (b) axial views of prostate TV and proximate OARs. Structures generated by contour tiling algorithm.*

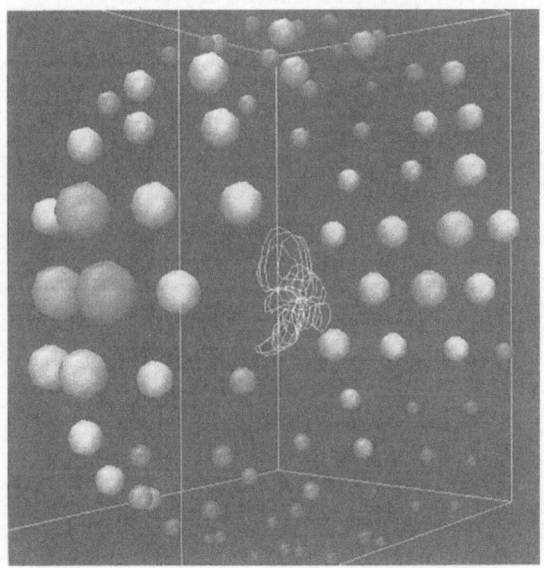

FIG. 18. *Rho-metric ranking of 84 physically accessible beams for prostate tumor treatment (highest ranked beams in red).*

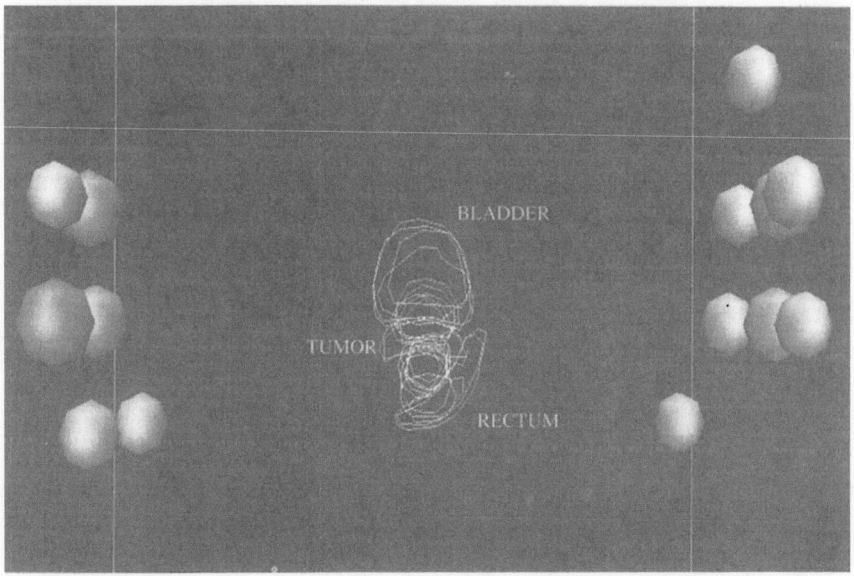

FIG. 19. *Rho-metric ranking of beam orientations for MGH prostate tumor. Fourteen highest ranked beams.*

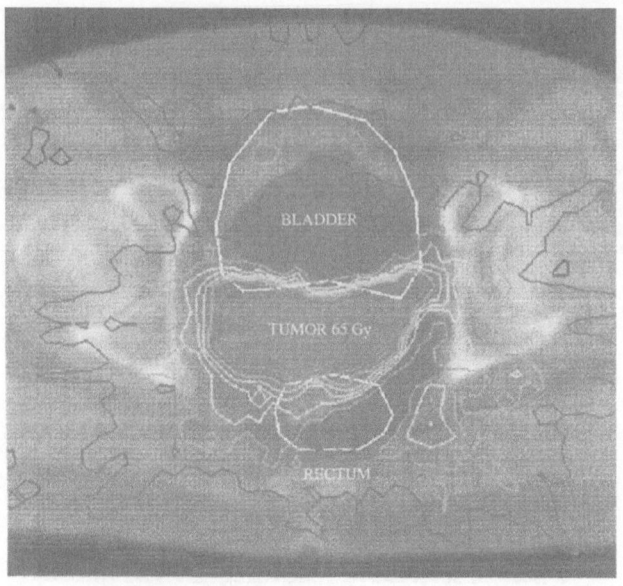

FIG. 20. *Ten percent delivered dose contours in axial slice for 84 beam MGH prostate tumor treatment. TV and OAR prescribed dose in solid color. Bladder and rectum contours included.*

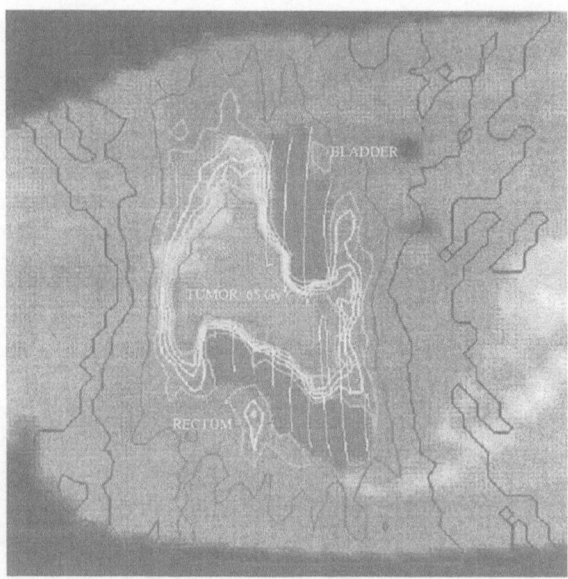

FIG. 21. *Ten percent isodose contours in sagittal slice for 84 beam MGH prostate tumor treatment. TV and OAR prescribed dose in solid color. Three-dimensional rectum and bladder contours included.*

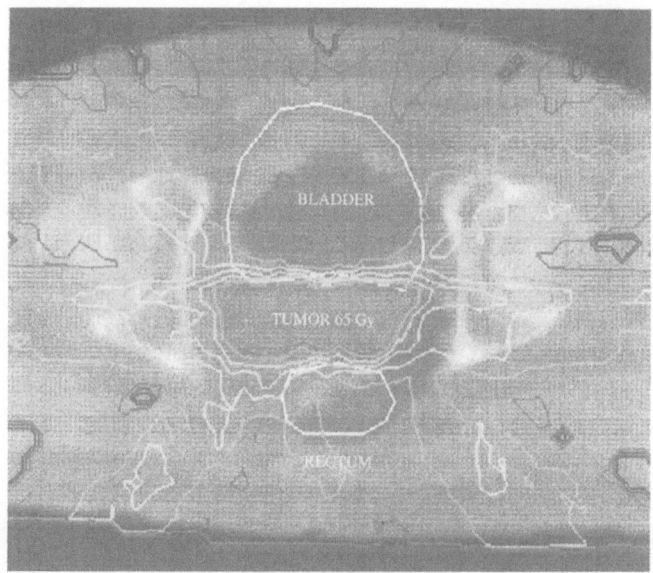

FIG. 22. *Ten percent delivered dose contours in axial slice for 24 beam MGH prostate tumor treatment. TV and OAR prescribed dose in solid color. Bladder and rectum contours included.*

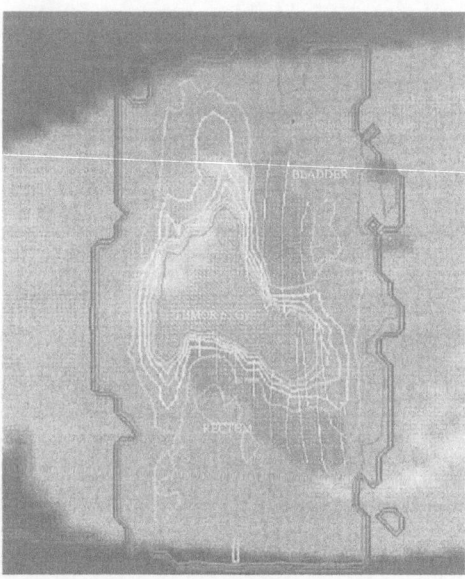

FIG. 23. *Ten percent isodose contours in sagittal slice for 24 beam MGH prostate tumor treatment. TV and OAR prescribed dosages in solid color. Three-dimensional rectum and bladder contours included.*

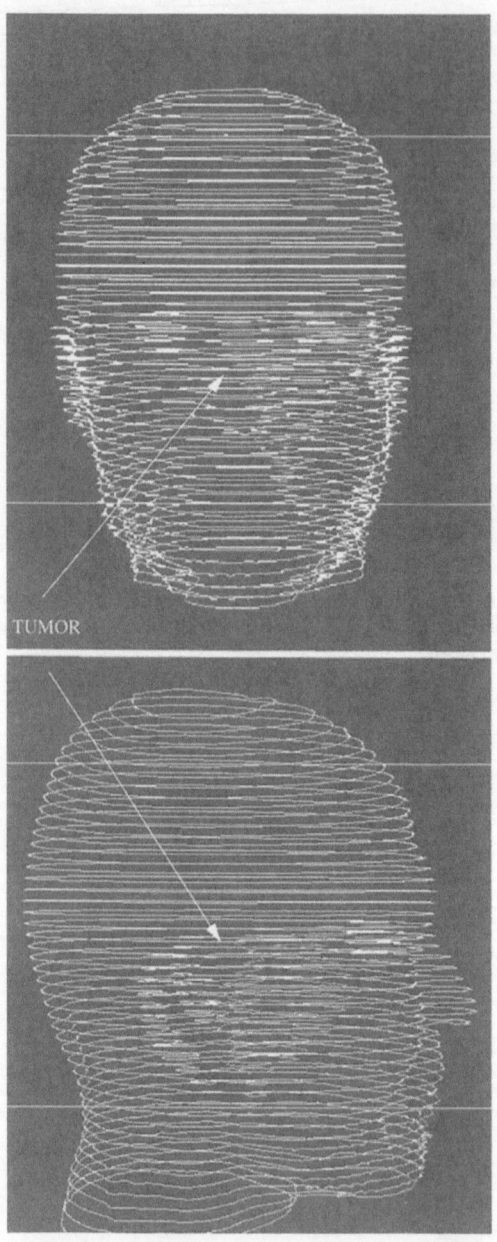

FIG. 24. *TV and OAR contours for UMASS fore-brain tumor data set.*

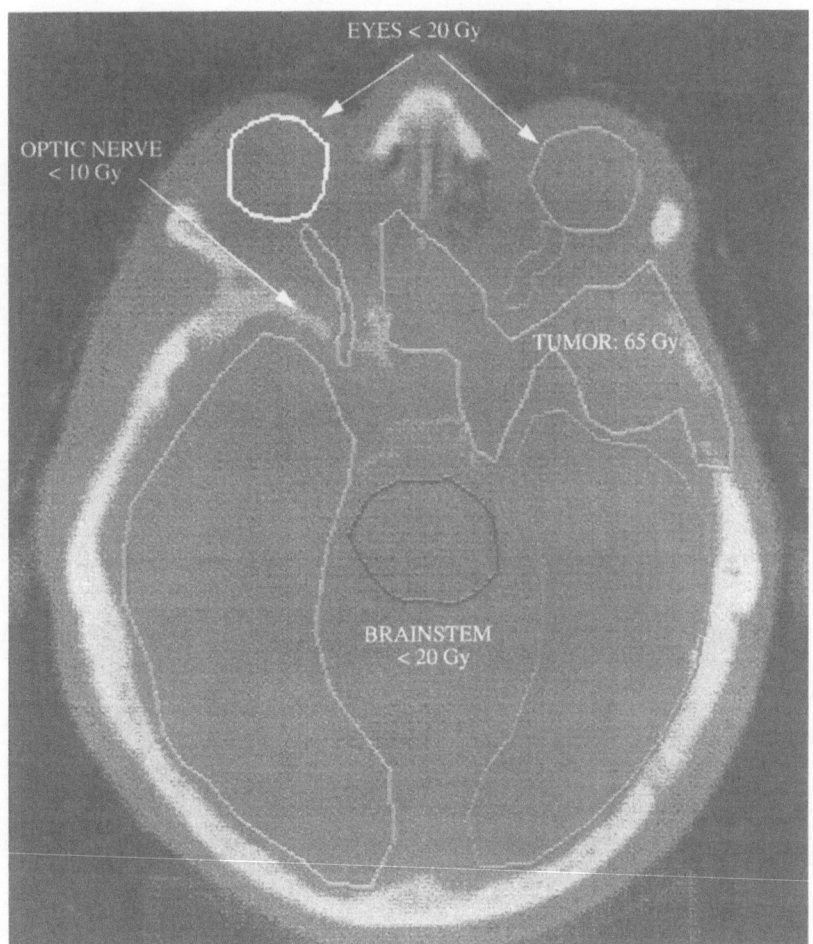

FIG. 25. *UMASS fore-brain TV and OAR contours in axial slice with prescribed dosage.*

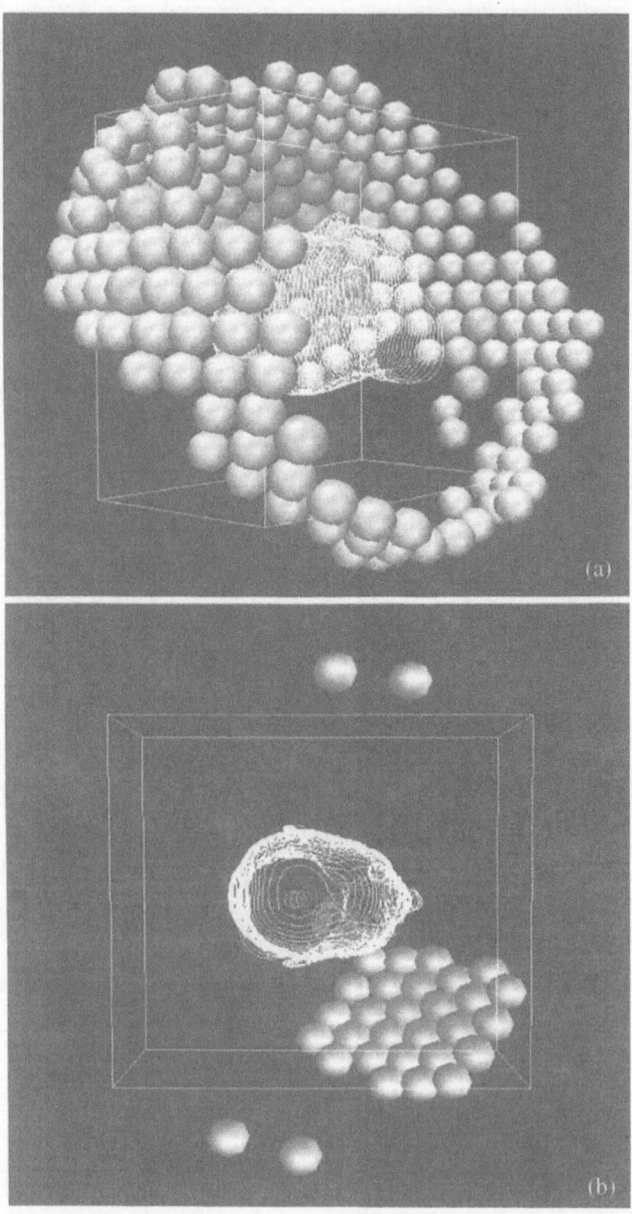

FIG. 26. *Rho-metric ranking for beam orientation for UMASS fore-brain tumor treatment. Iterative inversion with 481 candidate beams. Sagittal (a) and axial (b) views of highest ranked beams.*

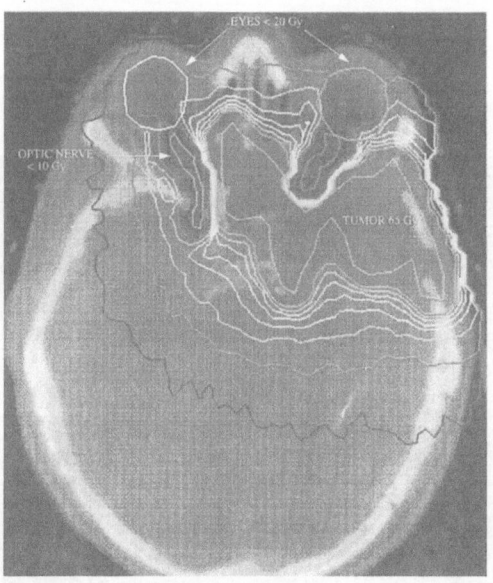

FIG. 27. *Ten percent delivered dose contours in axial slice for UMASS fore-brain tumor treatment with 481 beams.*

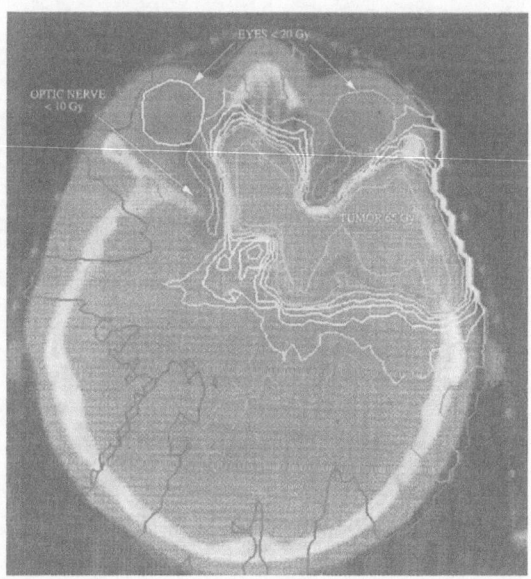

FIG. 28. *Ten percent delivered dose contours in axial slice for UMASS fore-brain tumor treatment with 33 beams.*

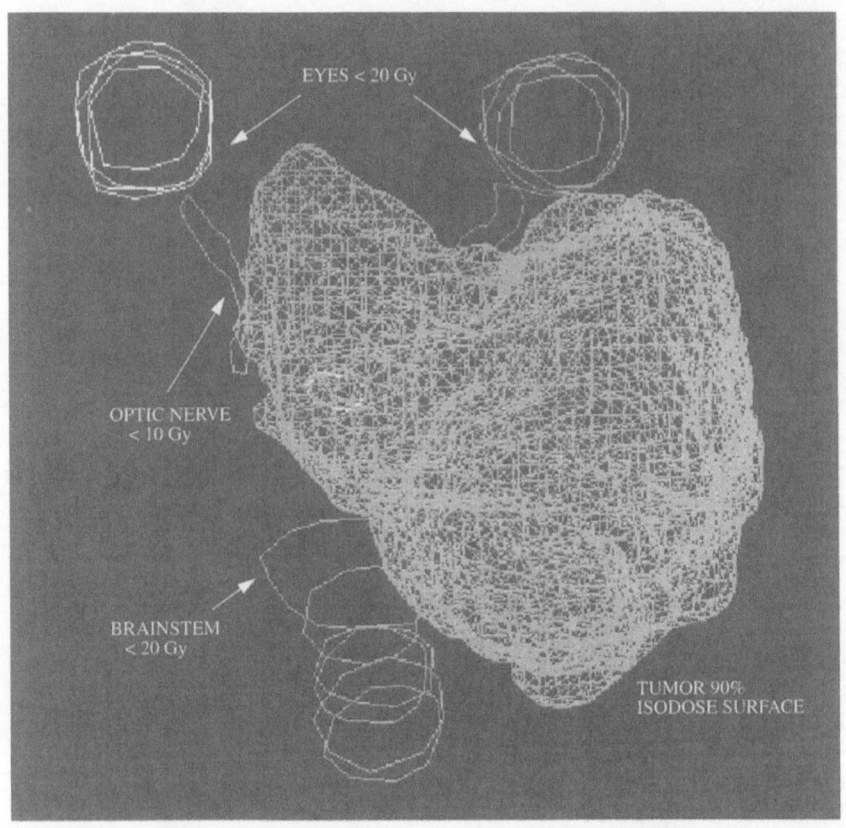

FIG. 29. *Ninety percent delivered isodose surface for UMASS fore-brain tumor treatment with 123 beams.*

APPENDIX

A. Multiple beam calibration algorithm. In this appendix we derive a calibration algorithm to convert relative beam fluence in Eq. (15) to absolute monitor unit (MU) values through the central ray of each beam. The definitions and notation for beam calibration can be found in Ref. [70]. Assume that the machine is calibrated to deliver dosage $\mathcal{D}$ to depth d_0 through a surface field size r_0 and source-to-calibration- point-distance (SCD) at a monitor unit setting of H. Let $x_i I_0$ be the ith beam central ray fluence, where x_i is computed using the iterative inversion algorithm, and I_0 is an arbitrary uniform MU factor for all beams. The dose to depth d_0 in the *calibration* set-up defined in Figure A1 from the ith beam central ray is given by $\frac{x_i I_0 \mathcal{D}}{H}$. Assume that, in the *treatment* set-up (Figure A2), an axis field size r_i^c at a source-to-axis-distance SAD is used for the ith beam central ray. Then the dose to isocenter, with the patient located with the isocenter at the calibration depth, results in a dosage

$$\text{(A1)} \qquad D_i^0 = \frac{x_i I_0 \mathcal{D}}{H} \left(\frac{SCD}{SAD}\right)^2 S_c(r_i^c) S_p(r_{0i}^p),$$

where $S_c(r_i^c)$ and $S_p(r_{0i}^p)$ are scatter factors for the collimator and phantom at the appropriate field sizes. In Eq. (A1) r_{0i}^p is the phantom surface field size. The tissue maximum ratios (TMRs) are applied to compute the dose to the treatment depth d_i in the patient from the central ray of each beam (Figure A3);

$$\text{(A2)} \qquad D_i = \frac{x_i I_0 \mathcal{D}}{H} \left(\frac{SCD}{SAD}\right)^2 S_c(r_i^c) S_p(r_i^p) TMR(d_i, r_i^c),$$

where the TMR is recorded at the treatment depth field size r_i^c, and r_i^p is the surface field size on the patient. The total dosage to the isocenter (treatment point) of all the central rays is then given by

$$\text{(A3)} \qquad D = \frac{I_0 \mathcal{D}}{H} \left(\frac{SCD}{SAD}\right)^2 \sum_{i=1}^{N} x_i S_c(r_i^c) S_p(r_i^p) TMR(d_i, r_i^c),$$

a relationship which determines the overall MU factor I_0 for N beams. For a multileaf collimator the scatter factor $S_c(r_i^c)$ corresponds to the leaf opening defining the ray (independent of the beam). In the case of compensating blocks for intensity modulation, the collimator scatter factor is estimated from the size of the beam opening.

Acknowledgement. We would like to thank Gregg Tracton for providing the NCI tool software and consultation, and Greg Moulton for assistance in data acquisition. Todd Quinto and Christoph Börgers are thanked for helpful discussions on the mathematics of the x-ray transform.

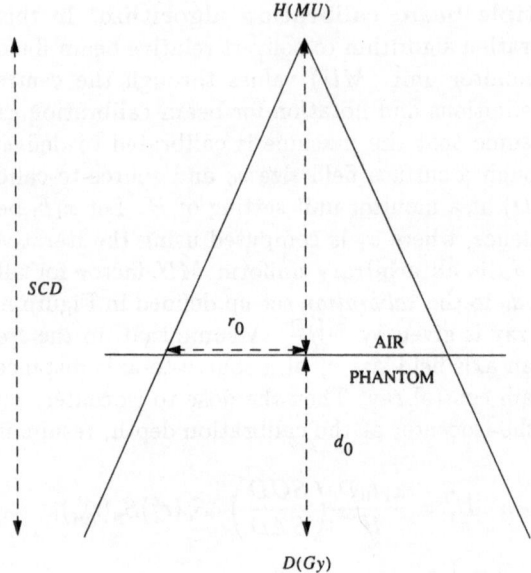

FIG. A1. *Calibration set-up for dose measurements at depth d_0.*

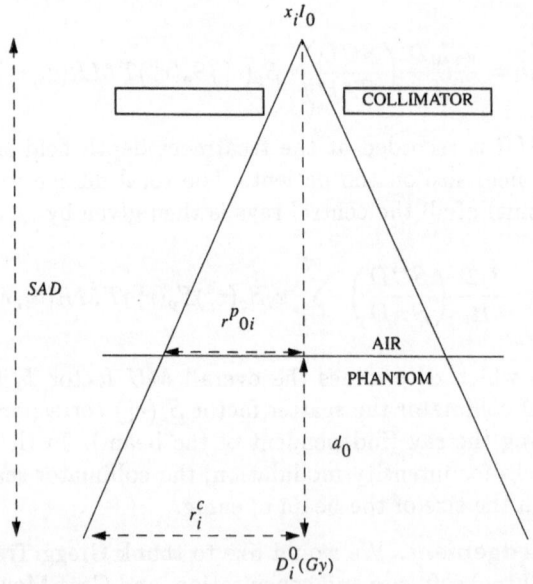

FIG. A2. *Treatment set-up for dose measurements at depth d_0.*

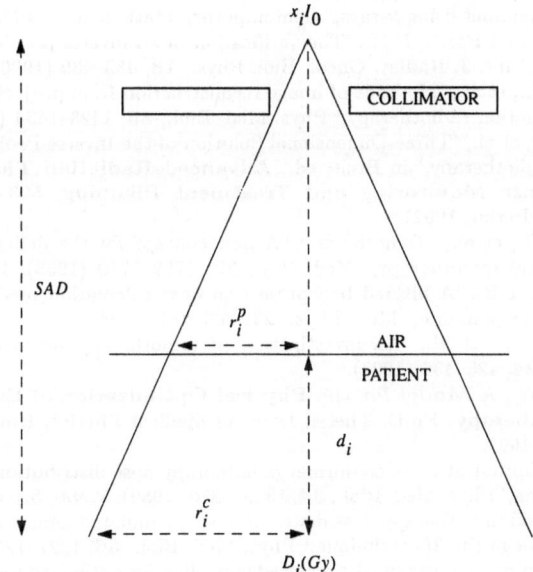

FIG. A3. *Treatment set-up for dose delivery at depth d_i.*

REFERENCES

[1] Natterer, F., **The Mathematics of Computerized Tomography** (J. Wiley and Sons, New York, 1986).

[2] Goitein, M., 'Three-dimensional Density Reconstruction from a Series of Two-dimensional Projections,' Nucl. Instr. Meth., **101**, 509–518 (1972).

[3] Brahme, A., et al., 'Solution of an integral equation encountered in rotation therapy,' Phys. Med. Biol., **27**, 1221–1229 (1982).

[4] Cormack, A.M., 'A Problem in Rotation Therapy with X-rays,' Int. J. Radiat. Oncol. Biol. Phys., **13**, 623–630 (1987). Cormack, A.M. and Cormack, R.A., 'A Problem in Rotation Therapy II: Dose Distributions with an Axis of Symmetry,' Int. J. Radiat. Oncol. Biol. Phys., **13**, 1921–1925 (1987).

[5] Cormack, A.M. and Quinto, E.T., 'On a Problem in Radiotherapy: Questions of Non-negativity,' Int. J. Imaging Systems and Technology, **1**, 120–124 (1989). Cormack, A.M. and Quinto, E.T., 'The Mathematics and Physics of Radiation Dose Planning using x-rays,' Contemporary Mathematics, **113**, 41–55 (1990).

[6] Kooy, H.M. and Barth, N.H., 'The verification of an inverse problem in radiation therapy,' Int. J. Radiat. Oncol. Biol. Phys., **18**, 433–439 (1990).

[7] Bortfeld, Th., et al., 'Methods of image reconstruction from projections applied to conformation radiotherapy,' Phys. Med. Biol., **35**, 1423–1434 (1990).

[8] Bortfeld, T., et al., 'Three-Dimensional Solution of the Inverse Problem in Conformal Radiotherapy,' in Breit, ed., **Advanced Radiation Therapy Tumor Response Monitoring and Treatment Planning**, 503–508 (Springer-Verlag, Berlin, 1992).

[9] Mackie, T.R., et al., 'Tomotherapy: A new concept for the delivery of dynamic conformal radiotherapy,' Med. Phy., **20**, 1709–1719 (1993). Holmes, T. and Mackie, T.R., 'A filtered backprojection dose calculation method for inverse treatment planning,' Med. Phys., **21**, 303–313 (1994). Yang, J.N., et al., 'An investigation of tomotherapy beam delivery,' Med. Phys., **24**, 425–436 (1997).

[10] Holmes, T.W., **A Model for the Physical Optimization of External Beam Radiotherapy**, Ph.D. Thesis, Dept. of Medical Physics, University of Wisconsin, 1993.

[11] Webb, S., 'Optimisation of conformal radiotherapy dose distributions by simulated annealing,' Phys. Med. Biol., **34**, 1349–1370 (1989). Webb, S., 'Optimisation of conformal radiotherapy dose distributions by simulated annealing: 2. Inclusion of scatter in the 2D technique,' Phys. Med. Biol., **36**, 1227–1237 (1991).

[12] Morrill, S., et al., 'Constrained simulated annealing for optimized radiation therapy treatment planning,' Computer Methods and Programs in Biomedicine, **33**, 135–144 (1990).

[13] Webb, S., 'Optimization by simulated annealing of three-dimensional conformal treatment planning for radiation fields defined by a multileaf collimator,' Phys. Med. Biol., **36**, 1201–1226 (1991).

[14] Webb, S., 'Optimization by simulated annealing of three-dimensional, conformal treatment planning for radiation fields defined by a multileaf collimator: II. Inclusion of two-dimensional modulation of the x-ray intensity,' Phys. Med. Biol., **37**, 1689–1704 (1992).

[15] Webb, S., 'Optimized Three-Dimensional Treatment Planning for Volumes with Concave Outlines, Using a Multileaf Collimator,' in Breit, ed., **Advanced Radiation Therapy Tumor Response Monitoring and Treatment Planning**, 495–502 (Springer-Verlag, Berlin, 1992).

[16] Mageras, G.S. and Mohan, R., 'Application of fast simulated annealing to optimization of conformal radiation treatments,' Med. Phys., **20**, 639–647 (1993).

[17] Censor, Y., et al., 'A Computational Solution of the Inverse Problem in Radiation-Therapy Treatment Planning,' Applied Mathematics and Computation, **25**, 57–87 (1988).

[18] Censor, Y., et al., 'On the use of Cimmino's simultaneous projections method for computing a solution of the inverse problem in radiation therapy treatment planning,' Inverse Problems, **4**, 607–623 (1988). Powlis, W.D., et al., 'Semi-automated radiotherapy treatment planning with a mathematical model to satisfy treatment goals,' Int. J. Radiat. Oncol. Biol. Phys., **16**, 271–276 (1989).

[19] Rosen, I.I., et al., 'Treatment plan optimization using linear programming,' Med. Phys., **18**, 141–152 (1991).

[20] Morrill, S., et al., 'Dose-volume considerations with linear programming optimization,' Med. Phys., **18**, 1201–1210 (1991).

[21] Neimierko, A., 'Random search algorithm (RONSC) for optimization of radiation therapy with both physical and biological end points and constraints,' Int. J. Radiation Oncology Biol. Phys., **23**, 89–98 (1992).

[22] Neimierko, A., et al., 'Optimization of 3D radiation therapy with both physical and biological end points and constraints,' Int. J. Radiation Oncology Biol. Phys., **23**, 99–108 (1992).

[23] Raphael, C.S., 'Mathematical modelling of objectives in radiation therapy treatment planning,' Phys. Med. Biol., **37**, 1293–1311 (1992). Raphael, C.S., 'Radiation therapy treatment planning: an L^2 approach,' Appl. Math. Comp., **52**, 251–277 (1992).

[24] Lind, B.K. and Kallman, P., 'Experimental verification of an algorithm for inverse radiation therapy planning,' Radiotherapy and Oncology, **17**, 359–368 (1990).

[25] Boyer, A.L., et al., 'Potential and limitations of invariant kernel conformal therapy,' Med. Phys., **18**, 703–712 (1991).

[26] Boyer, A.L., et al., 'Potential Applications of Invariant Kernel Conformal Therapy,' in Breit, ed., **Advanced Radiation Therapy Tumor Response Monitoring and Treatment Planning**, 471–477 (Springer-Verlag, Berlin, 1992).

[27] Lind, B.K. and Brahme, A., 'Photon field quantities and units for kernel based radiation therapy planning and treatment optimization,' Phys. Med. Biol., **37**, 891–909 (1992).

[28] Webb, S., **The Physics of Three-dimensional Radiation Therapy, Conformal Radiation Therapy, Radiosurgery and Treatment Planning** (IOP Publishing, Philadelphia, 1993).

[29] Boyer, A.L., 'Radiation Therapy Beam Modulation Techniques,' in Purdy, J.A. and Fraass, B.A., eds., **Syllabus: A Categorical Course in Physics: Three-dimensional Radiation Therapy Treatment Planning**, RSNA Publications, 1994.

[30] Convery, D.J. and Rosenbloom, M.E., 'The generation of intensity-modulated fields for conformal radiotherapy by dynamic collimation,' Phys. Med. Biol., **37**, 1359–1374 (1992).

[31] Carol, M., 'An automatic 3D treatment planning and implementation system for optimized conformal therapy by the NOMOS Corporation,' Proc. ASTRO, San Diego, CA., Nov., 1992.

[32] Carol, M.P., et al., '3-D Planning and Delivery System for Optimized Conformal Therapy,' Radiat. Oncol. Biol. Phys., **24**, Suppl. 1, p. 158 (1994).

[33] Bortfeld, T., et al., 'X-ray field compensation with multileaf collimators,' Int. J. Radiat. Oncol. Biol. Phys., **28**, 723–730 (1994).

[34] Spirou, S.V. and Chui, C.S., 'Generation of arbitrary fluence profiles by dynamic jaws or multileaf collimators,' Med. Phys., **21**, 1031–1041 (1994).

[35] Kallman, P., et al., 'Shaping of arbitrary dose distributions by dynamic multileaf collimation,' Phys. Med. Biol., **33**, 1291–1300 (1988).

[36] Zhu, Y., et al., 'Dose distributions of x-ray fields as shaped with multileaf collimators,' Phys. Med. Biol., **37**, 163–173 (1992).

[37] Bortfeld, T., et al., 'Realization and verification of three-dimensional conformal radiotherapy with modulated fields,' Int. J. Radiat. Oncol. Biol. Phys., **30**, 899–908 (1994).

[38] Chui, C.-S., et al., 'Dose calculation for photon beams with intensity modulation generated by dynamic jaw or multileaf collimations,' Med. Phys., **21**, 1237–1243 (1994).

[39] Wang, X., et al., 'Dosimetric verification of intensity-modulated fields,' Med. Phys., **23**, 317–327 (1996).

[40] Chui, C.-S., et al., 'Testing of dynamic multileaf collimation,' Med. Phys., **23**, 635–641 (1996).

[41] Boyer, A.L., et al., 'Modulated Beam Conformal Therapy for Head and Neck Tumors,' Int. J. Radiat. Oncol. Biol. Phys., **39**, 227–236, 1997.

[42] Ling, C.C., et al., 'Conformal radiation treatment of prostate cancer using inversely-planned intensity-modulated photon beams produced with dynamic multileaf collimation,' Int. J. Radiat. Oncol. Biol. Phys., **35**, 721–730 (1996). Reinstein, L.E., et al. 'A feasibility study of automated inverse treatment planning for cancer of the prostate,' Int. J. Radiat. Oncol. Biol. Phys., **40**, 207–214 (1998).

[43] Leibel, S. A., et al., 'Three-dimensional conformal radiation therapy in locally advanced carcinoma of the prostate: preliminary results of a phase I dose-escalation study,' Int. J. Radiat. Oncol. Biol. Phys., **28**, 55–65 (1993).

[44] Zelefsky, M.J., et al., 'The Feasibility of Dose Escalation with Three-Dimensional Conformal Radiotherapy in Patients with Prostatic Carcinoma,' Cancer J. Sci. Am., **1**, 142–150, 1995.

[45] Urie, M.M., et. al., 'Implementation and experimental verification of 3D optimized intensity-modulated radiotherapy,' Med. Phys., **23**, p. 1073 (1996).

[46] Smyczynski, M.S., et al., 'Implementation and experimental verification of individually optimized three-dimensional intensity-modulated radiotherapy,' Proc. ASTRO, Los Angeles, CA., October, 1996.

[47] Sherouse, G.W., 'A mathematical basis for selection of wedge angle and orientation,' Med. Phys., **20**, 1211–1218 (1993).

[48] Stein J., et al., 'Number and Orientation of Beams in Intensity-Modulated Radiation Treatments,' Med. Phys., **24**, 149–160 (1997).

[49] Soderstrom, S. and Brahme, A., 'Optimization of the dose delivery in a few field techniques using radiobiological objective functions,' Med. Phys., **20**, 1201–1210 (1993).

[50] E.A. Gregerson, et al., 'Beam Sampling and Selection for 3D Conformal Radiotherapy,' Med. Phys., **22**, 920–921 (1995).

[51] Niemierko, A., 'Optimization of Intensity Modulated Beams: Local or Global Optimum?,' Med. Phys., **23**, p. 1072 (1996).

[52] Mohan, R., et al., 'The potential and limitations of the inverse radiotherapy technique,' Radiotherapy and Oncology, **32**, 232–248 (1994).

[53] Goitein, M., 'The Inverse Problem,' Int. J. Radiat. Oncol. Biol. Phys., **18**, 489–491 (1990).

[54] Goitein, M. and Schultheiss, T., 'Strategies for treating possible tumor extension: some theoretical considerations,' Int. J. Radiat. Oncol. Biol. Phys., **11**, 1519–1528 (1985).

[55] Kutcher, G. and Burman, C., 'Calculation of complication probability factors for non-uniform normal tissue irradiation: the effective volume method,' Int. J. Radiat. Oncol. Biol. Phys., **16**, 1623–1630 (1989).

[56] Burman, C., et al., 'Fitting of normal tissue tolerance data to an analytic function,' Int. J. Radiat. Oncol. Biol. Phys., **21**, 123–135 (1991).

[57] Neimierko, A. and Goitein, M., 'Calculation of normal tissue complication probability and dose-volume histogram reduction schemes for tissues with a critical element architecture,' Radiotherapy and Oncology, **20**, 166–176 (1991).

[58] Neimierko, A. and Goitein, M., 'Modeling of Normal Tissue Response to Radiation: The Critical Volume Model,' Int. J. Radiat. Oncol. Biol. Phys., **25**, 135–145 (1992).

[59] Kallman, P., et al., 'An algorithm for maximizing the probability of complication-free tumour control in radiation therapy,' Phys. Med. Biol., **37**, 871–890 (1992).

[60] Crowther, R.A., et al., 'The reconstruction of a three-dimensional structure from projections and its applications to electron microscopy,' Proc. Roy. Soc. Lond., **A317**, 319–340 (1970).

[61] Medoff, B.P., 'Image Reconstruction from Limited Data: Theory and Applications in Computerized Tomography,' in Stark, H., ed., **Image Recovery: Theory and Application**, (Academic Press, New York, 1987).

[62] Lingren, A.G. and Rattey, P.A., 'The Inverse Discrete Radon Transform with Applications to Tomographic Imaging Using Projection Data,' Advanced in Electronics and Electron Physics, **56**, 359–410 (1981).

[63] The suggestion that a convex tissue-air interface can be compensated for in the definition of the beam profiles was made to us by E.T. Quinto in March, 1995.

[64] Jenkins, T.M., et al., eds., **Monte Carlo Transport of Electrons and Photons** (Plenum Press, New York, 1988).

[65] Tretiak, O. and Metz, C., 'The exponential Radon transform,' SIAM J. Appl. Math., **39**, 341–354 (1980). Tretiak, O.J., 'Attenuated and exponential Radon transforms,' Proc. Symposia in Applied Mathematics, **27**, 25–33 (1982).

[66] Clack, R., 'Toward a complete description of three-dimensional filtered backprojection,' Phys. Med. Biol., **37**, 645–660 (1992). Defrise, M., et al., 'Three-dimensional image reconstruction from complete projections,' Phys. Med. Biol., **34**, 573–587 (1989).

[67] Deans, S.R., **The Radon Transform and Its Applications**, (J. Wiley and Sons, New York, 1983).

[68] Drzymala, R.E., et al., **NCI Plan Evaluation Tools: Users Manual, Version 3.0**, February, 1994, Mallinckrodt Institute of Radiology, Washington University School of Medicine, St. Louis, MO., NCI Contract NO1-CM-97564.

[69] Drzymala, R., et al., 'Dose-volume histograms,' Int. J. Radiat. Oncol. Biol. Phys., **21**, 71–78 (1991).

[70] Khan, F.M., **The Physics of Radiation Therapy** (Williams and Wilkins, Baltimore, 1994).

OPTIMIZATION METHODS FOR RADIATION THERAPY PLANS*

WELDON A. LODWICK[†], STEVE MCCOURT[‡], FRANCIS NEWMAN[§], AND
STEPHEN HUMPHRIES[¶]

Abstract. We develop and apply a variety of optimization methods to obtain, in three dimensions, optimal beam angles and intensities in radiation treatment of benign and malignant tumors. Thus, flexible and fast algorithms, graphics, and human interaction using computed tomography scans in three dimensional treatment problems are incorporated in the development of the methods and in the computational experiments that are reported. We use MATLAB as a means to demonstrate the viability of our optimization methods to aid oncology clinics in three dimensional treatment plans.

1. Introduction. The radiation therapy problem, to which our optimization methods are applied, is to obtain, for a radiation machine being used to treat tumors, a set of beam angles and beam intensities at these angles so that the delivered dosage to the tumor and other tissue meets the physician-specified dosage criteria. Mathematically, this is the inverse of the computed tomography (CT) problem (see; for example, [7], [8], [14]). In this study, we use the discrete version of the inverse of the computed tomography problem since the methods developed here are simply a way to identify a set of optimal beam angles and beam weights in the sense of optimizing an objective function or a set of objective functions as we will see below. The identification of a set of beam angles and weights that yield the radiation dose distribution as specified by a physician is called a **treatment plan**. In this study, it is assumed that once a treatment plan is obtained via mathematical optimization, it becomes the input to a medically approved software system that computes the actual dosage by standard approved means. Thus, our methods are run first and then the treatment plan is verified. The treatment plans that the discrete transforms find are, as will be seen from the initial experiments we report, correct and fast and therefore useful in its discrete form. We continue to research how to implement a more robust radiation deposition model (see [15], [16]). However, since our objective is to develop a system that is useful in day-to-day identification of treatment plans for a radiation oncology clinic, we will report our findings implementing the discrete transform.

The discretized transform associated with the computed tomography

*This research was sponsored, in part, by Computerized Medical Systems Inc., 1195 Corporate Lake Drive, Saint Louis, Missouri 63132.

†Department of Mathematics, University of Colorado at Denver, E-mail: weldon.lodwick@cudenver.edu

‡University of Colorado Health Sciences Center, E-mail: steve.mccourt@uchsc.edu

§University of Colorado Health Sciences Center, E-mail: francis.newman@uchsc.edu

¶University of Colorado Health Sciences Center, E-mail: stephen.humphries@uchsc.edu

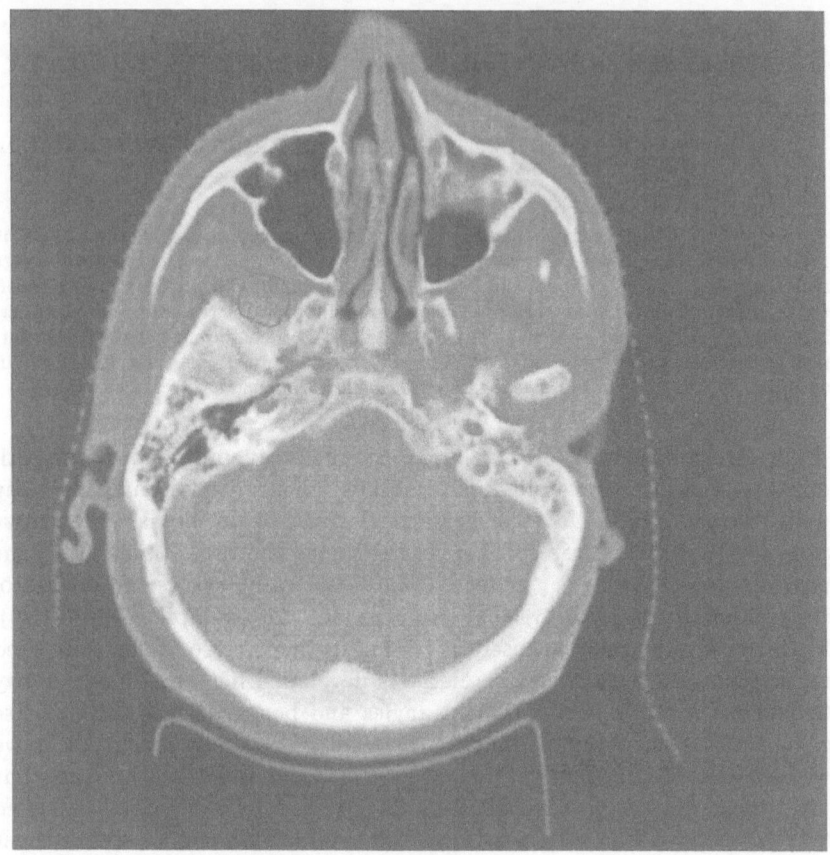

problem first considers a continuous beam as being subdivided into a number of distinct pencils passing through an $N \times N \times M$ grid of the patient region that intercept the grid cells or **voxels**, in a way that can be described
geometrically. The discrete model in this setting that approximates the
continuous Radon transform, is (see [8]),

$$(1.1) \qquad\qquad A\vec{z} = \vec{x},$$

where A is the matrix performing a "ray-sum" (the line integral in the continuous case) through the patient image space, $\vec{z}$ is the vectorized image,
and $\vec{x}$ is the vector of pencil beam weights whose value is the radiation
intensity for a source angle. The vectorized image $\vec{z}$ is the vector of voxels with components z_1 through z_{N^2M} that are stacked to create a N^2M
column vector.

One may explicitly show (see; for example, [7], [8], [23], [24]) that the

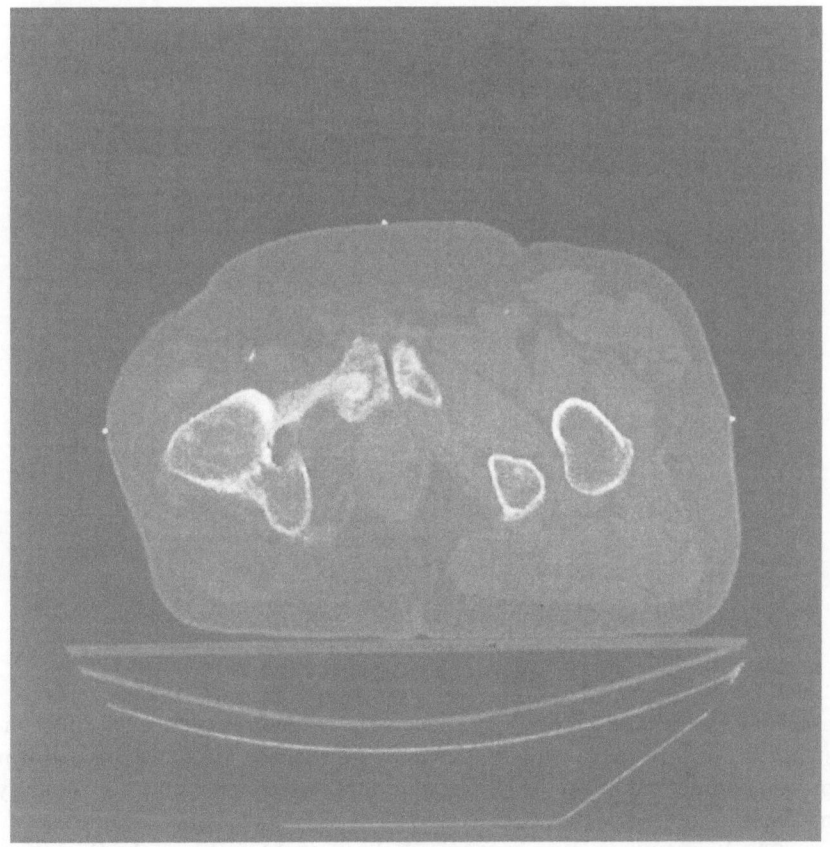

FIG. 2.

discrete radiation therapy problem is modeled as,

$$(1.2) \qquad A^T \vec{x} = \vec{z}$$
$$(1.3) \qquad subject\ to\ \vec{x} \geq \vec{0}$$

where row i of A^T gives the contribution of beam intensity x to the ith voxel. We call this operator, A^T, the **attenuation matrix** for reason that will be apparent. It should be emphasized that in (1.2), $\vec{x} \geq \vec{0}$ since radiation pencil beam weights must be non-negative. The vector $\vec{z}$, for the inverse problem, is the set of radiation values desired at each of the voxels as specified by an oncological physician. Thus (1.2) is the discrete dose deposition operator for the radiation therapy problem. The general mathematical problem is to solve (1.2) for $\vec{x}$ under the condition that $\vec{x} \geq \vec{0}$. This is called the inverse problem to the computed tomography problem (1.1). The optimization problem presented here is to find an optimal $\vec{x} \geq \vec{0}$ subject to a set of inequality constraints associated with the delivered dose at

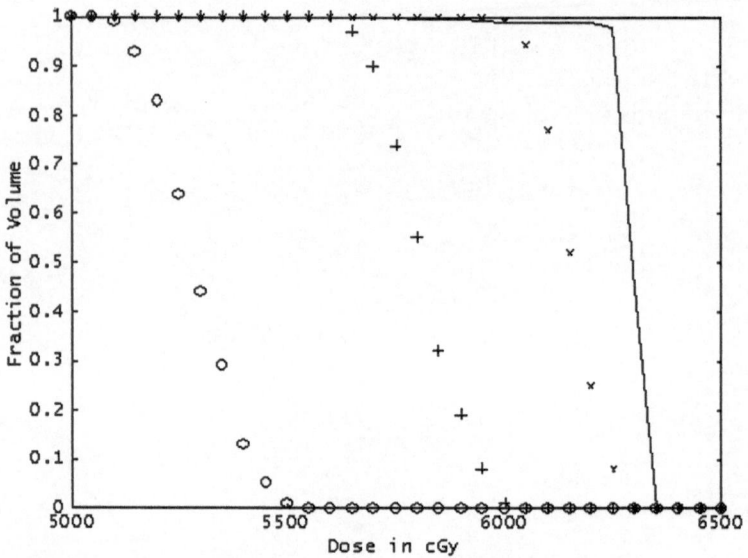

FIG. 3.

each voxel as computed by (1.2). There are several ways to solve this constrained optimization problem and four approaches are developed: linear programming, nonlinear programming, simulated annealing and penalty methods.

The linear programming approach developed here most closely resembles that of Yair Censor and his colleagues [1], [2], [4], [5], [20], [21]. Other approaches that are similar to ours are [3], [13], [18], [19], [25], [26], [27], [30], [31], [33] and [34]. Compared to Censor, Altschuler, and Powlis' approach, our method uses the simplex and other optimization methods directly, which means that our methods impose an objective function to the problem. Censor, Altschuler and Powlis do not use objective functions, they compute one feasible solution. Moreover, we do specify a set of possible gantry positions *a priori* whereas Censor, Altschuler, and Powlis allow the algorithm to select the angle. That is, the position variable is a continuous variable whereas our approach discretizes the beam positions and selects among the pre-selected discrete positions.

Two foci of this study are: (i) the analysis of various optimization methods, some of which are new to the literature, applied to radiation therapy planning, and (ii) the development of a computer system that uses these optimization methods in the day-to-day operating mode of a typical radiation oncology clinic. The methods that are new in their use in radiation therapy problems are the penalty methods as well as the use of linear programming as a pre-processor. Linear programming is used as

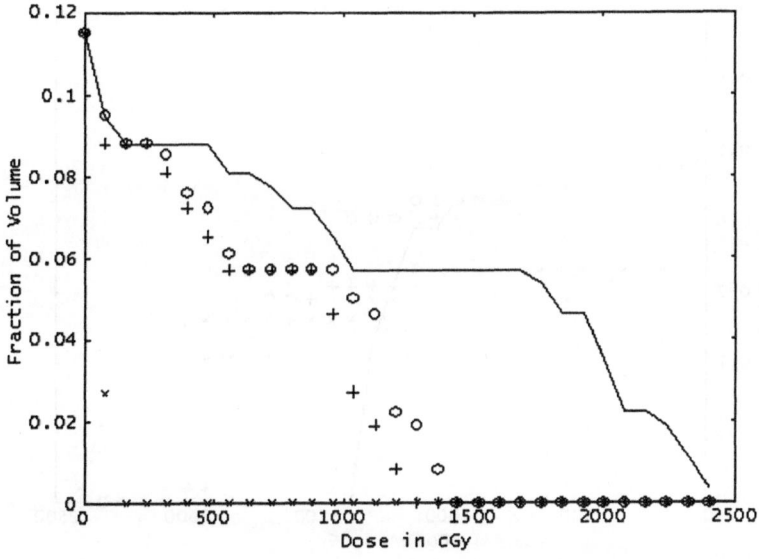

FIG. 4.

a pre-processor to identify a starting set of feasible beam angles. This is done by using one single (broad) beam (open field) per *a priori* chosen angle and allowing the linear program to identify the feasible angles. The feasible angles are those corresponding to positive beam weights. Only the angles with positive beam weights identified by the initial application of the linear programming model are used. The beams corresponding to these angles are then subdivided into pencils and become the variables for subsequent linear programming, nonlinear programming, simulated annealing and penalty function models. Another pre-processor that we use is forward scanning where forward scanning is a way to take as starting vectors only those that intersect the tumor without intersecting critical tissues (spinal cord, liver for example) and prioritizing these vectors *a priori*. This research compares performance as measured by execution times and closeness to physician specified dosage tolerances. The latter is measured using dose-volume histograms (see figures 3-14).

Image-processing capabilities that transform computed tomograph scans (so called CT scans) into constraint equations as well as display-ing solutions back onto the actual CT scans are required. Moreover, it is necessary to have the ability to quickly change constraints, objective functions and methods in addition to have access to high-quality optimiza-tion algorithms. MATLAB was chosen as the system for this project since it has image-processing, generalized user interface (GUI) and program-ming capabilities. The optimization algorithms are tailored to make use

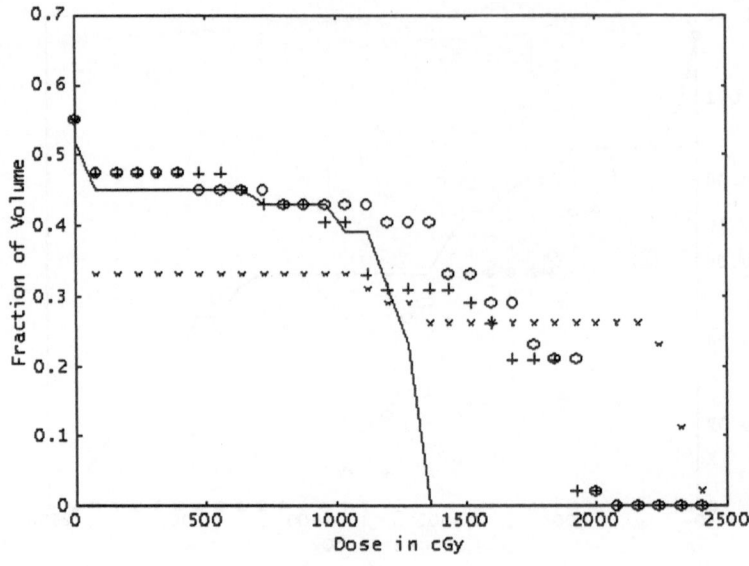

FIG. 5.

of the strengths of MATLAB, its vectorization for example, and to avoid its weakness, its slowness in execution of loops for example. In particular, tailored versions of the simplex, simulated annealing, and penalty function method were used for this study. The MATLAB optimization toolbox routine for constrained nonlinear least squares was used. An ideal approach would be to use an optimization package such as GAMS, C-Plex, or MINOS together with C, C++, or FORTRAN routines in conjunction with an image-processing system that handles CT scans and graphical displays. Experiments were run using GAMS (see [15]). For this study, we report only on the system developed using MATLAB.

This paper is organized in the following way. Section 2 shows how the attenuation matrix (1.2) is constructed. Section 3 delineates the optimization models. Section 4 contains results and discussion of numerical experiments that were run. In the last section are found the conclusions.

2. The attenuation matrix. In this section, the method to compute the attenuation matrix (1.2) is given. The process involves, (i) selecting a grid space for the patient volume, (ii) selecting a set of beam angles along with their associated subdivision into pencils and (iii) integrating (i) and (ii) with patient specific CT scans.

The CT scans are used to obtain a set of equations and inequalities that represent the way radiation from a linear accelerator is deposited into a particular patient's body. An algorithm is required to translate the CT

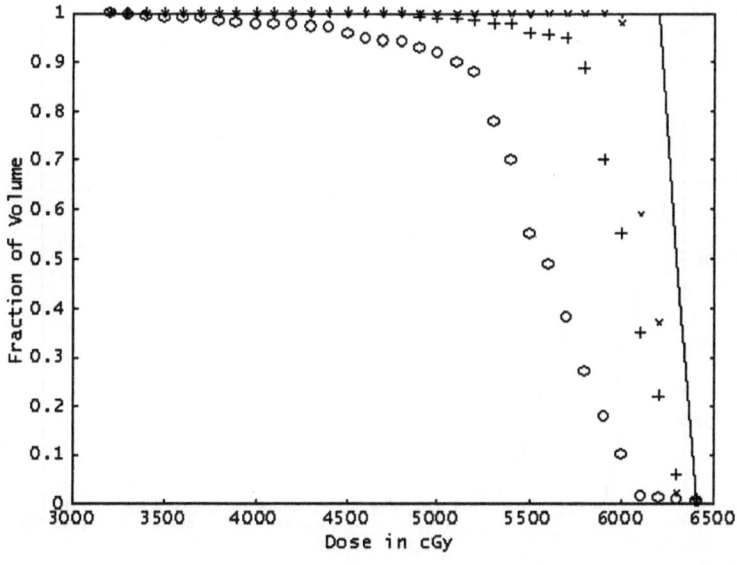

scans' numerical values (denoting distinct tissue types) into a set of constraint inequalities. Whatever format the CT scans are in, the distinct tissues, including the tumor, need to be identified; that is, segmentation of the CT scans is assumed to have been done and is part of the input (see [15], [16]).

A mathematical model to accurately calculate the dosages delivered to each voxel from a particular linear accelerator for a specified number of gantry angle positions is created next. This results in the attenuation matrix A^T. Two assumptions are made in computing the attenuation matrix:
a) A beam is made up of a set of divergent pencils beams,
b) Attenuation at a voxel along a pencil is obtained via a standard radiation attenuation table.

The standard attenuation table is constructed from empirical data measured at a specific photon intensity from a given linear accelerator and is part of a normal linear accelerator installation procedure. In particular, voxel depth and beam width are used to find an attenuation factor. Inhomogeneous attenuation could be incorporated into the models yielding more accurate dose deposition at a cost of additional computational complexity.

The grid system of three dimensional voxels is developed as follows. Let M be the number of patient CT scans that are available to use. The experiments use a grid system whose dimension is $96 \times 96 \times M$ where 96×96, in our case, is the resolution used in the optimization and in the

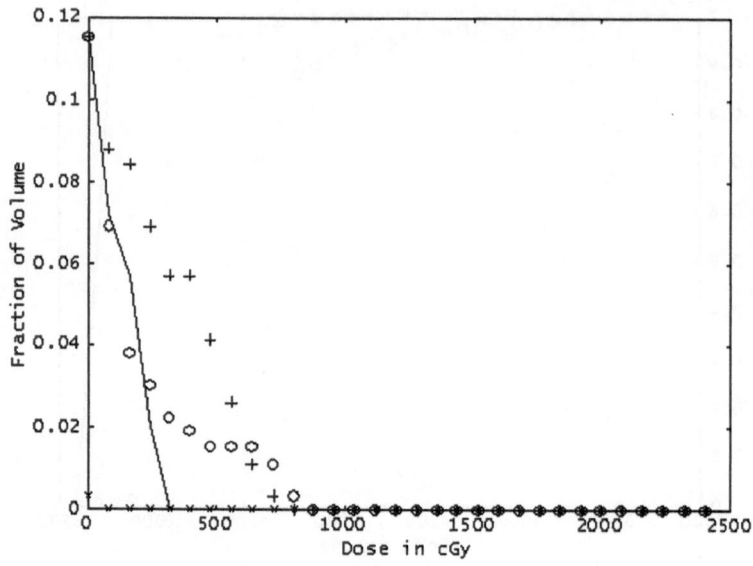

FIG. 7.

display of the results. Typically, CT scans have a 512 × 512 resolution. This is transformed into a 96 × 96 image for use in the optimization models. A radiation beam emanating from a point source (angle) is modeled as a divergent beam composed of pencils where the number of pencils per rectangular beam is equal to the dimensionality of the image which, for our experiments, is ninety-six. Of course, pencils that do not hit the tumor are masked off, but not at the initial stage since the particular patient information is not input at this point. Thus, the actual beam width will vary depending on the width of the tumor as viewed from a particular source, the so called beam's eye view. The equally spaced beam angle positions are chosen at this point and for our experiments, seventy-two (coplanar) angles were chosen, one every five degrees. These can be further minimized by applying a penalty to the objective function (see the penalized linear programming objective function given in section 3.3).

The second phase of the attenuation model takes into account the particular patient CT scans, the location and shape of the boundary of the body, critical tissues, and tumor. These outlines and seventy-two sets of angles each with ninety-six pencil beams are fit onto the 96 × 96 × M patient volume. At this stage, pencils that do not deposit primary dose to the tumor are masked off and the rest are attenuated. Attenuation is calculated by table look-up of empirical beam data. Among other advantages to this approach, the use of empirical data is superior to using straight exponential attenuation as it accounts for the dose build-up region. The

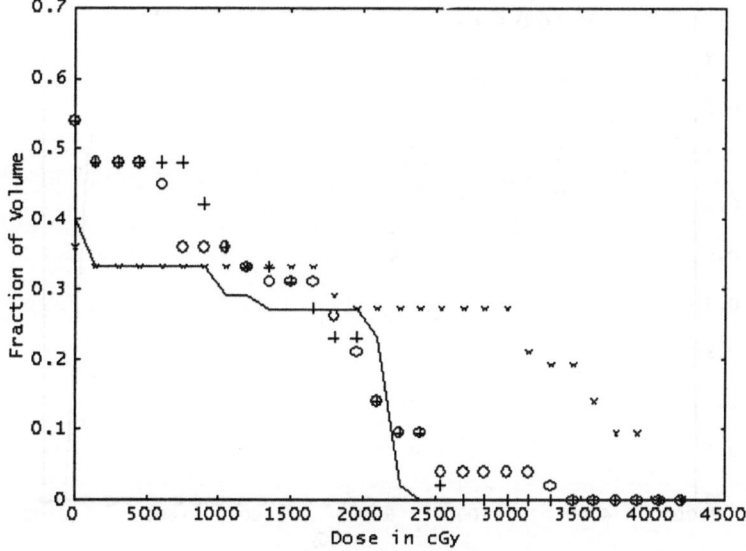

FIG. 8.

distance along a pencil from the body's surface to any given voxel in the image is computed. This distance and the width of the beam are then used to find the attenuation from the table look-up of measured data. A highly structured and sparse matrix is obtained. The structure and sparsity are exploited for computer storage and processing. If the full set of pencil beam intensities is denoted by $\vec{x}$, then

$$A^T \vec{x} \equiv delivered\ dosage\ at\ each\ voxel$$

where A^T is the attenuation matrix.

There are two approaches to three dimensional attenuation models. The first is an extension from a two dimensional model and assumes that the gantry positions (beam angle positions) lie in the same plane as the CT scans. This is called **coplanar** treatment planning. The second version allows beam angles to be placed anywhere without regard to the orientation of the two dimensional CT scans and is called **non-coplanar** treatment planning. The coplanar model is used in this study. The non-coplanar model was not implemented for this study but its theory is developed in [16].

The coplanar model uses each of the two dimensional patient CT scans to create an attenuation matrix. Only those CT scans that contain the tumor are used. Since the beams are assumed to be coplanar to the CT scans, the effect of one beam on each image is in fact a constraint on each CT scan; that is, the total amount of radiation that is deposited by

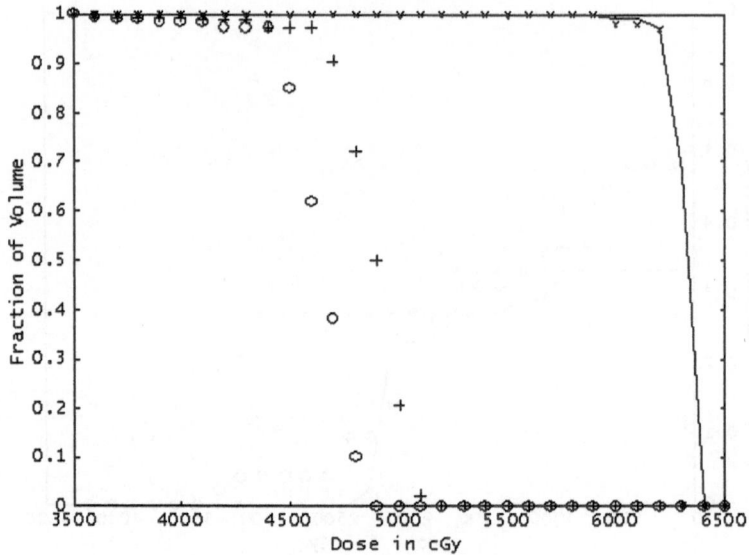

FIG. 9.

a particular pencil beam at each voxel on each CT scan is computed by
a the matrix vector multiplication of corresponding CT scan attenuation
matrix and the pencil beam intensity vector. Thus, each CT scan will
have individual constraints specified by the physician. These are "stacked"
vertically as constraint rows of the set of constraint equations corresponding
to the associated attenuation matrix. Thus, the coplanar model has an
attenuation matrix of the following form:

$$A^T = \begin{bmatrix} A_1^T \\ A_2^T \\ \vdots \\ A_M^T \end{bmatrix},$$

where A_m^T, $m = 1,.., M$ (M is the actual number of CT scans containing
the tumor) is the attenuation matrix corresponding to the mth CT scan.

3. Optimization models. Once the particular patient images have
been processed and converted into the attenuation matrix, a variety of optimization routines may be used. We have experimented with linear programming, constrained nonlinear programming (quadratic programming
and non-negative least squares), simulated annealing, and penalty function methods. The basic mathematical form of the linear and nonlinear
programming problem is:

(3.1) $opt \ z = f(\vec{x})$

subject to:

(3.2) $\vec{\alpha}_t^T A^T \vec{x} \geq \vec{T}_{\min} \equiv$ minimum tumor dosage,

(3.3) $\vec{\alpha}_t^T A^T \vec{x} \leq \vec{T}_{\max} \equiv$ maximum tumor dosage,

(3.4) $\vec{\alpha}_{c_k}^T A^T \vec{x} \leq \vec{d}_k \equiv$ critical structure $k = 1, ..., K$,

(3.5) $$\vec{x} \geq \vec{0}.$$

In the above, the subscript t denotes the index of tumor voxels, and the subscript c_k denotes the index of voxels associated with the kth critical structure (for example, optic nerve, spinal cord, bone, liver). Moreover, A^T is the attenuation matrix and $\vec{\alpha}^T$ selects a particular set of voxels. Its definition and properties are discussed below. The appropriate objective function (3.1) is chosen as linear if the model is a linear programming model and nonlinear if the model is a nonlinear programming model. The constraints associated with the discrete dose deposition operator (1.2) are the right-hand sides of (3.2) - (3.5) where $\vec{T}_{\min}$, $\vec{T}_{\max}$, and $\vec{d}_k$ are the physician specified radiation tolerances for the corresponding voxels.

The form of the quadratic programming, non-negative least squares, simulated annealing, and penalty function model is (3.1) and (3.5) where the objective function (3.1) is; for example, a Euclidean norm of a weighted discrepancy between the dosage specified by the physician for the tumor and critical organs, the right-hand sides of (3.2)-(3.4), and the delivered radiation, the left-hand sides of (3.2)-(3.4). The particular forms of these objective functions are given below.

3.1. Constraint sets. There are two types of constraints that are constructed using the $\vec{\alpha}$ of (3.2), (3.3) and (3.4). The dimension of $\vec{\alpha}$ is equal to the number of voxels contained in the patient volume and is composed of zeroes and ones. If a particular component of $\vec{\alpha}$ is one, then the corresponding voxel is "pulled." If more than one component of $\vec{\alpha}$ is one, then a group of voxels is pulled and added. All of the models presented here have voxel by voxel constraints for the tumor so that $\vec{\alpha}_t^T = (0, 0, ..., 0, 1, 0, ..., 0)$ where the one is the location of a particular tumor voxel, t, and t indexes all tumor voxels. For critical structures, the same type of vector, $\vec{\alpha}$, could and was used for some experiments. Constraints so constructed are called **voxel by voxel constraints**. In addition, all or a subset of voxels, corresponding to a tissue type, could be grouped into one or more **super-voxel(s)**. A constraint on such a super-voxel has the effect of limiting the average radiation dose to the corresponding region instead of dose at a single voxel. Super-voxel constraints reduce problem complexity and are called **integral constraints**.

Body voxels (those that are not tumors or critical structures) are generally not explicitly used as constraints in the model since excellent results were obtained with these constraints. Occasionally, however, so called hot-spots can arise outside of specified critical regions. These are dealt with by creating a body super-voxel constraint or folding it into the objective function via penalties.

3.2. Variables. The unknown is the vector $\vec{x}$ whose values, component-wise, correspond to the pencil beam intensity at a specified beam (gantry) angle. For example, x_{19} is the intensity of the 19th pencil, which, if there were 16 pencils per beam, would be the 3rd pencil of the 2nd beam. In addition to pencil beam intensities at specified angles, wedges to preferentially attenuate and shape beams were used. Wedges are like triangular metal doorstops that come in a variety of angles. Nine different triangular wedge angles were used in the numerical experiments, $0°, \pm 15°, \pm 30°, \pm 45°, \pm 60°$ where $0°$ means that no wedge is used. One can model wedges within the optimization model or as a post-optimization analysis.

Wedges modeled within the optimization, which is the way the numerical experiments were performed, compute an attenuation matrix, $A_0^T,...,$ A_8^T, for each of the wedge positions. The composite attenuation matrix is formed by concatenating column-wise (horizontally) each of the nine wedge attenuation matrices as follows:

$$A^T = \left[A_0^T | A_1^T | ... | A_8^T \right].$$

Post-optimization uses the solution obtained by an optimization run for each pencil within one beam and computes a least squares fit with a wedged attenuation of a unit intensity of a pencil going through a corresponding position on the wedge. That is, if the ith optimal solution vector component is x_i, and $W_{i,j}$ is the effect that triangular wedge j has on a unit intensity of pencil i, then the non-negative least squares minimization becomes:

$$\min z = \|W\vec{y} - \vec{x}\|_2^2$$
$$\text{subject to : } \vec{y} \geq \vec{0}$$

Thus, $\vec{y}$ is the linear combination of wedges required to most closely, for the $2 - norm$, reproduce the profile $\vec{x}$ where the values of the $\vec{y}$ are the intensities for a particular beam and a particular wedge. A report on post-optimization computations for wedges, using both triangular and B-spline wedges, is found in [16].

In addition, two different photon energies were used, 6 MV photons and 18 MV photons; future implementations will add electron beams. To model multiple energy beams, each of the beam types (6 MV photons, 18 MV photons) has its own corresponding attenuation matrix, A_{6MV}^T and A_{18MV}^T, so that the overall attenuation matrix is obtained by concatenating horizontally each of the individual attenuation matrices in the same way wedges are handled within an optimization model. Thus,

$$A^T = \left[A_{6MV}^T \mid A_{18MV}^T \right].$$

Models that use a single beam per gantry angle; that is, beams that are not subdivided into pencils (a broad beam), are called **open field** models. Models that subdivide a beam into more than one pencil and each intensity within a beam is allowed to be a different non-negative value are called **intensity modulation** models. In practice, intensity modulation is achieved with compensating filters or dynamic multi-leaved collimation. A wedge is not used on an intensity modulated beam. Both open field and intensity modulation models are used in the numerical experiments.

3.3. Objective functions and problem statement. There are three types of objective functions used in this study: linear, quadratic or least squares, and penalty. Their explicit forms and associated constraints are the following.

1. Linear Programming

a) Maximization of minimum tumor dose:

$$\max z$$

$$\text{subject to :}$$
$$z - \vec{\alpha}_t^T A^T \vec{x} \leq \vec{0}$$
$$\vec{\alpha}_{c_k}^T A^T \vec{x} \leq \vec{d}_k, \; k = 1, ... K$$
$$\vec{x} \geq \vec{0}$$

b) Penalized Linear Programming

$$\max \; z - \vec{\lambda}_{penalty}^T \vec{x}$$

$$\text{subject to :}$$
$$z - \vec{\alpha}_t^T A^T \vec{x} \leq \vec{0}$$
$$\vec{\alpha}_{c_k}^T A^T \vec{x} \leq \vec{d}_k, \; k = 1, ... K$$
$$\vec{x} \geq \vec{0}$$

where $\vec{\lambda}_{penalty}$ imposes penalties for the non-preferred beam elements (pencils).

2. Quadratic Programming

a) Constrained Least Squares

$$\min \; z = \| \vec{\alpha}_t^T A^T \vec{x} - \vec{T} \|_2^2$$
$$\text{subject to :}$$
$$\vec{\alpha}_{c_k}^T A^T \vec{x} \leq \vec{d}_k, \; k = 1, ... K$$
$$\vec{x} \geq \vec{0}$$

b) Penalized Constrained Least Squares

$$\min \; z = \|\vec{\alpha}_t^T A^T \vec{x} - \vec{T}\|_2^2 + \lambda_k^T \frac{\vec{\alpha}_{c_k}^T A^T \vec{x}}{\vec{d}_k}, \; k = 1, \ldots K$$

subject to :
$$\vec{x} \geq \vec{0}$$

where the vector division is understood to be component-wise; that is, by dividing by the right-hand side, each constraint right-hand side value is normalized to one. This component-wise division makes the right-hand sides of the constraint equal to one so that the components of the penalties reflect the "cost" of per unit violations.

3. Simulated Annealing

a) Simple Simulated Annealing

$$\min \; z = \|\vec{a} + \vec{b}\|_2^2$$

subject to : $\vec{x} \geq \vec{0}$

where (using component-wise computations),

$$\vec{a} = \begin{cases} 0 \text{ if } \vec{T}_{\min} \leq \vec{\alpha}_t^T A^T \vec{x} \leq \vec{T}_{\max} \\ |\vec{T}_{\min} - \vec{\alpha}_t^T A^T \vec{x}| \text{ if } \vec{\alpha}_t^T A^T \vec{x} < \vec{T}_{\min}, \text{ or } |\vec{T}_{\max} - \vec{\alpha}_t^T A^T \vec{x}| \text{ if} \\ \vec{\alpha}_t^T A^T \vec{x} > \vec{T}_{\max} \end{cases}$$

$$\vec{b} = \begin{cases} 0 \text{ if } \vec{\alpha}_{c_k}^T A^T \vec{x} \leq \vec{d}_k \\ \vec{\alpha}_{c_k}^T A^T \vec{x} - \vec{d}_k \text{ if } \vec{\alpha}_{c_k}^T A^T \vec{x} > \vec{d}_k \end{cases}$$

for the critical structures $k = 1, \ldots, K$.

b) Penalized Simulated Annealing

$$\min \; z = \|\vec{a} + \vec{b} + \vec{\lambda}_{penalty}^T \vec{x}\|_2^2$$

subject to : $\vec{x} \geq \vec{0}$

where (using component-wise computations),

$$\vec{a} = \begin{cases} 0 \text{ if } \vec{T}_{\min} \leq \vec{\alpha}_t^T A^T \vec{x} \leq \vec{T}_{\max} \\ \lambda_t |\vec{T}_{\min} - \vec{\alpha}_t^T A^T \vec{x}| \text{ if} \vec{\alpha}_t^T A^T \vec{x} < \vec{T}_{\min}, \text{ or } \lambda_t |\vec{T}_{\max} - \vec{\alpha}_t^T A^T \vec{x}| \text{ if} \\ \vec{\alpha}_t^T A^T \vec{x} > \vec{T}_{\max} \end{cases}$$

$$\vec{b} = \begin{cases} 0 \text{ if } \vec{\alpha}_{c_k}^T A^T \vec{x} \leq \vec{d}_k \\ \lambda_{c_k}(\vec{\alpha}_{c_k}^T A^T \vec{x} - \vec{d}_k) \text{ if } \vec{\alpha}_{c_k}^T A^T \vec{x} > \vec{d}_k \end{cases}$$

TABLE 1
Problem Complexity.

HEAD tumor	Case 1: LP	Case 2: LP	Case 3: SA	Case 4: QP
Beam subdivisions per angle	1	1	96	96
Beam angles	72	72	72	72
Wedges	1	9	1	1
Beam energies	1	2	1	1
Maximum columns (variables)	72	1,296	6,912	6,912
Maximum rows (constraints)	700	700	2,583*	700
Maximum rows×columns	50,400	907,200	17,853,696	4,838,400
Actual rows×columns	50,400	75,600	139,482	37,800

and λ_t, λ_{c_k} are penalty terms that force the minimization in preferred directions for the tumor or critical structures $k = 1, ..., K$. If λ_t is larger than any of the λ_{c_k}, then the minimization delivers target dose at the expense of critical dose. The penalty term $\vec{\lambda}_{penalty}$ has the effect of reducing the number of angles and moving the minimization toward preferred angles.

4. Numerical experiments. Two tumors were used in the numerical experiments: a head tumor (see Figure 1), and a prostate tumor (see Figure 2). The head tumor had fifteen CT scans and the prostate tumor had nine CT scans and both were translated into a 96 × 96 resolution. Seventy-two evenly spaced (every five degrees) source beams, nine wedges and two energy types were used. Four optimization model experiments are reported and the way the optimization models are applied to obtain optimal beam weights is explained in detail in the Appendix. Table 1 summarizes problem complexity for the head tumor for the four optimization types used. The prostate problem complexity is similar. Case 1 is a linear programming (LP) model with open field. Case 2 is a LP with wedges and two energy types. Case 3 is a simulated annealing (SA) model with intensity modulation and case 4 is a quadratic programming (QP) model with intensity modulation. For cases 1, 2 and 4, integral constraints (supervoxels) are used for the critical structures whereas for case 3, voxel by voxel constraints are used.

Problem complexity is measured in this study by the number of constraints (rows) multiplied by the number of variables (columns) since the discrete inverse problem leads to linear constraints. There is, of course, algorithmic complexity associated with the optimization algorithm used. For an indication of algorithmic complexity, run times are used (see Table 2). Since the problems are pre-processed (by forward scanning and running a linear programming model to identify a feasible set of broad beam angles), the run time measure is simply a measure of whether or not our methods are practical from the point of view of being implementable in day-to-day operations of a radiation oncology clinic and should not be construed as an absolute indicator of superiority of one algorithm over another. There are, however, noticeable differences in the performances of the algorithms; a summary is found in Table 2.

TABLE 2
Algorithm times - head and prostate tumors.

	Case 1: LP	Case 2: SA	Case 3: QP	Case 4: NNLS
Head - Open field	6.34	0.39	3.14	3.00
Head - Modulated	1.76	13.59	7.16	58.93
Head - Wedges	4.57	N/A	N/A	N/A
Prostate - Open field	2.52	3.00	3.40	5.40
Prostate - Modulated	1.00	9.14	6.20	23.3
Prostate - Wedges	1.06	N/A	N/A	N/A

The total number of voxels in fifteen CT scans whose resolution is 96×96 is $138,240 = 96^2 \times 15$ voxels. However, of these, the actual number of voxels that are "seen" (intersected) by the 6,912 intensity modulated pencils is 2,583 in this case. Without masking, the actual maximum total row×column is $955,514,880 = 96^2 \times 15 \times 6,912$.

The reason for the reduction in the number associated with actual problem complexity even after masking pencils that are not in the path of the tumor, is that the problem is pre-processed by running a linear programming optimization. The pre-processing linear programming optimization uses seventy-two open field angles to identify the feasible angles. The feasible beam angles (those whose corresponding intensities were positive) were then the only beam angles that were used in all other subsequent optimization runs. Intensity modulation models subdivided each beam into 96 pencils per beam angle and the resulting pencils were masked again with respect to whether or not they intersected the tumor.

The timing results, reported below in Table 2, are from implementations on a Pentium Pro PC using MATLAB. The dose-volume histograms of the results associated with the times given in Table 2 are found in Figures 3-14. As stated before, the model that is being used is a the three-dimensional coplanar model using actual patient CT scans at a 96×96 resolution on prototype code using an interpretive language, MATLAB. The units of time are seconds as returned by MATLAB.

5. Conclusions. It has been demonstrated that optimization models are well suited to solve the three dimensional radiation therapy problem on actual patient tumors. The run times are certainly encouraging and point to the viability of these approaches. Moreover, the solutions generated are applicable to use in a typical radiation oncology clinic. In particular, the best method, as measured in terms of the dose-volume histograms (Figures 3-14) and run times, from our experience so far, is open field linear programming as a pre-processor to reduce the number of variables (pencil beam intensities) followed by simulated annealing or possibly quadratic programming. The advantage of simulated annealing is that an assortment of constraints or objective functions can be easily constructed and used. For example, preferentially selecting various visibly promising beam angles or constraining the number of angles can be easily created. The advantage of linear programming is that it can find a solution with few beams. The combination of these two is powerful.

Simulated annealing did not fare as well as linear programming when it was given the full set of possible source angles with seventy-two pencils per angle and the full set of constraints (voxel by voxel constraints). This is because simulated annealing has many different parameters to adjust (CT scan resolution, cooling schedule, weights on penalties), which in combination with a large number of variables, will cause the run times to be excessive and the resulting quality of the solutions to be poor.

Our research continues to explore various optimization models such as goal programming, fuzzy programming, and tabu search (see [16]) as well as the hybrid of linear followed by nonlinear programming such as a fast QP with simple bound constraints.

Appendix: Head tumor and prostate tumor.

The input image series for the head tumor was made up of seventeen transverse CT slices through the central area of the head. The target volume was long and narrow, located on fifteen of the slices to the right orbit and right optic nerve. A characteristic slice of this series, with manually segmented regions, is given in Figure 1.

An initial variable set of seventy-two open-field beams (6 MV photons) was the input to the linear programming routine as follows:
minimize right orbit dose

    subject to: 6,000 cGy < each tumor voxel < 6,300 cGy

    average spinal cord dose < 500 cGy

    average right optic nerve dose < 1,000 cGy

    all other regional (critical structure) doses < 2,000 cGy

    intensity ≥ 0

where the "other" regions are the right lens, maxilla, mandible, and the sinuses. Of the initial variables, the linear programming system returned six angles that satisfied the constraints with an optimal objective value (dose) of 6,000 cGy. These six angles (115°, 130°, 140°, 180°, 190°, 320°) were then used as the only variables in subsequent optimizations since these are the feasible angles identified by the pre-processing stage for open field and intensity modulation models.

The same problem was then solved again using the various optimization routines that are described above. All of the above constraints were used with the addition of an average dose of less than 10 cGy for the right orbit as this was attained in the solution of the linear programming minimization. A comparison of the final radiation intensities, measured in units of cGy, for each of the six angles is given in Table 3 below. The dose-volume histograms for the head tumor using open -field models are given in Figures 3-5. The next models subdivided each angle into ninety-six pencils and allowed each pencil to have its own intensity. All routines, including

TABLE 3
Final open-field weights for head tumor.

	115°	130°	140°	180°	190°	320°
LP	76	69	1,171	3,252	49	1,383
SA	1,117	245	1,150	1,154	1,901	433
NNLS	1,402	210	832	801	2,755	0
CONLS	1,740	0	0	0	1,656	2,604

linear programming, were restricted to the same six angles. The final average pencil weights for each angle are given in Table 4. The corresponding dose-volume histograms are found in Figures 6-8.

TABLE 4
Final average modulated weights for head tumor.

	115°	130°	140°	180°	190°	320°
LP	1,172	408	573	3,637	40	160
SA	1,514	349	592	1,415	1,840	290
NNLS	1,352	518	436	1,430	2,193	71
CONLS	129	2,103	338	2,077	39	1,314

The input image series for the prostate tumor was made up of nine transverse CT slices through the pelvis. The same method as described in the head case was used. The initial set of seventy-two open-field beams (18 MV photons) were reduced by the linear programming model with the following constraints:
minimize bladder dose
    subject to: 6,000 cGy < each prostate voxel < 6,300 cGy
    average rectum dose < 1,000 cGy
    average right/left femur dose < 1,000 cGy
    intensity ≥ 0
The optimal objective value of 6,000 cGy was achieved with 4 beams (at 25°, 205°, 220°, 230°). These four angles, identified in the pre-processing phase, were then used as the only four angles allowed for the subsequent optimization routines. The final solution weights for the open-beam models are given in Table 5 and the corresponding dose-volume histograms are given in Figures 9-11. The average pencil weights for each angle for the intensity modulation models are given in Table 6. Figures 12-14 show the dose-volume histograms for the intensity modulation.

TABLE 5
Final open-field weights for prostate tumor.

	25°	205°	220°	230°
LP	3,222	1,237	500	1,040
SA	0	1,536	3,379	1,235
NNLS	0	1,713	3,258	1,179
CONLS	3,321	1,896	0	933

TABLE 6
Final average modulated weights for prostate tumor.

	25°	205°	220°	230°
LP	1,327	4,432	218	22
SA	301	1,809	2,209	1,681
NNLS	464	1,491	2,436	1,608
CONLS	3,458	2,529	12	0

REFERENCES

[1] M. ALTSCHULER, W. POWLIS, AND Y. CENSOR, *Teletherapy treatment planning with physician requirements included in calculations: I. Concepts and methodology,* in Optimization of Cancer Radiotherapy (edited by R.R. Paliwal, D.E. Herbert and C.G. Orton), American Institute of Physics, New York, 1985, pp. 443-452.

[2] M. ALTSCHULER, Y. CENSOR, W. POWLIS, *Feasibility and optimization methods in teletherapy planning,* in Advances in Radiation Oncology Physics: Dosimetry, Treatment Planning and Brachytherapy (edited by J. Purdy), 1992, 1022-1057.

[3] G. BAHR, J. KEREIAKES, H. HORWITZ, R. FINNEY, J. GALVIN, AND K. GOODE, *The method of linear programming applied to radiation treatment planning,* Radiology, 91 (1968), pp. 686-693.

[4] Y. CENSOR, M. ALTSCHULER, AND W. POWLIS, *On the use of Cimmino's simultaneous projections method for computing a solution of the inverse problem in radiation therapy treatment planning,* Inverse Problems, 4 (1988), pp.607-623.

[5] Y. CENSOR, M. ALTSCHULER, AND W. POWLIS, *A computational solution of the inverse problem in radiation-therapy treatment planning,* Applied Mathematics and Computation, 25 (1988), pp. 57-87.

[6] R. COOPER, *A gradient method of optimizing external-beam radiotherapy treatment plans,* Radiology, 128 (July 1978), pp. 235-243.

[7] A. M. CORMACK, *Some early radiotherapy optimization work,* International Journal of Imaging Systems and Technology, 6 (1995), pp. 2-5.

[8] A. M. CORMACK AND E. T. QUINTO, *The mathematics and physics of radiation dose planning using x-rays,* Comtemporary Mathematics, 113 (1990), pp. 41-55.

[9] G. EZZELL, *Genetic and geometric optimization of three-dimensional radiation therapy treatment planning,* Medical Physics, 23:3 (March 1996), pp. 293-305.

[10] C. HOPE, J. LAURIE, AND K. HALNAN, *Optimization of X-ray treatment planning by computer judgement,* Physics of Medical Biology, 12:4 (1967), pp. 531-542.

[11] M. LANGER AND J. LEONG, *Optimization of beam weights under dose-volume restrictions,* International Journal of Radiation Oncology Biology Physics, 13:8 (August 1987), pp. 1255-1260.

[12] M. LANGER, R. BROWN, M. URIE, J. LEONG, M. STRACHER, AND J. SHAPIRO,

Large scale optimization of beam weights under dose-volume restrictions, International Journal of Radiation Oncology Biology Physics, 18:4 (April 1990), pp. 887-893

[13] J. LEGRAS, B. LEGRAS, AND J. LAMBERT, *Software for linear and nonlinear optimization in external radiotherapy*, Computer Programming in Biomedicine, 15 (1982), pp. 233-242.

[14] B. LIND AND A. BRAHME, *Development of treatment techniques for radiotherapy optimization*, International Journal of Imaging Systems and Technology, 6 (1995), pp. 33-42.

[15] W. LODWICK (DIRECTOR), *Final Report of the Mathematics Clinic: Algorithms for Radiation Therapy Treatment Planning*, University of Colorado at Denver, Department of Mathematics, MC96F001, Fall 1996.

[16] W. LODWICK (DIRECTOR), *Final Report of the Mathematics Clinic: Algorithms for Radiation Therapy Treatment Planning*, University of Colorado at Denver, Department of Mathematics, MC97S001, Spring 1997.

[17] S. MCDONALD AND P. RUBIN, *Optimization of external beam radiation therapy*, International Journal of Radiation Oncology Biology Physics, 2:3-4 (March-April 1977), pp. 307-317.

[18] S. MORRILL, I. ROSEN, R. LANE, AND J. BELLI, *The influence of dose constraint point placement on optimized radiation therapy treatment planning*, International Journal of Radiation Oncology Biology Physics, 19 (1990), pp. 129-141.

[19] S. MORRILL, R. LANE, G. JACOBSON, AND I. ROSEN, *Treatment planning optimization using constrained simulated annealing*, Physics and Medical Biology, 36:10 (1991), pp. 1341-1361.

[20] W. POWLIS, M. ALTSCHULER, AND Y. CENSOR, *Teletherapy treatment planning with physician requirements included in the calculations: II. Clinical applications*, in Optimization of Cancer Radiotherapy (edited by R.R. Paliwal, D.E. Herbert and C.G. Orton), American Institute of Physics, New York, 1985, pp. 453-461.

[21] W. POWLIS, M. ALTSCHULER, Y. CENSOR, AND E. BUHLE, *Semi-automatic radiotherapy treatment planning with a mathematical model to satisfy treatment goals*, International Journal of Radiation Oncology Biology Physics, 16:1 (January 1989), pp. 271-276.

[22] C. RAPHAEL, *Mathematics in radiation therapy treatment planning*, Ph.D. Thesis, Department of Applied Mathematics at Brown University, May 1991.

[23] C. RAPHAEL, *Mathematical modelling of objectives in radiation therapy treatment planning*, Physics and Medical Biology, 37:6 (1992), pp. 1293-1311.

[24] C. RAPHAEL, *Radiation therapy treatment planning: An L^2 approach*, Applied Mathematics and Computation, 52 (1992), pp. 251-277.

[25] A. REDPATH, B. VICKERY, AND D. WRIGHT, *A set of Fortran subroutines for optimizing radiotherapy plans*, Computer Programming in Biomedicine, 5 (1975), pp. 158-164.

[26] A. REDPATH, B. VICKERY, AND D. WRIGHT, *A new technique for radiotherapy planning using quadratic programming*, Physics and Medical Biology, 21:5 (1976), pp. 781-791.

[27] I. ROSEN, R. LANE, S. MORRILL, AND J. BELLI, *Treatment plan optimization using linear programming*, Medical Physics, 18:2 (March-April 1991), pp. 141-152.

[28] W. SANDHAM, Y. YUAN, AND T. DURRANI, *Conformal therapy using maximum entropy optimization*, International Journal of Imaging Systems and Technology, 16 (1995), pp. 80-90.

[29] S. SHALEV, D. VIGGARS, M. CAREY, AND P. HAHN, *The objective evaluation of alternative treatment plans: II. Score functions*, International Journal of Radiation Oncology Biology Physics, 19 (November1990), pp. 120-126.

[30] D. SONDERMAN, AND P. ABRAHAMSON, *Radiation treatment design using mathematical programming models*, Operations Research, 33:4 (July-August 1985), pp. 705-725.

[31] G. STARKSCHALL, *A constrained least-squares optimization method for external beam radiation therapy treatment planning*, Medical Physics, 11:5 (1984), pp. 659-665.

[32] V. TULOVSKY, M. RINGOR, AND L. PAPIEZ, *Optimization of rotational radiotherapy treatment planning*, International Journal of Radiation Oncology Biology Physics, 32:4 (1995), pp. 1205-1214.

[33] S. WEBB, *Optimisation of conformal radiotherapy dose distributions by simulated annealing*, Physics of Medical Biology, 34:10 (1989), pp. 1349-1370.

[34] S. WEBB, *Optimizing radiation therapy inverse treatment planning using the simulated annealing technique*, International Journal of Imaging Systems and Technology, 16 (1995), pp. 71-79.

[35] A. WOLBARST, L. CHIN, AND G. SVENSSON, *Optimization of radiation therapy: Integral-response of a model biological system*, International Journal of Radiation Oncology Biology Physics, 8:10 (1982), pp. 1761-1769.

[36] A. WOLBARST, *Optimization of radiation therapy II: The critical-voxel model*, International Journal of Radiation Oncology Biology Physics, 10:5 (May 1984), pp. 741-745.

[37] L. XING, AND G. CHEN, *Iterative methods for inverse treatment planning*, Physics of Medical Biology, 41 (1996), pp. 2107-2123.

FULLY THREE-DIMENSIONAL RECONSTRUCTION IN ELECTRON MICROSCOPY

ROBERTO MARABINI*, GABOR T. HERMAN†, AND JOSÉ M. CARAZO*

Abstract. Algebraic reconstruction techniques (ART) are iterative procedures for solving systems of linear equations. They have been used in tomography to recover objects from their projections. In this work we apply an ART approach in which the basis functions used to describe the objects are not based on voxels, but are much smoother functions named "blobs." The data collection studied in this work follows the so-called "conical tilt geometry" that is commonly used in many applications of three-dimensional electron microscopy of biological macromolecules. The performance of ART with blobs is carefully compared with a currently well known 3D reconstruction algorithm (weighted backprojection or WBP) using a methodology which assigns a level of statistical significance to a claim of relative superiority of one algorithm over another for a particular task. The conclusion we reach is that ART with blobs produces high quality reconstructions and is, in particular, superior to WBP in recovering features along the "vertical" direction. For the exact implementation recommended in this paper, the computational costs of ART are almost an order of magnitude smaller than those of WBP, though ART has to solve an order of three million noisy equations in an order of 50,000 unknowns

Key words. Three-dimensional reconstruction, image processing.

1. Introduction. Traditionally, three-dimensional image reconstruction from projections in radiology (and in many other fields of application) has been done slice by slice as a sequence of two-dimensional reconstructions. Many instruments collect data in a truly three-dimensional fashion; a prime example in radiology is positron emission tomography (PET) and the same is true for certain modes of data collection using a transmission electron microscope (TEM). Reconstructing from such data directly (i.e., without reducing the problem to a sequence of two-dimensional problems by interpolation and/or by ignoring some of the data) is referred to as fully three-dimensional image reconstruction. In this chapter we explain some of the approaches that have been proposed for this purpose and illustrate that among these the so-called "ART with blobs" appears to be the most efficacious. This claim is supported by experimental evidence from two very different modalities: PET in radiology and the conical tilt data collection geometry in transmission electron microscopy.

The problem of reconstructing an object from a set of its projections arises in a great number of fields including medicine, molecular biology and non-destructive industrial testing. It is therefore not surprising that the number of approaches proposed over the years has been large [1–3]. In this work we concentrate on the field of structural biology, focusing our

---

*Centro Nacional de Biotecnología, Universidad Autónoma de Madrid, 28049 Madrid, Spain.

†Department of Radiology, University of Pennsylvania, Blockley Hall, 4th Floor, 423 Guardian Drive, Philadelphia, PA 19104-6021, USA.

attention on the experimental situations that arise when a TEM is used to obtain projection images of biological macromolecules which are not ordered into crystals. This is referred to as "three-dimensional electron microscopy of single particles," where the term "single particle" stands for isolated, unordered particles with, in principle, identical structure. A recent list of biological macromolecules whose 3D structure has been determined by this approach can be found in [3].

Conceptually, the task of reconstructing the volume of a macromolecule from a set of its transmission electron microscopic projections is very similar to the one of obtaining the distribution in a human body of injected or inhaled compounds labeled with positron emitting isotopes in PET (positron emission tomography; see, e.g., [4]). An emitted positron will interact with an electron very near the location of its emission and this results in two photons travelling in (very nearly) opposite directions. If both these photons are detected, then we know that a positron emission has taken place somewhere near the line joining the two sites of photon detection. By considering tubes of constant width in a large number of locations and orientations (typically of the order of ten million), we can estimate the total number of positron emissions that have taken place within each tube.

This number is approximately proportional to the total concentration of the isotope in the tube: the task of reconstruction is to estimate the relative local concentrations of the isotope from these measurements. A difficulty with PET (which it shares with electron microscopy) is that the data are very noisy: the total count for some of the tubes may be less than ten and, since we are dealing with a Poisson process, this provides a rather unreliable estimate of the total concentration in the tube. This is the reason why we do not just use a subset of the tubes which contains those tubes which are all orthogonal to a fixed axis in space; while such a data collection allows in principle the reconstruction of the 3D distribution of the concentration slice by slice (using only those tubes which are in the same slice perpendicular to the fixed axis), such reconstructions would be unreliable due to the noisiness of the measurements. The data collected for the additional tubes allows us to reduce the noise in the reconstructions.

In the area of fully three-dimensional (i.e., not slice by slice) PET, it has been found that the recently proposed "ART with blobs" approach produces results which are superior from a number of points of view to those produced by the standard method of the field, the so called FBP (filtered backprojection) method [5–7]. Since FBP is mathematically very similar to the method commonly used in TEM with conical tilt geometry (namely, weighted backprojection or WBP; see [8]), it is reasonable to test if ART with blobs may also be an improvement on the current standard for TEM with conical tilt geometry.

In a typical application we use over 800 copies of the particle, each providing us with a 64 × 64 projection image, resulting in a total of over

$3,000,000$ projection measurements. From these we try to estimate the values of the particle at $250,000$ individual points. The first number is an order of magnitude larger than the second one, since our measurements are very noisy and so we need an "overdetermined" system to somewhat compensate for the noise in the individual measurements. In any case, we end up with a computationally large problem; it is therefore significant that the solution by ART requires less than one seventh of the computer time required for the solution by WBP. For our test case, the exact CPU time per reconstruction on a SGI Power Challenge 194 mHz workstation is 24.16 minutes for WBP and 3.34 minutes for ART with blobs.

A preliminary experiment using real TEM data of the DnaB helicase [9] revealed that ART with blobs and WBP produce very similar reconstructions, except that in the one produced by ART the particle appears to be shorter in the "vertical" direction (as explained in the next section, by vertical we understand the direction perpendicular to the specimen support). This indicates a need for a careful comparative evaluation of the accuracy of the two methods, with special attention paid to the vertical resolution. This paper reports on such a comparison.

The paper is divided into two main parts. In the first one (Sections 2–5), TEM with conical tilt geometry is explained and our implementations of the fully three-dimensional reconstruction algorithms are presented together with some resulting reconstructions from real TEM data; in the second part (Sections 6–9), the objective approach followed in this work to compare reconstruction methods is described and an extensive comparison between ART and WBP is performed. In the end we present our conclusions, which indicate that ART with blobs may well be superior to WBP in three-dimensional reconstruction of volumes from a large set of TEM images of single particles obtained following a conical tilt geometry. This reinforces the corresponding results reported in [5–7] on the efficacy of ART with blobs in fully three-dimensional PET.

2. Data collection in TEM with the conical tilt geometry. To recover a three-dimensional macromolecular structure from TEM images, projection data of the specimen need to be collected for a large number of different orientations of the specimen relative to the microscope. The only way to collect a sufficient amount of data for the three-dimensional reconstruction of a single specimen (while keeping the electron dose low, so as not to destroy the specimen while data are still being collected) is to combine images from a large number of particles. Amongst the different techniques of TEM projection data collection for the purpose of three-dimensional reconstruction of single particles, the one that is most widely used is the so-called *conical tilt geometry*, developed in 1987 by Radermacher and coworkers [8]. In this section we explain and illustrate this mode of data collection.

The underlying assumption is that we can attach a coordinate system (x, y, z) to the particle to be reconstructed and that we have the means of insuring that when identical particles are placed on the same specimen support, then the z-axes which are attached to the various particles are all (almost) parallel to each other (and orthogonal to the plane of the specimen support). On the other hand the same is not true for the x- and y-axes that are attached to the particles; in fact, as we shall see, this would not be desirable. A schematic illustration of what such an arrangement of particles looks like from the direction of their common z-axis is shown in fig. 1(a). Part of an electron micrograph of multiple particles with all their z-axis aligned with the direction of the electron beam is shown in fig. 2(a). Typically a single electron micrograph will contain an order of a few hundreds such particles.

If we now tilt the specimen support around the indicated broken line (this is a line in a plane perpendicular to the z-axes), then (looking at it from the same direction as before) the resulting appearance would be as indicated schematically in fig. 1(b). Since the (x, y, z)-axes are attached to the particle, the x- and y-axes are now foreshortened and the z-axes become visible. Taking an electron micrograph of the preparation after tilting the specimen support results in the image in fig. 2(b). The images of the identical particles in this tilted case are what provide us with the projection data of the structure of the particle in various orientations. Each particle provides projection data for a single orientation.

To explain this better, we attach to every particle a second coordinate system (x', y', z') with the same origin as the coordinate system (x, y, z). The z'-axis is selected to be the same as the z-axis prior to tilting. The y'-axis is selected parallel to the line around which we tilt and the x'-axis is uniquely determined by the requirement that (x', y', z') is a right-handed rectangular coordinate system. An important difference between the two coordinate systems is that (x', y', z') is not attached to the particle. While the origin will move with the particle, the orientation of the axes in this coordinate system does not change as a result of the tilt: the z'-axis remains parallel to the electron beam. We may express this by saying that the (x', y', z') coordinate system is attached to the microscope.

The function f that we wish to reconstruct is defined over the (x, y, z) coordinate system: it represents the structure of the particle. After tilting, the location of z' in the (x, y, z) coordinate system varies from particle to particle (see fig. 3) and so the different particles in the tilted electron micrograph provide us with projection data of the structure of the particle for different orientations. The whole process is summarized and the technical expressions associated with the process are introduced in the caption of fig. 3.

The reason why this data collection mode is referred to as the conical tilt geometry is clear from fig. 3(c): the z' for the various particles in the tilted electron micrograph all lie on the surface of a fixed cone. This has an important consequence on the nature of the information available in the projection data. A particular particle in the tilted electron micrograph provides us with the projection data (estimates of the line integrals) of the structure f to be reconstructed along lines parallel to its particular z'. Using the projection theorem (see, e.g., [1], p. 149), such data allow us to calculate the three-dimensional Fourier transform of f on a plane through the origin and perpendicular to z'. If we have many different such z' (all lying on the same cone) we will be able to accurately estimate the three-dimensional Fourier transform of f at all points outside a double-conical region (whose axis is parallel to z), but generally not inside this double-conical region. This observation is summarized by saying that the conical tilt geometry leads to a "missing cone" problem. Assigning the value zero to the Fourier transform of f in the missing cone results in a blurring of f in the vertical (z) direction. Any reconstruction method from the data will (in some sense) fill the missing cone, since the three-dimensional Fourier transform of the reconstruction will have values in the missing cone; a difference between reconstruction techniques is how accurately they do this for structures of various kind.

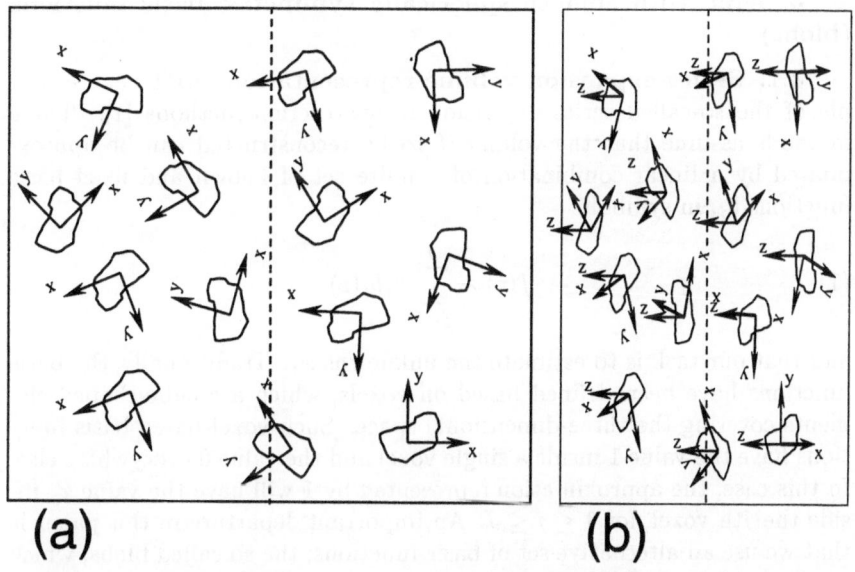

(a)   **(b)**

FIG. 1. *Illustration of the basic principle of the conical tilt geometry of data collection. (a) A view of an idealized untilted electron micrograph containing several identical specimens with their attached coordinates indicated. The z-axis is not visible since it points in the direction of the electron beam. (b) An idealized electron micrograph of the same specimens after tilting around the broken line. All three axes in the attached coordinate system become visible.*

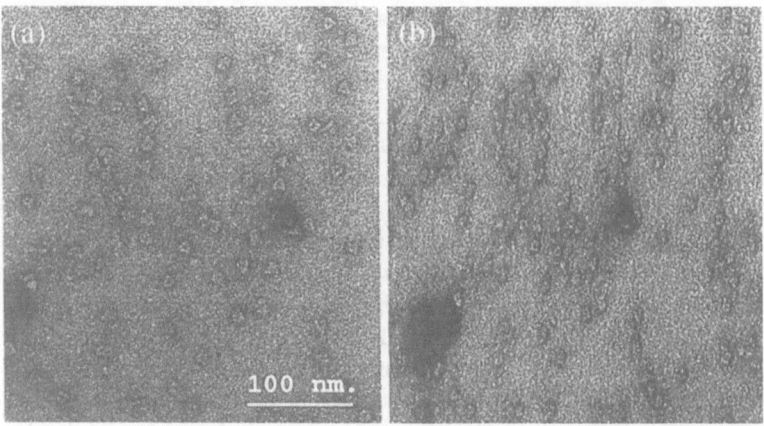

FIG. 2. *A pair of electron micrographs from a field of negatively stained DnaB particles. (a) An untilted electron micrograph containing several identical specimens. (b) An electron micrograph of the same specimens after a 55° tilt. (From [9], Academic Press Inc 1995, reproduced with permission.)*

3. ART with smooth spherically symmetric basis functions (blobs).

3.1. Series expansion volume representation. ART is an example of the so-called series expansion reconstruction methods [1]. These methods assume that the volume f to be reconstructed can be approximated by a linear combination of a finite set of known and fixed basis functions b_j, in symbols

$$(1) \qquad f(\mathbf{r}) \approx \sum_{j=1}^{J} x_j b_j(\mathbf{r})$$

and that our task is to estimate the unknowns x_j. Traditionally, the basis functions have been defined based on voxels, which are cube-shaped elements covering the three-dimensional space. Such voxel-based basis functions have the value 1 inside a single voxel and the value 0 everywhere else. In this case, the approximation represented by 1 will have the value x_j inside the jth voxel, for $1 \leq j \leq J$. An important departure in this paper is that we use an alternative set of basis functions; the so called blobs, which we discuss below. If we assume, as it is usually done in TEM [10], that the measurements depend linearly on the object to be reconstructed and that we know (at least approximately) what the measurements would be if the object to be reconstructed was one of the basis functions (we use $l_{i,j}$ to denote the value of the ith measurement of the jth basis function), then we can conclude [1] that the ith of our measurements, y_i, of f satisfies the approximation

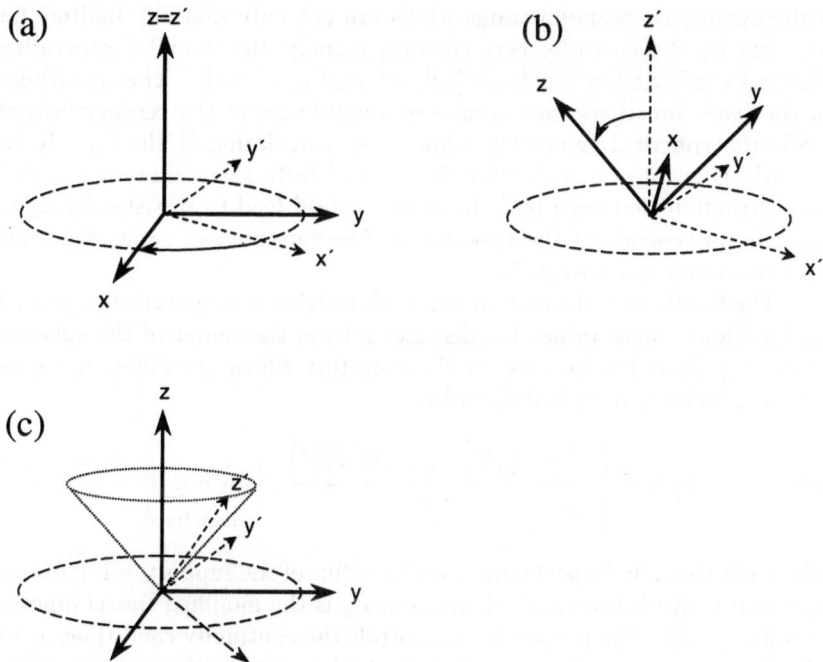

FIG. 3. *In these drawings there are two coordinate systems: (x, y, z) which is thought of as being attached to the particle and (x', y', z') which is thought of as being attached to the electron microscope, with the direction of the z'-axis being parallel to the electron beam.*

In (a) we indicate the situation when the specimen support is not tilted (i.e., its plane is orthogonal to the z'-axis). Since we can prepare the specimen in such a way that its z-axis is (at least approximately) perpendicular to the support, in this case $z' = z$. On the other hand, the other two axes are not aligned, the angle between x' and x (and between y' and y) is the "rotation angle."

In (b) we indicate the situation after the specimen support has been tilted around the y'-axis (which is what we do prior to data collection). The resulting angle between z' and z is the "tilt angle." The z-axis is still orthogonal to the specimen support.

In (c) we depict the same situation, but from the point of view of the particle to be reconstructed; that is, in the coordinate system of the particle, it is the electron microscope that appears to be tilting. Thus in a fixed (x, y, z) coordinate system, the location of z' after the tilt will lie on a cone; the shape of this cone is determined by the tilt angle and the location of the z' on it is determined by the rotation angle. The volume f is reconstructed as a function over (x, y, z), as depicted in this figure. We therefore refer to the direction parallel to the z-axis as "vertical" and when we use words such as "elongation" and "height" we mean them in the direction of the z-axis.

$$(2) \qquad\qquad y_i \approx \sum_{j=1}^{J} l_{i,j} x_j$$

Our problem is then to estimate the x_j from such a system of approximate equations.

Lewitt [11] proposed to use for reconstruction from projections spherically symmetric basis functions which are not only spatially limited, but also can be chosen to be very smooth, namely, the so-called generalized Kaiser-Bessel window functions [12], referred to as *blobs*. The smoothness of the basis functions then results in smoothness of the reconstructions, while the spherical symmetry allows easy calculation of the $l_{i,j}$. It has recently been demonstrated for the case of fully three-dimensional PET reconstruction that such basis functions indeed lead to statistically significant improvements in the task-oriented performance of series expansion reconstruction methods [5–7].

The family of blobs used in our work is defined as spherically symmetric functions whose value at a distance r from the center of the spherical symmetry (it is the location of these centers which are different for the different blobs b_j in 1) is defined by:

$$(3) \qquad b(r) = \begin{cases} \dfrac{\left[\sqrt{1-\left(\frac{r}{a}\right)^2}\,\right]^m I_m\left[\alpha\sqrt{1-\left(\frac{r}{a}\right)^2}\right]}{I_m[\alpha]}, & \text{for } r \le a, \\ 0, & \text{elsewhere,} \end{cases}$$

where r is the radial coordinate, a is the radius of the support, α is a parameter that controls the blob's shape, and I_m is the modified Bessel function of order m [12]. The parameter m controls the continuity conditions at the blob boundary: $m = 0$ means that the blob is not continuous at radius a and, if $m > 0$, then the function b has $m - 1$ continuous derivatives. It is desirable to use $m \ge 2$, so that the basis function and its first derivative are continuous at any point. The basis functions b_j in 1 are shifted versions of such a single blob. The arrangement of the centers of these shifted versions is referred to as a *grid*; the choice of the grid (as well as of the parameters a, m, α) will influence the quality of the blob-based reconstructions. These choices are discussed in great detail by Matej and Lewitt [7,13] for PET. We point out that they found that grids others than a cubic one are preferable; they advocate the use of the so called body-centered cubic grid. Since the principles involved in selecting particular types of blobs as good basis functions by Matej and Lewitt [7] were quite general, we have decided to use the basis functions advocated by them in our work for this paper (that is, $a = 2$, $m = 2$, $\alpha = 3.6$, and a relatively sparsely spaced body-centered cubic grid).

3.2. ART for 3D reconstruction. The algebraic reconstruction techniques (ART) form a large family of iterative reconstruction algorithms [1]. The particular ART algorithm that we use produces a sequence $x^{(0)}$,

$x^{(1)}$, $x^{(2)}$, ... of vectors which converges to a solution of the system provided that the system has a solution. When there are several solutions, the sequence converges to the one which is nearest to $x^{(0)}$ [1].

In the ART version we use in this work, the kth iterative step makes use of the $i(k)$th measurement and is described by

$$(4) \qquad x_j^{(k+1)} = x_j^{(k)} + \lambda \frac{\left(y_{i(k)} - \sum_{j'=1}^{J} l_{i(k),j'} x_{j'}^{(k)} \right)}{\sum_{j'=1}^{J} (l_{i(k),j'})^2} l_{i(k),j} ,$$

where λ is a real number (called the *relaxation parameter*). This particular version presents two interesting properties: first, it permits a very easy updating of $x^{(k)}$ by the addition of a multiple of a vector $l_{i(k)}$ and, second, most of the elements $l_{i(k),j}$ are zero-valued and the location and magnitude of the nonzero elements can be easily computed. With the relaxation parameter set to 1, this approach was first proposed in the literature by Kaczmarz [14] as a general method for solving systems of linear equations.

4 is the core of our algorithm. For its final implementation we need to specify some details. The initial estimate $x^{(0)}$ is chosen to be a vector of zeros. The data-access-ordering function, $i(k)$, needs to be defined, as well as the relaxation parameter. Finally, we have to decide when to stop the iterative process.

The order in which the data are accessed by the algorithm can have a significant effect on its practical performance [1]. The underlying intuitive principle is that, in a subsequence of iterative steps, we wish the changes introduced in a given step to be as independent as possible from the previous changes. This can be brought about by having the vectors $l_{i(k)}$ successively presented to the algorithm as orthogonal as possible (in the sense that, if we take the sum of the successive dot products of a vector with the vectors presented after it, then the result will be near zero for many successive vectors). Our implementation, based on [5, 15], selects a projection so that its angle is as different as possible from the angles of the previously selected projections and, within the projection, the measurements are ordered so that in an iterative step the recently updated x_j are not changed again.

The relaxation parameter λ controls the magnitude of the update in each step of the reconstruction process. Its practical importance has been repeatedly demonstrated [1]. Our approach to choosing the value of λ is presented in Section 8.1. The optimum value of λ depends not only on the projection quality and data collection geometry (maximum resolution, misalignment, noise, etc.), but also on the specific type of information that we are aiming to obtain from the reconstructed volume. It should be stressed that, in general, the "best" λ value will be different for different tasks. As an example, the "best" λ for the task of detecting the presence or absence of a given structure will be different from the "best" λ for estimating the integral of f over a region of interest [5].

As is commonly done in the literature discussing ART (and other techniques in which a single iterative step makes use of only one of the measurements), we refer to the number of iterations which results in each measurement having been used once and only once as "one cycle through the data." In case the total number of measurements is I, the reconstruction after one cycle through the data is $x^{(I)}$. Since we have a fixed data access ordering, this terminology allows us to talk about fractional cycles through the data: thus the reconstruction after r cycles through the data (where r is a real number such as 1.07 or 0.5) is defined as $x^{([rI])}$, where $[rI]$ is the nearest integer to rI.

It has been demonstrated that in the PET application ART with blobs produces high quality reconstructions after cycling through the measurements approximately once [5–7]. Our experience with the data sets treated in this work is also that there is not a significant improvement by cycling through the data more than once (this aspect is addressed in detail in Section 8.1). For this reason, the experimental results reported below are based on $x^{(I)}$, where I is the total number of measurements.

4. WBP (weighted backprojection). For this work, WBP has been implemented as described in [16]. In this section we give a brief summary of the approach; for the exact reasoning behind it and for computational details the reader should consult [16].

Consider fig. 3(c). Lines along which data are collected are parallel to the z'-axis. The projection data for this copy of the particle can be conceived of as a function p of the two coordinates in the (x', y')-plane. If we choose a positive real number a, then *the backprojection of p with radius a* is defined as the function $[B_a p]$ of three variables satisfying

$$(5) \qquad [B_a p](x, y, z) = \begin{cases} p(x', y'), & \text{if } -a \leq z' \leq a, \\ 0, & \text{otherwise,} \end{cases}$$

where (x', y', z') are the coordinates in the system (x', y', z') of the point whose coordinates in the system (x, y, z) are (x, y, z). Now if we choose a to be large enough so that a sphere of radius a centered at the origin is sure to enclose the particle, then the sum of the functions $[B_a p]$ over all the different orientations for which we have collected projection data, will be a rough representation of the object we wish to reconstruct. (For implementational details, see A.1. on p. 112 of [16].) This is a version of the so-called backprojection reconstruction method; see, e.g., [1]. To approximate the original object, such a reconstruction needs to be normalized; but even after normalization it will be a very blurry approximation. The typical approach to deblurring it (and one can bundle the normalization with the deblurring) is to use some filtering process to compensate for the blurring.

The method that we have adopted (its computational details can be found in A.2.1. on p. 113 of [16]) uses filtering on the projection data prior to backprojecting. The exact functions to be used for the filtering are arrived at in [16] by following reasoning.

Assume for now that the object to be reconstructed is an impulse at the origin (a three-dimensional Dirac δ function). Then the three-dimensional Fourier transform of $[B_a p]$ is the function

$$(6) \qquad S(X, Y, Z) = \frac{\sin(2\pi a Z')}{\pi Z'},$$

where Z' is the distance of the point (X, Y, Z) from the (X', Y')-plane; that is from the plane through the origin in Fourier space which is parallel to the (x', y')-plane. Hence the three-dimensional Fourier transform of the backprojection reconstruction of an impulse at the origin is the sum T, over all the different orientations for which we have collected projection data, of such sinc functions S. However, we know that the three-dimensional Fourier transform of our true object (the impulse) is 1 everywhere. This is the observation that is used to produce the filter to be applied to the physically-collected projection data. To be more precise, if we take the two-dimensional Fourier transform of the physically-collected projection data $p(x', y')$, then we get a function $P(X', Y')$. The arguments (X', Y') of this function correspond to the points with coordinates $(X', Y', 0)$ in the system (X', Y', Z'). Hence, for every one of them we can work out the corresponding coordinates (X, Y, Z) in the system (X, Y, Z). Our proposed filtering step should now be achieved by a division by $T(X, Y, Z)$. However, it is possible that $T(X, Y, Z) = 0$ and, even if it is not, it may be so small that division by it would result in an unacceptable amplification of noise in the projection data. For this reason we introduce the idea of a *threshold* $t > 0$ and define

$$(7) \qquad H(X', Y') = \begin{cases} T(X, Y, Z), & \text{if } |T(X, Y, Z)| \geq t, \\ t & \text{otherwise}, \end{cases}$$

where the relationship between the coordinates (X', Y') and (X, Y, Z) is as defined above. Finally, the filtering is completed by taking the inverse two-dimensional Fourier transform of $\frac{P(X', Y')}{H(X', Y')}$. It is this filtered projection data that gets backprojected to obtain the WBP reconstruction. (Note that the standard ways used to avoid noise amplification in medical imaging [1] and in electron microscopy with conical tilt geometry [16] are different: in the former it is customary to use some window of an appropriate shape, in the latter noise amplification is prevented through the use of the just introduced threshold.)

As can be seen from this discussion, WBP has two parameters which need to be determined in order to get an unambiguous reconstruction method: the radius (denoted by a above) and the threshold (denoted by

t above). These should be optimized for the particular situation to which the method is being applied.

5. A preliminary experiment with real data. For our initial comparison of the general qualitative performance of the new ART with blobs method versus WBP in a real case, the DnaB helicase [9] was used as test specimen. Eight hundred and thirty-seven negative staining projections of 64×64 pixels were obtained following the conical tilt geometry (tilt angle $= 55°$) and used as input projections. For this preliminary study, the free parameters of the reconstruction algorithms (λ for ART, threshold and radius for WBP) were selected by visual inspection of the reconstructions. The results are shown in fig. 4, in which corresponding horizontal slices of reconstructions obtained with WBP (rows 1 and 3) and ART with blobs (rows 2 and 4) are presented.

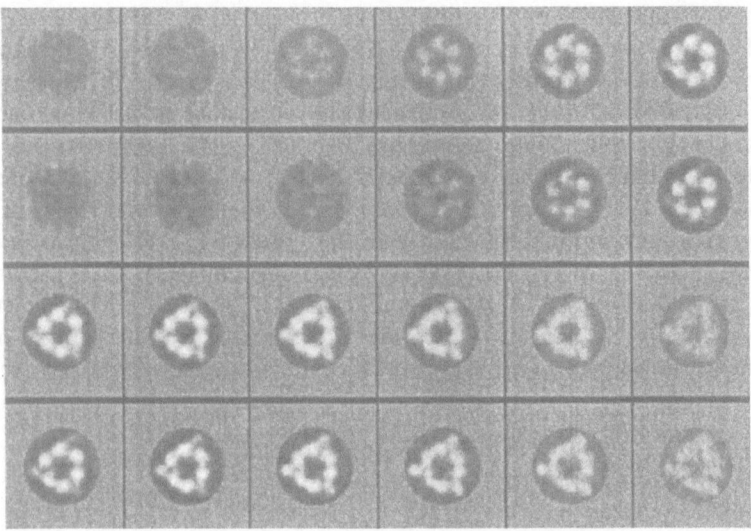

FIG. 4. *Horizontal slices of a reconstruction of the DnaB protein obtained from 837 projections using two different reconstruction algorithms: WBP (rows 1 and 3) and ART with blobs after cycling through the data just once (rows 2 and 4).*

The results obtained with the two algorithms are similar, but not identical. One of the most noticeable differences is the height of the reconstructed particle: *it seems that* ART with blobs has reconstructed a shorter particle than WBP (see fig. 4). This fact, if it is general, could have a significant impact on the selection of the algorithm for 3D reconstruction of a macromolecule. The main point of the rest of this work is the description of an objective approach to comparing different reconstruction algorithms and its application to our particular area of interest. It will be demonstrated that this approach allows us to go from a comparison using statements such as "*it seems that...*" to a comparison using statements such as "*we claim with a statistical significance of xx that...*"

6. A general methodology for objective comparison of reconstruction methods. The analysis of the relative performance of two different algorithms is a key task in image processing. Such an analysis is typically not trivial. In fact, its outcome depends on the purpose for which the processed images are to be used. A serious comparison cannot rely on visual inspection of a few images, especially if the tested methods produce similar results. We now present a methodology which allows us to experimentally compare the efficacy of two different reconstruction algorithms for a particular task. The approach assigns a statistical significance level to the claim of superiority of one method over the other.

The proposed methodology, using numerical observers, compares the reconstructions obtained from artificially generated projections with the ideal volumes [13, 17]. Briefly it consists of:

1. Generation of random samples from a statistically described set of *phantoms*. These phantoms must resemble the real objects of interest in some specific way.

2. Generation of the corresponding noisy *projections*, simulating the instrument used for collecting the data.

3. Optimization of free parameters of the reconstruction algorithms on a *training set* of data.

4. Reconstructions by the methods to be compared on a *testing set* of projection data (statistically independent from the training set).

5. Assignment of *figures of merit* (FOMs) to each reconstructed volume. The FOMs should be measures of quality for solving specified tasks. Our convention is that FOMs are defined so that if the reconstructed volume coincides with the ideal volume, then the FOM has the value 1.

6. Calculation of the *statistical significance*, based on the FOMs for all the reconstructions in the testing set, at which we can reject the null hypothesis that the algorithms perform equally well in favor of the alternative hypothesis that the method with the higher average FOM is better. This is done using the *t*-test for paired data [17, 18]. (It is assumed by the *t*-test that certain underlying distributions are gaussian; the gaussianness of corresponding distributions has been confirmed by [17, 19].)

7. Calculation of the *relevance* of the superiority of one algorithm over the other [20]. This magnitude is defined as:

$$(8) \qquad \frac{f_2 - f_1}{1 - f_1} \times 100\%$$

where f_1 and f_2 are the average values of the FOM for the reconstructions made with the two algorithms respectively, selected in such a way that $f_2 \geq f_1$.

In the following sections we discuss each of the above points for our particular problem of comparing ART with blobs with WBP in the area of conical tilt TEM.

7. Application of the comparison methodology to TEM reconstruction methods.

7.1. Phantom generation.
As the main difference observed in the reconstructions using real data is the apparent particle height, we have designed an ensemble of phantoms with the aim of comparing the fidelity of ART with blobs versus WBP for the measurement of features along the vertical axis. Our ultimate purpose is to answer the practical question: which method yields a more realistic measurement of the particle height?

Each phantom of the ensemble is composed of nine pairs of cylinders, each pair arranged vertically (fig. 5(a)). Their diameter is always 10 length units (one length unit is the size of a pixel side). The height of the cylinders and the gap between them are always integer and odd values, selected from a uniform distribution from 3 to 9 and from 3 to 7 length units, respectively. Each of the two cylinders belonging to a pair has the same height and is symmetrically placed with respect to the central horizontal plane. The "density" value (in arbitrary units) inside the cylinders is 1 and outside the cylinders is 0.

The phantom shape has been selected using the above stated criterion of similarity with real data. There are many biological systems with two structures (two cylinders in our case) with one stacked on the other; the chaperonine GroEL is one of the best known examples of this. We have placed a gap between the two structures since we are mostly interested in measuring the blur across such a gap (i.e., in the vertical direction).

The phantoms are mathematically defined; that is, the file that stores them contains the collection of parameters which geometrically describes each of the cylinders. However, in order to compare a phantom with its reconstruction, for each cubic grid point at which the reconstruction is evaluated (using eq. 1) the phantom is also evaluated as the average density within the small cubic region (voxel) around that grid point. We refer to these values as "voxel values."

7.2. Projections.
Image formation in the electron microscope is a quite complex process. For this work it was assumed, as is usually done in high resolution electron microscopy, that micrographs can be interpreted as projections, that is, as sets of line integrals over the specimen. The effect of the contrast transfer function has not been taken into account and will be addressed in a separate work. The conical tilt geometry was simulated (see fig. 3), with the tilt angle equal to 55°. This value is quite common in TEM, where technical (related to the microscope) and fundamental problems (related to the process of image formation) restrict the maximum tilt angle to around 55–60 degrees. These limitations in the data collection pro-

(a)

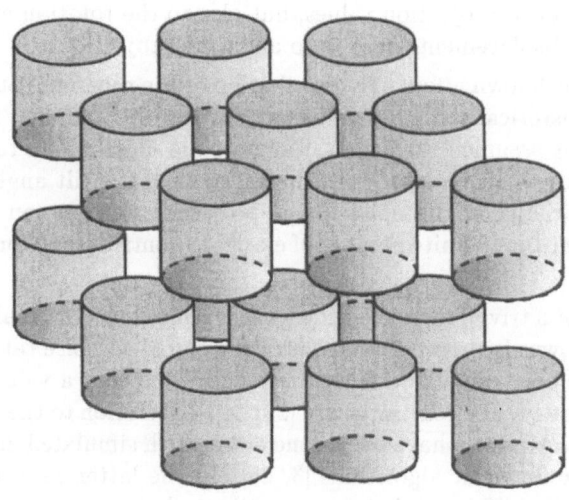

(b)

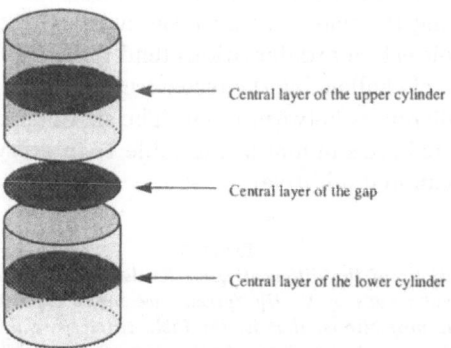

Central layer of the upper cylinder

Central layer of the gap

Central layer of the lower cylinder

FIG. 5. *(a) Phantom of cylinders, composed of nine pairs of cylinders with a gap between them. Although this is not reflected in the figure, the cylinder height and the gap length are random values. The phantom is symmetric with respect to the central horizontal plane.*
(b) Layers involved in computing the vrFOM.

cedure are responsible for a reduction of the total information available on the specimen structure. When formulated in Fourier space, the frequencies for which data are not available fall in a vertically oriented cone-shaped region, usually referred to as the "missing cone" [2].

The ideal projections were computed analytically based on the intersection length of each cylinder with the projector ray. Special care was taken in simulating realistic noise by considering the addition of noise not only to the ideal projection values, but also to the rotation and tilt angles and to the displacements needed to align the projections.

Little is known about the distribution of the different sources of noise, although historically they have been simulated by gaussian functions. In general, it is assumed that the variance of the noise added to the rotation angle and the shift is lower than the variance of the tilt angle [3, 21], and that the variance of the noise in the projection images can be estimated from the fact that a finite number of electrons contribute to form the images [22].

It is not a trivial task to decide what levels of noise are realistic. Several attempts towards the measurement of the signal-to-noise ratio of a micrograph have been made and it has been suggested that a value around 0.55 is representative of a real data case [23, 24]. In relation to the tilt angle, the only studies available have been done either with simulated data [21] or are related to refinement algorithms [3, 25]. In the latter case, the measured quantity is the difference between the tilt angle before and after the alignment. In any case, a variance of 0.09 rad seems to be representative. There is even less data regarding the noise related to in-plane shifts and rotations. For our work their standard deviations were chosen by testing some trial values and selecting the ones that gave the most realistic-looking results. In general, our noise standard deviations tend to be slightly overestimated with the intention of challenging the reconstruction algorithms and thereby amplifying the differences between them. The six different noise cases that we have investigated are summarized in table 1. In every case, the noise is gaussian, zero mean and additive.

TABLE 1

Standard deviations of the different noise levels added to the projections and the corresponding optimal values of λ. By "pixels" we mean the values of the individual measurements in the projections; that is, the table entries are the standard deviations of the noise added to projection values. Angles are expressed in radians and shifts in length units. The standard deviation of a typical projection (prior to adding noise) is approximately 4.3 for the cylinder phantoms of fig. 5.

	Pixels	Rot angle	Tilt angle	Shirt	Opt λ
Case 1	15	0.00	0.00	0.00	0.0200
Case 2	30	0.00	0.00	0.00	0.0125
Case 3	15	0.09	0.00	0.00	0.0400
Case 4	15	0.00	0.00	0.75	0.0400
Case 5	15	0.00	0.16	0.00	0.0300
Case 6	9	0.02	0.08	0.30	0.0600

The projection data generation can be summarized in the following steps. First, 64×64 projection values are computed as path integrals of the volume using analytical integration of the geometrically described phantom. This procedure is repeated for 800 images. The projections are generated for random rotation angles although a not totally correct (noisy) angle is supplied to the 3D reconstruction algorithm. A similar, but not identical, scheme is followed with the tilt angle. In this case the projections are generated at random values (with different deviations, see table 1) around 55°, which is the value assumed by the algorithm. Second, gaussian zero mean noise (with different deviations, see table 1) is added to the projection values and to the parameters needed to make the translational alignment of the set of images.

7.3. Figures of merit. A figure of merit (FOM) is a measure of the quality of the reconstruction from a specific point of view; it compares an easily quantifiable similitude of the phantom with the reconstructed volume. Following Matej et al. [20], the FOMs are defined so that for a perfect reconstruction the FOM value is 1. Four FOMs have been used in the main body of this work. The first three, eFOM, tFOM and saFOM, are different measures of the error between the reconstruction and the phantom. The fourth FOM, vrFOM, is related to the resolution of the reconstruction parallel to the vertical axis.

Generally speaking, for this kind of evaluation one should avoid global measures of similarity (such as these based on the mean square error taken over all points), since these may be highly influenced by the accuracy of the reconstructions in regions away from the ones of interest and, thus, will not reflect the practical usefulness of the reconstructions. (For example, see the discussion regarding Figs. 8.10 and 8.11 and Table 8.3 on pp. 140–144 in [1].) Our measures tFOM and saFOM are designed to indicate the accuracy of reconstruction for our regions of interest (namely the cylinders), both at individual points and when considered as whole features, respectively. Nevertheless, since the global mean squared error is such a standard measure, we have decided to report also on performance according to this measure which we call eFOM.

7.3.1. Error FOM (eFOM). The error FOM is defined as

$$(9) \qquad\qquad \text{eFOM} = 1 - s,$$

where s is the mean squared error between the reconstructed volume and the phantom averaged over the whole volume.

7.3.2. Training FOM (tFOM). We define the training FOM as

$$(10) \qquad\qquad \text{tFOM} = 1 - \frac{1}{F} \sum_{f=1}^{F} e_f,$$

where F is the number of cylinders (in our case always 18) and e_f is:

$$e_f = \sum_{i \in \mathcal{J}_f} |x_i^p - x_i^r|,$$

where $i \in \mathcal{J}_f$ means all the voxels belonging to the fth cylinder and the superscripts p and r refer to the phantom and the reconstruction respectively.

7.3.3. Structural accuracy FOM (saFOM). Structural accuracy is defined as:

$$(11) \qquad \text{saFOM} = 1 - \frac{1}{F} \sum_{f=1}^{F} |m_f - \mu_f|,$$

(where F is the number of cylinders, m_f is the average of voxel values within a particular cylinder in the reconstruction, and μ_f is the average of voxel values within the same cylinder in the phantom.

7.3.4. Vertical resolution FOM (vrFOM). This FOM is defined as the average over the nine pairs of cylinders of the quantity:

$$(12) \qquad \frac{m_1 + m_2 - 2m_3}{\sqrt{\nu_1 + \nu_2 + \nu_3}} \left/ \left(\frac{\mu_1 + \mu_2 - 2\mu_3}{\sqrt{\omega_1 + \omega_2 + \omega_3}} \right) \right.,$$

where m_1, m_2 and m_3 (μ_1, μ_2 and μ_3) are the means of the voxel values in the reconstruction (phantom) in the central layer of the upper cylinder, of the lower cylinder and of the gap between them, respectively, while the ν_i (ω_i) are the corresponding variances (see fig. 5(b)). The denominator of eq. 12 is a normalization factor (constant in our case) that guarantees that the FOM is 1 for a perfect reconstruction. With increasing blurring in the vertical direction, the values of m_1 and m_2 get smaller and the value of m_3 increases, resulting in a loss of contrast between the cylinders and the gap separating them. The observability of such a contrast by the user of the images also depends on the variance within the individual slices used to determine the contrast; as this variance increases, the observability of the contrast is reduced. Thus the vrFOM is a measure of the contrast that remains observable between the cylinders and the gap separating them in spite of the blurring in the vertical direction.

In tFOM and saFOM, we consider that a voxel belongs to a cylinder when its center is inside the cylinder. In vrFOM, we have relaxed this condition so that any voxel that has at least one edge in the indicated layer of a cylinder or a gap (see fig. 5) is considered to belong to it. In this way, the variances for the phantom (ω_i) are positive and the normalization factor $\frac{\mu_1 + \mu_2 - 2\mu_3}{\sqrt{\omega_1 + \omega_2 + \omega_3}}$ is finite.

Another quantity that *a priori* could have been a very interesting FOM is *resolution*, as defined, for example, by Frank et al. [26]. Unfortunately resolution measures in 3D electron microscopy of single particles only check the consistency between two reconstructions, but not necessarily their absolute accuracy. An extreme case could be an algorithm which always assigns zero to every voxel; such an algorithm would come out as perfect as judged by any such "consistency criterion." More seriously, let us suppose that WBP consistently elongates the specimen in the vertical-direction and that ART is always more accurate, but less consistent. Should that really be in favor of WBP? This is a familiar discussion in other fields as well. In medicine one talks about reproducibility and accuracy, in statistics about variance and bias. As a general principle, accuracy and lack of bias is more important than lack of reproducibility and variance; problems with the latter can be overcome by doing more experiments; problems with the former are not affected by doing more experiments, which will just confirm the wrong results. The real situation with resolution could be even worse because resolution, as measured in this field [26], presents a huge variance (see [27]). These are the reasons why we have not included in our study any reproducibility-criterion-based FOMs.

8. Results.

8.1. Training. Before starting the reconstruction process itself, the different parameters used in each algorithm must be tuned. In view of the choices we have made in Sections 3 and 4, two parameters need to be optimized for ART (λ and iteration number) and another two for WBP (threshold and radius).

Six phantoms and their projections were randomly generated for each noise case (see Section 7), to be used for the optimization of the parameters. The optimization process uses the training FOM as the measure of reconstruction fidelity, neglecting the information provided by the other FOMs. (At first it may appear reasonable to repeat the training process with the vrFOM and to compare the results obtained training on the tFOM and on the vrFOM. However, the vrFOM, as all signal-to-noise measures of fidelity, has the property that the phantom itself does not optimize it. For example, by adding 1 to each phantom voxel value in the central layers of the cylinders, but not of the gaps, we obtain an object whose vrFOM is 2.)

To determine when to stop the parameter refinement process we follow the 5% criterion proposed by Matej et al. [5]: let f_{max} be the maximum FOM currently obtained and let $f*_{max}$ be the largest possible FOM as projected based on the neighboring points, then the search is stopped when the relevance of the possible improvement, $100 \times (f^*_{max} - f_{max})/(1 - f_{max})$, is less than 5 (see fig. 6).

The first parameter to be optimized in ART is the relaxation parameter λ. Fixing the number of iterations so that we cycle though the data exactly once, the best results were obtained with small values around $\lambda = 0.02$ when the noise added to the pixels had a standard deviation 15 (see plot 1(a))

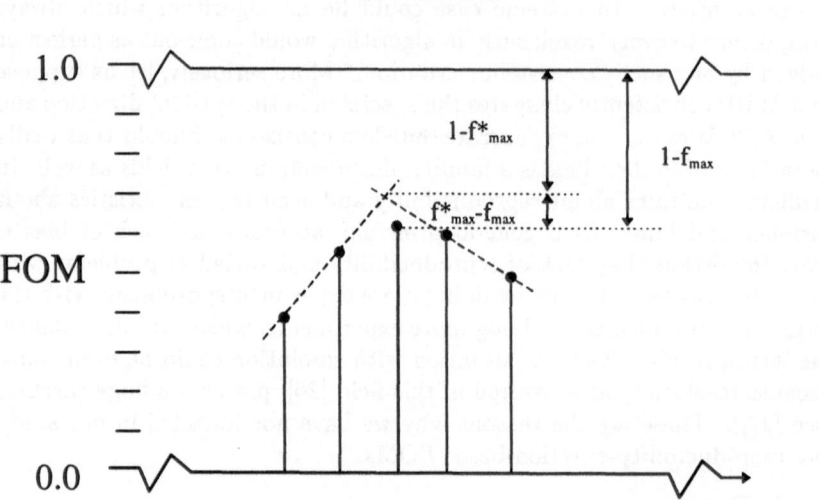

FIG. 6. *Graphical illustration of the stopping criterion in training. When the relevance of possible improvement $(\frac{f^*_{max}-f_{max}}{1-f_{max}} \times 100)$ is less than 5, the parameter search stops.*

and around $\lambda = 0.0125$ when the standard deviation was increased up to 30 (see plot 1(b)). If other sources of noise are added, the plot of tFOM versus λ becomes flatter and, at the same time, the peak moves slightly towards the right (see table 1).

The second parameter to be optimized is the number of iterations. 3D reconstructions from sets of $200, 400, 800, 1600, 2400$ or 3200 projections were obtained. As we have only created 800 different projections, a set of 1600 projections is the same as two cycles through the data. Plot 1(c) shows the tFOM versus the number of cycles through the data for one particular noise case (with λ fixed at its optimal value as found from independent training sets for each of the different number of iterations). The value of the tFOM obtained after one cycle through the data is similar to the maximum value; in fact, using the 5% criterion we conclude that the difference is not relevant and therefore any number of cycles equal to or greater than 1 is an acceptable choice.

The parameters optimized for WBP are the threshold and the radius. As shown in plot 1(d) for a fixed radius, any value of the threshold that is below 8 can be considered optimal. This result surprised us, because it is well demonstrated in the field that a very small threshold can lead to noise amplification. Further investigations showed that for a smaller number of projections, for example 100, a threshold about 12 gives the best visual results; but this is definitely not the case for a large number of projections, such as 800, provided that the tilt angle assumed by WBP is the same for all the projections.

In order to optimize the radius, the 64×64 projections were padded into a 128×128 frame and then reconstructed with different diameters from 64 to 128. The results (plot 1(e)) show that the tFOM hardly changes with the radius size and that smaller radii give slightly better reconstructions.

8.2. Testing. Twelve different realizations of the cylinder phantom have been used. Table 2 summarizes the comparison between ART with blobs and WBP for each FOM. In the table we report the relevance of the improvement (eq. 8) together with the level of statistical significance at which we can reject the null hypothesis that both methods are equally good in favor of the alternative hypothesis that ART performs better.

PLOT 1. *Parameter optimization.*
(a), (b) Optimization of the relaxation parameter λ using the tFOM for noise cases 1 and 2 (see table 1).
(c) Optimization of the iteration number.
(d) Optimization of the threshold. The graph suggests, for this data set, that the best reconstructions are obtained when the threshold is smaller than 8.
(e) Optimization of the radius.

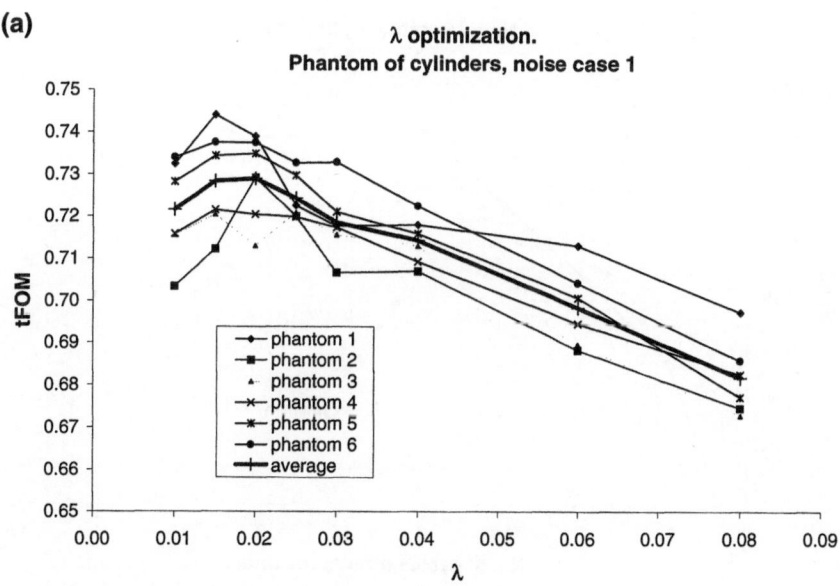

1(a)

(b)

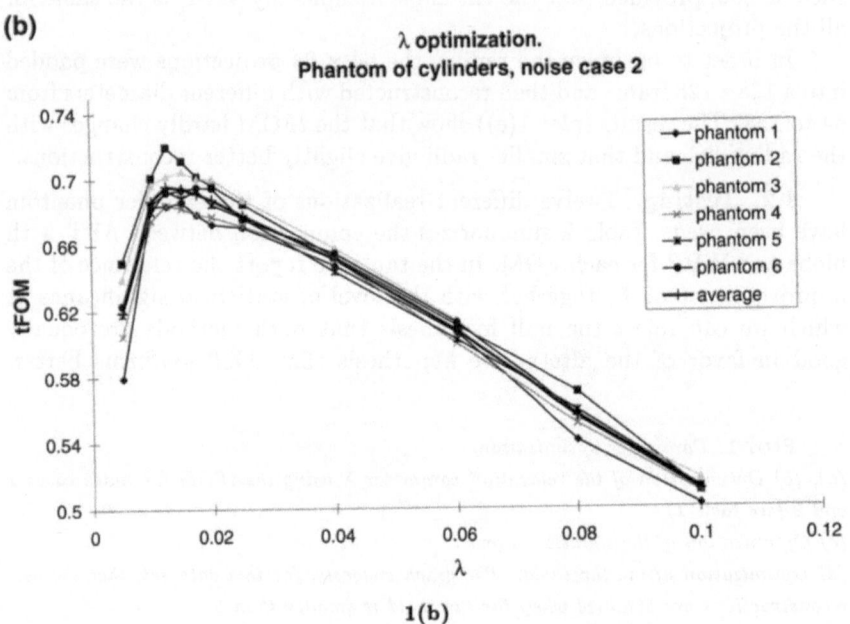

1(b)

(c)

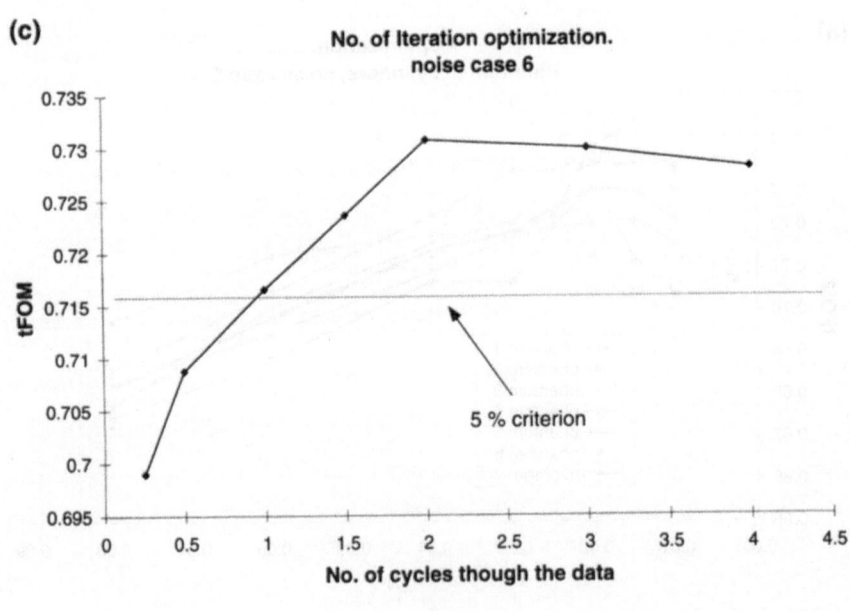

1(c)

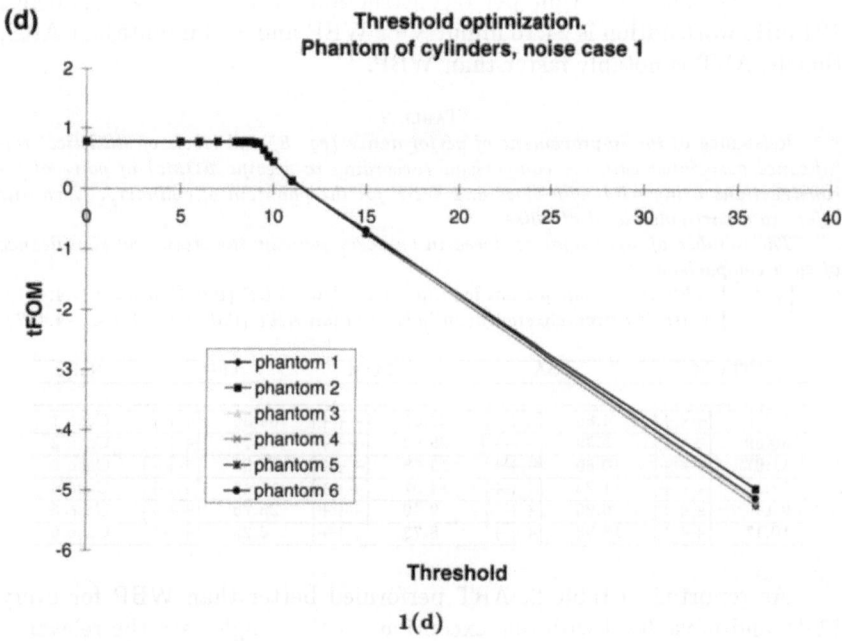

(d) Threshold optimization.
Phantom of cylinders, noise case 1

1(d)

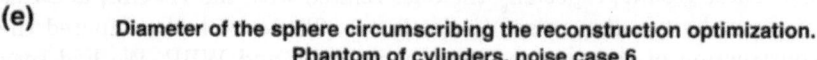

(e) Diameter of the sphere circumscribing the reconstruction optimization.
Phantom of cylinders, noise case 6

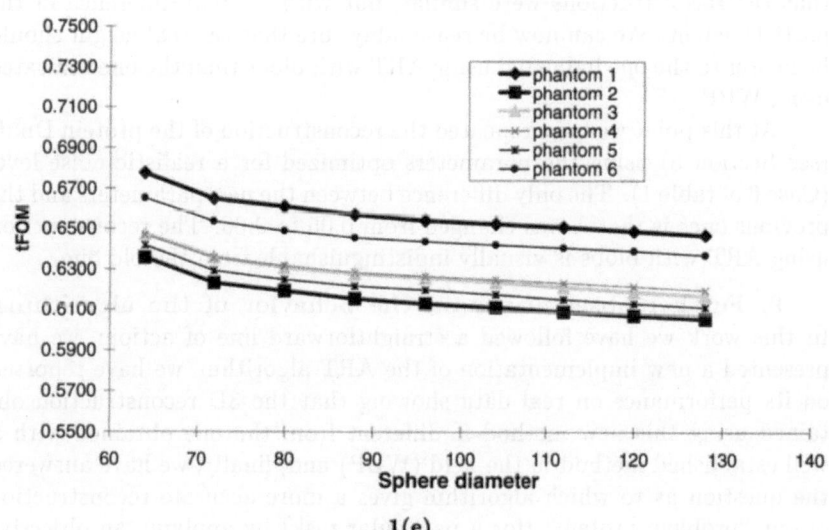

1(e)

The statistical significance is indicated by the number of plus (or minus) signs: three indicates that the level of statistical significance is better than 0.0005 [18]. The CPU time per reconstruction on a SGI Power Challenge 194 mHz workstation is 24.16 minutes for WBP and 3.34 minutes for ART; that is, ART is notably faster than WBP.

TABLE 2

Relevance of the improvement of performance (eq. 8) and levels of statistical significance associated with the comparison (according to specific FOMs) of pairs of reconstructions using ART with blobs and WBP for the phantom of cylinders. Each line refers to a particular level of noise.

The number of $+/-$ signs enclosed in brackets quantify the statistical significance of each comparison:

$[+++]$ *ART is extremely significantly better than WBP (0.05% level, $t > 4.437$);*
$[---]$ *WBP is extremely significantly better than ART (0.05% level, $t < -4.437$).*

TFOM		SSA		vrFOM		eFOM		Noise
11.75	[+++]	4.85	[+++]	22.65	[+++]	34.35	[+++]	Case 1
30.59	[+++]	5.39	[---]	25.62	[+++]	54.77	[+++]	Case 2
11.97	[+++]	10.80	[+++]	13.68	[+++]	19.36	[+++]	Case 3
12.10	[+++]	11.22	[+++]	14.07	[+++]	19.42	[+++]	Case 4
9.14	[+++]	6.26	[+++]	9.40	[+++]	23.75	[+++]	Case 5
10.17	[+++]	13.53	[+++]	8.78	[+++]	6.25	[+++]	Case 6

As reported in table 2, ART performed better than WBP for every FOM and noise level with one exception (in this single case the relevance of improvement by WBP over ART is only slightly over 5%, in most other cases the relevance of improvement by ART over WBP is much greater than this). These results, especially the ones related with the vrFOM, allow us to answer the question raised in Section 5. There we had compared the reconstruction of the protein DnaB using ART and WBP. We had seen that the reconstructions were similar, but with a small difference in the particle height. We can now be reasonably sure that the real height should be nearer to the one indicated using ART with blobs than the one indicated using WBP.

At this point we have repeated the reconstruction of the protein DnaB (see Section 5) using the parameters optimized for a realistic noise level (Case 6 of table 1). The only difference between the new parameters and the previous ones is that λ was changed from 0.05 to 0.06. The reconstruction using ART with blobs is visually indistinguishable from the old one.

9. Further observations on the behavior of the algorithms. In this work we have followed a straightforward line of action: we have presented a new implementation of the ART algorithm, we have reported on its performance on real data showing that the 3D reconstruction obtained using this new method is different from the one obtained with a well established method in the field (WBP) and, finally, we have answered the question as to which algorithm gives a more accurate reconstruction of our "problem protein" (for a particular task) by applying an objective methodology which allows us to assign statistical significance to the claim of superiority of one method over the other. This section presents extra information that may help the reader to better understand the behavior of the ART versus the WBP algorithm. In particular we discuss two points: first, we test if the reconstruction is differently influenced by the first-presented projections than by the last-presented ones and, second, we examine the quality of a reconstructed feature when we take into consideration its position in the volume.

9.1. Relative influence of the first and last presented projections. The specialized literature [3, 28] has raised several general concerns related to the behavior of iterative methods (such as ART) versus analytic methods (such as WBP). One of the strongest causes of concern is that, as the volume is updated sequentially after each projection is presented, some projections may have a much greater influence on the reconstruction than others [28]. This may be especially so in our case, since we cycle through the data just once. In order to test this possibility we have plotted (plot 2) the error between input projections and projections of the reconstructed structure; a linear fit of the error (continuous line) is also plotted. (Error has been defined as the square root of the mean squared distance between pixel values of a projection used for the reconstruction and the corresponding projection of the reconstruction.) The noise case 6 was used (see table 1), so a perfect reconstruction (i.e., the phantom) would produce an error similar to the standard deviation of the noise added to the pixels, that is, 9. The results indicate that even for ART there is no reason to worry about the relative influence of the projections at the end of one cycle though the data: the differences are negligible, since the slope of the interpolated line is -9.13×10^{-5}, while the deviation is much bigger: 0.122. It may help in understanding this result to recall that we are using ART with a very small relaxation parameter; this limits the influence of each individual projection. As expected, WBP gives almost an horizontal line (slope $= 6.67 \times 10^{-6}$, deviation $= 0.125$.)

9.2. Behavior of a reconstructed small sphere depending on its position in the volume. In this section we address the problem of the behavior of a reconstructed point depending on its position in the volume. In general, and specially for linear and stationary systems, it is a good idea to know how a point is imaged by an optical device or how a point is reconstructed by a reconstruction algorithm. A point spread function approach is suitable for such studies. Since the behavior of the implemented reconstruction algorithms is not stationary, it is not easily described using such a formalism, so we have decided to follow the more experimental approach described below.

We have created a new phantom composed of ten small spheres placed at random voxel centers (see table 3). The sphere radius is 1.75 units (one unit is the edge of a voxel); choosing this radius results in a sphere which

PLOT 2. *Error between original projections and projections of the reconstructed structure. The projections have been ordered as they are presented to the algorithm.*

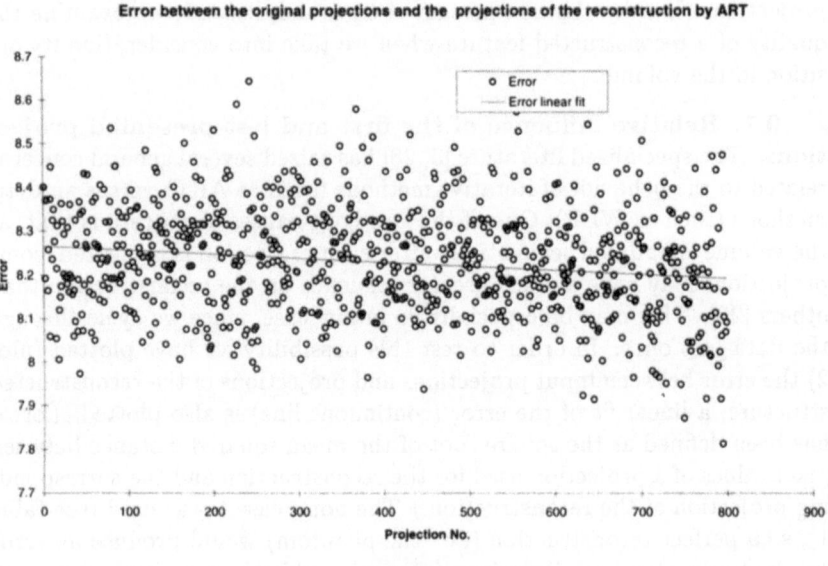

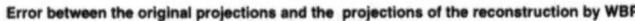

2(a)

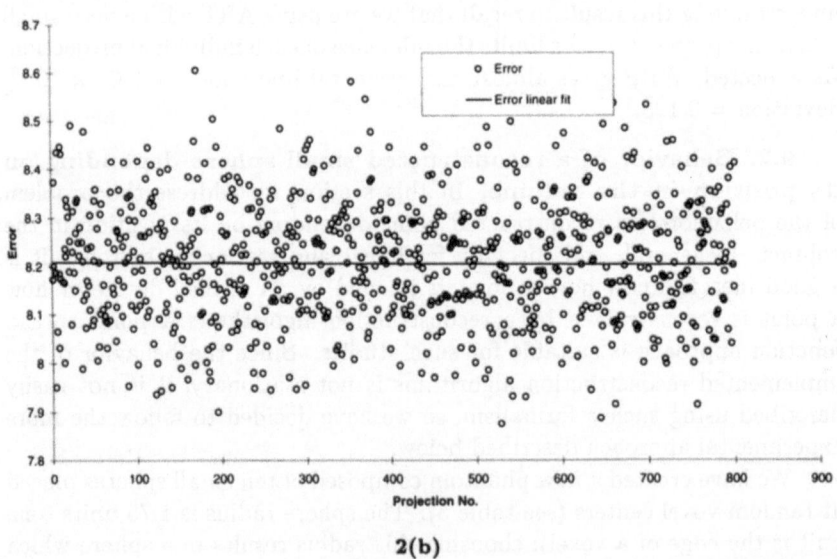

2(b)

encloses the voxel and its six face-neighbors. The rationale of this selection is as follows: we want to measure how a "point" is reconstructed by the algorithms, so spheres as small as possible should be used. But, at the same time, images in TEM never present a one voxel resolution, so a realistic "point" size should be greater than 1 unit.

TABLE 3

Positions of the centers of the spheres that form the little spheres phantom. The origin of coordinates is at the volume center.

	x	y	z
1	-2.5	7.5	12.5
2	6.5	-4.5	-18.5
3	6.5	0.5	8.5
4	-18.5	13.5	-17.5
5	6.5	8.5	0.5
6	-4.5	-3.5	2.5
7	5.5	1.5	13.5
8	-12.5	0.5	-0.5
9	-0.5	10.5	-3.5
10	-0.5	15.5	8.5

Twenty-five statistically independent sets of 800 projections of the same phantom were calculated using the noise case 1 (table 1) and reconstructions were obtained from all data sets by both methods. Two new FOMs related to point detectability along the planes parallel and perpendicular to the specimen support were calculated. The first one, called $FOMxy^i$, is defined as:

$$(13) \qquad FOMxy^i = \frac{a^i - b^i}{s^i}.$$

The variables a^i, b^i, and s^i are all obtained from reconstructed voxel values for voxels whose centers lie in the same horizontal plane as the center of the ith little sphere of the phantom; a^i is the mean value for those voxels whose center is within a distance of 1.75 of the center of the ith little sphere (we think of this as the signal), b^i is the mean value for those voxels whose center is at a distance between 3.5 and 5.25 from the center of the ith little sphere (we think of this as the background) and s^i is the standard deviation of the second set of voxel values. The second FOM called, $FOMzy^i$, is defined similarly, but using a vertical plane containing the center of the ith little sphere instead of the horizontal one.

Following this scheme, each reconstruction provides us with 20 FOMs, two for each of the ten little spheres that form a phantom: one for a horizontal plane and the other for a vertical plane. The statistical evaluation is the same as with all the other FOMs. The results are presented in table 4.

Table 5 presents the levels of statistical significance at which we can reject the null hypothesis that the two methods are equally good in favor of the alternative hypothesis that the one with the higher average FOM is better.

TABLE 4

FOMxy^i and FOMzy^i FOMs for the little spheres phantom. For each little sphere we get two contributions by ART and another two by WBP. The last row contains the averages of the columns.

i	ART		WBP	
	FOMxy$i$	FOMzy^i	FOM	FOMz
1	6.19	6.27	5.27	5.86
2	6.68	7.04	5.51	5.99
3	6.33	3.33	5.32	3.36
4	6.35	7.75	6.20	6.96
5	7.05	7.73	5.49	6.09
6	6.60	5.40	5.14	5.39
7	4.84	4.38	5.19	3.39
8	6.33	6.08	5.55	5.37
9	6.66	6.50	5.69	5.49
10	6.39	6.26	5.58	5.82
mean	6.34	6.07	5.50	5.37

We do not observe any pattern between the FOM values and the position of the spheres, neither for ART nor for WBP. However, we observe that in general ART presents a significantly better detectability than WBP. As expected, for both techniques the average FOM (last row of table 4) is better for the horizontal planes than for the vertical planes, but even the vertical plane average for ART is better than the horizontal plane average for WBP.

10. Discussion. We have introduced in TEM the algorithm called "ART with blobs" for 3D reconstruction of single particles. We have also adapted a methodology developed for PET by Furuie et al. [17] to evaluate the performance of TEM reconstruction algorithms. Concentrating our study on the case of the conical tilt geometry involving large data sets, we have found that, for the tested FOMs, ART performs better than WBP for almost all tested cases and, therefore, is a very promising alternative to WBP. In particular, ART gives more accurate measurements for specimens presenting elongated structures perpendicular to the support level. A possible explanation of this behavior is the way in which ART and WBP fill the "missing cone". WBP does essentially nothing to interpolate into the missing cone. (There are some accidental interpolation effects due to sampling and finite size.) On other hand, ART will try to fit a combination of blobs to the data which acts as an implicit interpolator.

TABLE 5

Levels of statistical significance associated with the comparison (according to specific FOMs) of pairs of reconstructions using ART with blobs and WBP for the phantom of little spheres. Each line refers to a particular sphere.

The number of +/− signs enclosed in brackets quantify the level of statistical significance of each comparison.:

[+ + +] *ART is extremely significantly better than WBP (0.05% level, $t > 3.745$);*
[++] *ART is very significantly better than WBP (0.5% level, $t > 2.797$);*
[+] *ART is significantly better than WBP (5% level, $t > 1.711$);*
[   ] *the difference in performance is not significant ($|t| < 1.711$);*
[−] *WBP is significantly better than ART (5% level, $t < −1.711$).*

	FOMxy^i	FOMzy^i
1	[++]	[+]
2	[+ + +]	[++]
3	[+ + +]	[ ]
4	[ ]	[+]
5	[+ + +]	[+ + +]
6	[+ + +]	[ ]
7	[−]	[+ + +]
8	[++]	[++]
9	[+ + +]	[+ + +]
10	[+]	[+]

Regarding future developments it is interesting to note that ART lends itself very easily to the incorporation of *a priori* spatial information on the specimen, while this kind of information cannot be easily incorporated into a WBP scheme. In this way we can incorporate in a very straightforward way information on high resolution "surface reliefs" of macromolecules as provided either by high resolution shadowing or by other microscopies, allowing for much improved structural information.

Acknowledgments. We acknowledge partial financial support from the Spanish Comisión Interministerial de Ciencia y Tecnología through grant BIO95-0768, NATO through grant CRG 960070 and NIH through grants HL28438 and CA54356. The initial steps of this work started with partial financial help of the Spanish Dirección General de Investigación Científica y Técnica as a sabbatical grant to G.T. Herman (grant number SAB94-0298). We acknowledge fruitful discussions with and suggestions from S. Matej, R.M. Lewitt and T.K. Narayan. The programming work was greatly facilitated by a set of routines originally developed for PET by Furuie et al. [29]. This work contains data previously published in "3D reconstruction in electron microscopy using ART with ..." Ultramicroscopy, Vol 72, No 1-2, pp. 53–65, 1998 (with permission from Elrevier Science - NL).

REFERENCES

[1] HERMAN, G.T., *Image Reconstruction From Projections: The Fundamentals of Computerized Tomography* (Academic Press, New York 1980).

[2] CARAZO, J.M., *The fidelity of 3D reconstructions from incomplete data and the use of restoration methods*, In *Electron Tomography* (Frank J., editor. Plenum Press, New York 1992).

[3] FRANK, J., *Three-Dimensional Electron Microscopy of Macromolecular Assemblies* (Academic Press, San Diego 1996).

[4] REIVICH, M. AND A. ALAVI, *Positron Emission Tomography* (Alan R. Liss, Inc., New York 1985).

[5] MATEJ, S., G.T. HERMAN, T.K. NARAYAN, S.S. FURUIE, R.M. LEWITT AND P. KINAHAN, *Evaluation of task-oriented performance of several fully 3-D PET reconstruction algorithms*, Phys. Med. Biol. 39:355–367, 1994.

[6] KINAHAN, P.E., S. MATEJ, J.S. KARP, G.T. HERMAN AND R.M. LEWITT, *A comparison of transform and iterative reconstruction techniques for a volume-imaging PET scanner with a large axial acceptance angle*, IEEE Trans. Nuclear Sci. 42:2281–2287, (1995).

[7] MATEJ, S. AND R.M. LEWITT, *Efficient 3D grids for image reconstruction using spherically symmetric volume elements*, IEEE Trans. Nuclear Sci. 42:1361–1370, (1995).

[8] RADERMACHER, M., T. WAGENKNECHT, A. VERSCHOOR AND J. FRANK, *Three-dimensional reconstruction from a single-exposure, random conical tilt series applied to the 50S ribosomal subunit of Escherichia coli*, J. Microsc. 146:113–136 (1987).

[9] SAN MARTÍN, C., N.P.J STAMFORD, N. DAMMEROVA, N. DIXON AND J.M. CARAZO, *A structural model for the Escherichia coli DnaB helicase based on electron microscopy data*, J. Struc. Biol. 114: 167–176 (1995).

[10] HAWKES, P.T., *The electron microscope as a structure projector*, In *Electron Tomography* (J. Frank, editor. Plenum Press, New York 1992).

[11] LEWITT, R.M., *Alternatives to voxels for image representation in iterative reconstruction algorithms*, Phys. Med. Biol. 37:705–716 1992.

[12] LEWITT, R.M., *Multidimensional digital image representations using generalized Kaiser-Bessel window functions*, J. Opt. Soc. Am. A7:1834–1846, 1990.

[13] MATEJ, S. AND R.M. LEWITT, *Practical considerations for 3D image reconstruction using spherically-symmetric volume elements*, IEEE Trans. Med. Img. 15:68–78, 1996.

[14] KACZMARZ, S., *Angenährte Auflösung von Systemen linearer Gleichungen*, Bull. Int. Acad. Pol. Sci. Lett. A 35:355–357, 1937.

[15] HERMAN, G.T. AND L.B. MEYER, *Algebraic reconstruction techniques can be made computationally efficient*, IEEE Trans. Med. Img. 12:600–609, 1993.

[16] RADERMACHER, M., *Weighted backprojection methods*, In *Electron Tomography* (Frank J. editor. Plenum Press, New York. 1992).

[17] FURUIE, S.S., G.T. HERMAN, T.K. NARAYAN, P.E. KINAHAN, J.S. KARP, R.M. LEWITT AND S. MATEJ, *A methodology for testing for statistically significant differences between fully 3D PET reconstruction algorithms*, Phys. Med. Biol. 39:343–354, 1994.

[18] VARDEMAN, B.S., *Statistics for Engineering Problem Solving*, Academic Press. Boston, 1994.

[19] CHAN, M.T., G.T. HERMAN AND E. LEVITAN, *A Bayesian approach to PET reconstruction using image-modeling Gibbs priors: Implementation and comparison*, IEEE Trans. on Nuclear Sci. 44: 1347–1354, 1997.

[20] MATEJ, S., S.S. FURUIE AND G.T. HERMAN, *The relevance of statistically significant differences between reconstruction algorithms*, IEEE Trans. Img. Proc. 5:554–556, 1996.

[21] HARAUZ, G. AND F.P. OTTENSMEYER, *Direct three-dimensional reconstruction for macromolecular complexes from electron micrographs*, Ultramicroscopy 12:309–320, 1984.

[22] SAXTON, W.O., *Computer Techniques for Image Processing in Electron Microscopy* (Academic Press, New York 1978).

[23] FRANK, J., A. VERSCHOOR AND M. BOUBLIK, *Computer averaging of electron micrographs of 40S ribosomal subunits*, Science 214:1353–1355, 1981.

[24] HANICKE, W., J. FRANK AND H.P. ZINGSHEIM, *Statistical significance of molecule projections by single particle averaging*, J. Microsc. 133:223–238, 1984.

[25] PENCZEK, P., R.A. GRASSUCCI AND J. FRANK, *The ribosome at improved resolution: New techniques for merging and orientation refinement in 3D cryoelectron microscopy of biological particles*, Ultramicroscopy. 53:251–270, 1994.

[26] FRANK, J., M. RADERMACHER, T. WAGENKNECHT AND A. VERSCHOOR, *Studying ribosome structure by electron microscopy and computer image processing*, Methods Enzymol. 164: 3–35, 1988.

[27] DE LA FRAGA L.G., J. DOPAZO AND J.M. CARAZO, *Confidence limits for resolution estimation in image averaging by random subsampling*, Ultramicroscopy 60:385–391, 1995.

[28] GILBERT, P., *Iterative methods for the three-dimensional reconstruction of an object from projections*, J. Theor. Biol. 36:105–117, 1971.

[29] FURUIE, S.S., S. MATEJ, G.T. HERMAN, T.K. NARAYAN, R.M. LEWITT, P. KINAHAN AND J.S. KARP, *Programs for evaluation of 3D PET reconstruction algorithms*, (Technical report No. MIPG206, Department of Radiology, University of Pennsylvania, Philadelphia 1994).

IMA SUMMER PROGRAMS

1987 Robotics
1988 Signal Processing
1989 Robust Statistics and Diagnostics
1990 Radar and Sonar (June 18 - June 29)
      New Directions in Time Series Analysis (July 2 - July 27)
1991 Semiconductors
1992 Environmental Studies: Mathematical, Computational, and
         Statistical Analysis
1993 Modeling, Mesh Generation, and Adaptive Numerical Methods
         for Partial Differential Equations
1994 Molecular Biology
1995 Large Scale Optimizations with Applications to Inverse Problems,
         Optimal Control and Design, and Molecular and Structural
         Optimization
1996 Emerging Applications of Number Theory (July 15 – July 26)
      Theory of Random Sets (August 22 – August 24)
1997 Statistics in the Health Sciences
1998 Coding and Cryptography (July 6 – July 18)
      Mathematical Modeling in Industry (July 22 – July 31)
1999 Codes, Systems and Graphical Models

SPRINGER LECTURE NOTES FROM THE IMA:

The Mathematics and Physics of Disordered Media
    Editors: Barry Hughes and Barry Ninham
    (Lecture Notes in Math., Volume 1035, 1983)

Orienting Polymers
    Editor: J.L. Ericksen
    (Lecture Notes in Math., Volume 1063, 1984)

New Perspectives in Thermodynamics
    Editor: James Serrin
    (Springer-Verlag, 1986)

Models of Economic Dynamics
    Editor: Hugo Sonnenschein
    (Lecture Notes in Econ., Volume 264, 1986)

The IMA Volumes in Mathematics and its Applications

Current Volumes:

1   **Homogenization and Effective Moduli of Materials and Media**
     J. Ericksen, D. Kinderlehrer, R. Kohn, and J.-L. Lions (eds.)

2   **Oscillation Theory, Computation, and Methods of Compensated
     Compactness** C. Dafermos, J. Ericksen, D. Kinderlehrer,
     and M. Slemrod (eds.)

3   **Metastability and Incompletely Posed Problems**
     S. Antman, J. Ericksen, D. Kinderlehrer, and I. Muller (eds.)

4   **Dynamical Problems in Continuum Physics**
     J. Bona, C. Dafermos, J. Ericksen, and D. Kinderlehrer (eds.)

5   **Theory and Applications of Liquid Crystals**
     J. Ericksen and D. Kinderlehrer (eds.)

6   **Amorphous Polymers and Non-Newtonian Fluids**
     C. Dafermos, J. Ericksen, and D. Kinderlehrer (eds.)

7   **Random Media** G. Papanicolaou (ed.)

8   **Percolation Theory and Ergodic Theory of Infinite Particle
     Systems** H. Kesten (ed.)

9   **Hydrodynamic Behavior and Interacting Particle Systems**
     G. Papanicolaou (ed.)

10   **Stochastic Differential Systems, Stochastic Control Theory,
     and Applications** W. Fleming and P.-L. Lions (eds.)

11   **Numerical Simulation in Oil Recovery** M.F. Wheeler (ed.)

12   **Computational Fluid Dynamics and Reacting Gas Flows**
     B. Engquist, M. Luskin, and A. Majda (eds.)

13   **Numerical Algorithms for Parallel Computer Architectures**
     M.H. Schultz (ed.)

14   **Mathematical Aspects of Scientific Software** J.R. Rice (ed.)

15   **Mathematical Frontiers in Computational Chemical Physics**
     D. Truhlar (ed.)

16   **Mathematics in Industrial Problems** A. Friedman

17   **Applications of Combinatorics and Graph Theory to the Biological
     and Social Sciences** F. Roberts (ed.)

18   **q-Series and Partitions** D. Stanton (ed.)

19   **Invariant Theory and Tableaux** D. Stanton (ed.)

20   **Coding Theory and Design Theory Part I: Coding Theory**
     D. Ray-Chaudhuri (ed.)

21   **Coding Theory and Design Theory Part II: Design Theory**
     D. Ray-Chaudhuri (ed.)

22   **Signal Processing Part I: Signal Processing Theory**
     L. Auslander, F.A. Grünbaum, J.W. Helton, T. Kailath,
     P. Khargonekar, and S. Mitter (eds.)

23   **Signal Processing Part II: Control Theory and Applications of Signal Processing** L. Auslander, F.A. Grünbaum, J.W. Helton, T. Kailath, P. Khargonekar, and S. Mitter (eds.)

24   **Mathematics in Industrial Problems, Part 2** A. Friedman

25   **Solitons in Physics, Mathematics, and Nonlinear Optics** P.J. Olver and D.H. Sattinger (eds.)

26   **Two Phase Flows and Waves** D.D. Joseph and D.G. Schaeffer (eds.)

27   **Nonlinear Evolution Equations that Change Type** B.L. Keyfitz and M. Shearer (eds.)

28   **Computer Aided Proofs in Analysis** K. Meyer and D. Schmidt (eds.)

29   **Multidimensional Hyperbolic Problems and Computations** A. Majda and J. Glimm (eds.)

30   **Microlocal Analysis and Nonlinear Waves** M. Beals, R. Melrose, and J. Rauch (eds.)

31   **Mathematics in Industrial Problems, Part 3** A. Friedman

32   **Radar and Sonar, Part I** R. Blahut, W. Miller, Jr., and C. Wilcox

33   **Directions in Robust Statistics and Diagnostics: Part I** W.A. Stahel and S. Weisberg (eds.)

34   **Directions in Robust Statistics and Diagnostics: Part II** W.A. Stahel and S. Weisberg (eds.)

35   **Dynamical Issues in Combustion Theory** P. Fife, A. Liñán, and F.A. Williams (eds.)

36   **Computing and Graphics in Statistics** A. Buja and P. Tukey (eds.)

37   **Patterns and Dynamics in Reactive Media** H. Swinney, G. Aris, and D. Aronson (eds.)

38   **Mathematics in Industrial Problems, Part 4** A. Friedman

39   **Radar and Sonar, Part II** F.A. Grünbaum, M. Bernfeld, and R.E. Blahut (eds.)

40   **Nonlinear Phenomena in Atmospheric and Oceanic Sciences** G.F. Carnevale and R.T. Pierrehumbert (eds.)

41   **Chaotic Processes in the Geological Sciences** D.A. Yuen (ed.)

42   **Partial Differential Equations with Minimal Smoothness and Applications** B. Dahlberg, E. Fabes, R. Fefferman, D. Jerison, C. Kenig, and J. Pipher (eds.)

43   **On the Evolution of Phase Boundaries** M.E. Gurtin and G.B. McFadden

44   **Twist Mappings and Their Applications** R. McGehee and K.R. Meyer (eds.)

45   **New Directions in Time Series Analysis, Part I** D. Brillinger, P. Caines, J. Geweke, E. Parzen, M. Rosenblatt, and M.S. Taqqu (eds.)

46   **New Directions in Time Series Analysis, Part II**
D. Brillinger, P. Caines, J. Geweke, E. Parzen, M. Rosenblatt,
and M.S. Taqqu (eds.)

47   **Degenerate Diffusions**
W.-M. Ni, L.A. Peletier, and J.-L. Vazquez (eds.)

48   **Linear Algebra, Markov Chains, and Queueing Models**
C.D. Meyer and R.J. Plemmons (eds.)

49   **Mathematics in Industrial Problems, Part 5**   A. Friedman

50   **Combinatorial and Graph-Theoretic Problems in Linear Algebra**
R.A. Brualdi, S. Friedland, and V. Klee (eds.)

51   **Statistical Thermodynamics and Differential Geometry**
of Microstructured Materials
H.T. Davis and J.C.C. Nitsche (eds.)

52   **Shock Induced Transitions and Phase Structures in General**
Media   J.E. Dunn, R. Fosdick, and M. Slemrod (eds.)

53   **Variational and Free Boundary Problems**
A. Friedman and J. Spruck (eds.)

54   **Microstructure and Phase Transitions**
D. Kinderlehrer, R. James, M. Luskin, and J.L. Ericksen (eds.)

55   **Turbulence in Fluid Flows: A Dynamical Systems Approach**
G.R. Sell, C. Foias, and R. Temam (eds.)

56   **Graph Theory and Sparse Matrix Computation**
A. George, J.R. Gilbert, and J.W.H. Liu (eds.)

57   **Mathematics in Industrial Problems, Part 6**   A. Friedman

58   **Semiconductors, Part I**
W.M. Coughran, Jr., J. Cole, P. Lloyd, and J. White (eds.)

59   **Semiconductors, Part II**
W.M. Coughran, Jr., J. Cole, P. Lloyd, and J. White (eds.)

60   **Recent Advances in Iterative Methods**
G. Golub, A. Greenbaum, and M. Luskin (eds.)

61   **Free Boundaries in Viscous Flows**
R.A. Brown and S.H. Davis (eds.)

62   **Linear Algebra for Control Theory**
P. Van Dooren and B. Wyman (eds.)

63   **Hamiltonian Dynamical Systems: History, Theory,**
and Applications
H.S. Dumas, K.R. Meyer, and D.S. Schmidt (eds.)

64   **Systems and Control Theory for Power Systems**
J.H. Chow, P.V. Kokotovic, R.J. Thomas (eds.)

65   **Mathematical Finance**
M.H.A. Davis, D. Duffie, W.H. Fleming, and S.E. Shreve (eds.)

66   **Robust Control Theory**   B.A. Francis and P.P. Khargonekar (eds.)

67   **Mathematics in Industrial Problems, Part 7**   A. Friedman

68   **Flow Control**   M.D. Gunzburger (ed.)

69 **Linear Algebra for Signal Processing**
A. Bojanczyk and G. Cybenko (eds.)

70 **Control and Optimal Design of Distributed Parameter Systems**
J.E. Lagnese, D.L. Russell, and L.W. White (eds.)

71 **Stochastic Networks** F.P. Kelly and R.J. Williams (eds.)

72 **Discrete Probability and Algorithms**
D. Aldous, P. Diaconis, J. Spencer, and J.M. Steele (eds.)

73 **Discrete Event Systems, Manufacturing Systems,
and Communication Networks**
P.R. Kumar and P.P. Varaiya (eds.)

74 **Adaptive Control, Filtering, and Signal Processing**
K.J. Aström, G.C. Goodwin, and P.R. Kumar (eds.)

75 **Modeling, Mesh Generation, and Adaptive Numerical Methods
for Partial Differential Equations** I. Babuska, J.E. Flaherty,
W.D. Henshaw, J.E. Hopcroft, J.E. Oliger, and T. Tezduyar (eds.)

76 **Random Discrete Structures** D. Aldous and R. Pemantle (eds.)

77 **Nonlinear Stochastic PDEs: Hydrodynamic Limit and Burgers'
Turbulence** T. Funaki and W.A. Woyczynski (eds.)

78 **Nonsmooth Analysis and Geometric Methods in Deterministic
Optimal Control** B.S. Mordukhovich and H.J. Sussmann (eds.)

79 **Environmental Studies: Mathematical, Computational,
and Statistical Analysis** M.F. Wheeler (ed.)

80 **Image Models (and their Speech Model Cousins)**
S.E. Levinson and L. Shepp (eds.)

81 **Genetic Mapping and DNA Sequencing**
T. Speed and M.S. Waterman (eds.)

82 **Mathematical Approaches to Biomolecular Structure and Dynamics**
J.P. Mesirov, K. Schulten, and D. Sumners (eds.)

83 **Mathematics in Industrial Problems, Part 8** A. Friedman

84 **Classical and Modern Branching Processes**
K.B. Athreya and P. Jagers (eds.)

85 **Stochastic Models in Geosystems**
S.A. Molchanov and W.A. Woyczynski (eds.)

86 **Computational Wave Propagation**
B. Engquist and G.A. Kriegsmann (eds.)

87 **Progress in Population Genetics and Human Evolution**
P. Donnelly and S. Tavaré (eds.)

88 **Mathematics in Industrial Problems, Part 9** A. Friedman

89 **Multiparticle Quantum Scattering With Applications to Nuclear,
Atomic and Molecular Physics** D.G. Truhlar and B. Simon (eds.)

90 **Inverse Problems in Wave Propagation** G. Chavent, G. Papanicolau,
P. Sacks, and W.W. Symes (eds.)

91 **Singularities and Oscillations** J. Rauch and M. Taylor (eds.)

92   **Large-Scale Optimization with Applications, Part I:**
        Optimization in Inverse Problems and Design
        L.T. Biegler, T.F. Coleman, A.R. Conn, and F. Santosa (eds.)
93   **Large-Scale Optimization with Applications, Part II:**
        Optimal Design and Control
        L.T. Biegler, T.F. Coleman, A.R. Conn, and F. Santosa (eds.)
94   **Large-Scale Optimization with Applications, Part III:**
        Molecular Structure and Optimization
        L.T. Biegler, T.F. Coleman, A.R. Conn, and F. Santosa (eds.)
95   **Quasiclassical Methods**
        J. Rauch and B. Simon (eds.)
96   **Wave Propagation in Complex Media**
        G. Papanicolaou (ed.)
97   **Random Sets: Theory and Applications**
        J. Goutsias, R.P.S. Mahler, and H.T. Nguyen (eds.)
98   **Particulate Flows: Processing and Rheology**
        D.A. Drew, D.D. Joseph, and S.L. Passman (eds.)
99   **Mathematics of Multiscale Materials** K.M. Golden, G.R. Grimmett,
        R.D. James, G.W. Milton, and P.N. Sen (eds.)
100   **Mathematics in Industrial Problems, Part 10** A. Friedman
101   **Nonlinear Optical Materials** J.V. Moloney (ed.)
102   **Numerical Methods for Polymeric Systems** S.G. Whittington (ed.)
103   **Topology and Geometry in Polymer Science** S.G. Whittington,
        D. Sumners, and T. Lodge (eds.)
104   **Essays on Mathematical Robotics** J. Baillieul, S.S. Sastry,
        and H.J. Sussmann (eds.)
105   **Algorithms For Parallel Processing** M.T. Heath, A. Ranade,
        and R.S. Schreiber (eds.)
106   **Parallel Processing of Discrete Problems** P.M. Pardalos (ed.)
107   **The Mathematics of Information Coding, Extraction, and**
        Distribution G. Cybenko, D.P. O'Leary, and J. Rissanen (eds.)
108   **Rational Drug Design** D.G. Truhlar, W. Howe, A.J. Hopfinger,
        J. Blaney, and R.A. Dammkoehler (eds.)
109   **Emerging Applications of Number Theory** D.A. Hejhal, J. Friedman,
        M.C. Gutzwiller, and A.M. Odlyzko (eds.)
110   **Computational Radiology and Imaging: Therapy and Diagnostics** C.
        Börgers, F. Natterer (eds.)

FORTHCOMING VOLUMES

1992–1992: *Control Theory*
     Robotics

1996 Summer Program: *Emerging Applications of Number Theory*

1996–1997: *Mathematics in High Performance Computing*
     Algorithms for Parallel Processing
     Evolutionary Algorithms
     The Mathematics of Information Coding, Extraction and Distribution
     Structured Adaptive Mesh Refinement Grid Methods
     Computational Radiology and Imaging: Therapy and Diagnostics
     Mathematical and Computational Issues in Drug Design
     Rational Drug Design
     Grid Generation and Adaptive Algorithms
     Parallel Solution of Partial Differential Equations

1997 Summer Program: *Statistics in the Health Sciences*
     Week 1: Genetics
     Week 2: Imaging
     Week 3: Diagnosis and Prediction
     Weeks 4 and 5: Design and Analysis of Clinical Trials
     Week 6: Statistics and Epidemiology: Environment and Health

1997–1998: *Emerging Applications for Dynamical Systems*
     Numerical Methods for Bifurcation Problems
     Multiple-time-scale Dynamical Systems
     Dynamics of Algorithms